BEYOND THE GLOBAL CITY

Beyond the Global City

Understanding and Planning for the Diversity of Ontario

Edited by
GORDON NELSON

McGill-Queen's University Press
Montreal & Kingston • London • Ithaca

© McGill-Queen's University Press 2012

ISBN 978-0-7735-3985-3 (cloth)
ISBN 978-0-7735-3986-0 (paper)

Legal deposit second quarter 2012
Bibliothèque nationale du Québec

Printed in Canada on acid-free paper that is 100% ancient forest free (100% post-consumer recycled), processed chlorine free

This book has been published with the help of a grant from the University of Waterloo. Funding has also been received from the Carolinian Canada Coalition.

McGill-Queen's University Press acknowledges the support of the Canada Council for the Arts for our publishing program. We also acknowledge the financial support of the Government of Canada through the Canada Book Fund for our publishing activities.

Library and Archives Canada Cataloguing in Publication

Beyond the global city : understanding and planning for the diversity of Ontario / edited by Gordon Nelson.

Includes bibliographical references and index.
ISBN 978-0-7735-3985-3 (bound). – ISBN 978-0-7735-3986-0 (pbk.)

1. Regional planning – Ontario. 2. Ontario – Geography.
I. Nelson, J. G. (James Gordon), 1932–

HT395.C320522 2012   307.1'209713   C2012-900929-6

Typeset by Jay Tee Graphics Ltd. in 10.5/13 Sabon

Dedicated to the citizens of Ontario

Contents

Acknowledgments   ix
Introduction   *Gordon Nelson*   xi

1. The Origin of Ontario's Georegions: Pre-European Times to the 1850s   3
   Gordon Nelson and Michael Troughton

2. The Industrial Era and the Further Development of Ontario's Georegions: The Railroads, 1850s–1900s   27
   Gordon Nelson and Michael Troughton

3. Sweetwater Seas and Shores: Waters and Coasts of the Great Lakes Georegion   55
   Patrick L. Lawrence

4. Toronto: Putting the Georegion in Perspective   80
   Lucy M. Sportza

5. The Carolinian Canada Georegion: Farmland, Forests, and Freeways in Conflict   102
   Stewart Hilts

6. The Huron Georegion: Rurality in an Urbanizing Province   127
   Wayne Caldwell

7. Peterborough: A Georegion in Transition?   151
   Alison Bain and John Marsh

8. Georgian Bay, Muskoka, and Haliburton: More than Cottage Country?   169
   Nik Luka and Nina-Marie Lister

9 The Kingston Georegion: Going "Glocal" In a Globalizing World   201
   Brian S. Osborne

10 The Ottawa Valley Georegion   227
   Mark Seasons

11 Learning from the Past, Assessing the Future: A Geo-historical Overview of Northeastern Ontario   245
   Raynald Harvey Lemelin and Rhonda Koster

12 Northwestern Ontario and the Lakehead Georegion: A Region Apart?   277
   Margaret E. Johnston and R.J. Payne

13 The Intersection of Landscape, Legislation, and Local Perceptions in Constructing the Niagara Escarpment as a Distinct Georegion   309
   Susan Preston

14 The Changing Natural Heritage of Ontario in Broad Perspective   336
   Stephen D. Murphy

15 Changing Cultural Landscapes of Ontario   359
   Robert Shipley

16 Retrospect and Prospect   388
   Gordon Nelson

Contributors   433
Index   439

Acknowledgments

First I would like to thank all the authors of the chapters in *Beyond the Global City* for their long commitment to the completion of this book. It has been a challenging process to link all our efforts, and the dedication of the authors is much appreciated. I would also like to acknowledge the early influences of the late Michael Troughton, professor of geography, University of Western Ontario, in getting the work going and in contributing to the first two chapters. I am also grateful to Susan Preston for her comments on the last chapter and her role in helping complete the manuscript.

I would especially like to acknowledge the support of the Faculty of Environment (formerly Environmental Studies), University of Waterloo, its faculty and staff, during the six years in which this book was under way. Thanks are also extended to the Social Sciences and Humanities Research Council and other agencies for their financial and general support for land use and environmental studies by me and other authors over recent decades, support that has ultimately made this book possible. I cannot thank directly everyone who helped, although I am aware of many contributions in diverse ways for which I am very grateful. I have benefitted greatly from service as chair of the Carolinian Canada Coalition, the Sources of Knowledge and Public Advisory Committees for Bruce Peninsula National Park, and the Heritage Resources Centre, University of Waterloo, and as a member of Community Heritage Ontario, the Ontario Parks Board of Directors, and other relevant organizations. I gratefully acknowledge the financial contribution of Carolinian Canada in support of publication.

When the manuscript was essentially complete, the Environmental Commissioner of Ontario collaborated with the Carolinian Canada

Coalition in holding an informal workshop on land use in Ontario. The manuscript for this book was one of the major documents providing background for this workshop, and the financial and general support of the commissioner and his staff is gratefully acknowledged. I would also like to thank Professors Chad Day and George Francis of Simon Fraser and the University of Waterloo, respectively, for their very helpful reviews of an early version of the manuscript. Very special thanks to Shirley Nelson and Ashleigh Beyer for their vital work in preparing the manuscript. And much appreciation to Stacy Cooper from Cooper Admin for preparing the final manuscript and to Barry Levelly for preparing maps. Figures, tables, boxes, and other information taken directly from other sources are acknowledged according to the source. Those figures without explicit acknowledgment of their source were prepared by the chapter authors for this volume. Finally, I am very grateful to the staff of McGill-Queen's University Press: Professor Don Akenson, Mary-Lynne Ascough, Ryan Van Huijstee, Ron Curtis, Ryan Thom, Katie Heffring, and Jacqueline Michelle Davis for their efforts and support in carrying this manuscript to publication and arranging for its marketing.

Gordon Nelson

Introduction

GORDON NELSON

PURPOSES AND MOTIVATION

The title of this book is intended to highlight its two primary purposes. The first is to paint a very different image of Ontario from the one that has been dominant for several decades. The prevailing view is of a province centred on and driven by the great city of Toronto, one of a growing number of huge metropolises expanding vigorously into surrounding lands and waters in different parts of North America and the world. The continued growth of Toronto and other great global cities is generally associated with massive urban sprawl; traffic gridlock; rising service costs; water, air, and landscape pollution; loss of plant and animal species and biodiversity; encroachment on good agricultural land; decline in community well-being; and reductions in local planning and political powers. In this context, places beyond the big city are seen predominantly as sources of goods and services essential to metropolitan expansion. They are not generally viewed as places with distinctive historical, environmental, social, and land use qualities and opportunities that are being progressively eroded by what very likely is unsustainable growth.

In *Beyond the Global City* we aim to look beyond Toronto and give people, planners, and politicians a wider and richer view of Ontario, the places that make it up, the challenges and opportunities confronting them, and the potentials that would be foregone without greater understanding and consideration of their distinctive qualities in local, regional, and provincial planning and decision making. In analysing and making the case for a more diverse understanding and planning framework for Ontario, we identify what we call its georegions, whose qualities and modes of recognition are described later in this introduction. The result is eleven georegions ranging from the Great Lakes through Toronto; Southwestern Ontario, or Carolinian Canada; Huron; Peterborough; Kingston; Ottawa; Muskoka and Georgian Bay; Northeastern and Northwestern Ontario; and the Niagara Escarpment.

This takes us to the second major reason for the title and the writing of this book. Toronto is frequently understood and planned as a global city. The idea of a global city is closely tied to the concept of globalization, a way of thinking about the most desirable route to local, regional, national, and international development that has been salient in geography, economics, planning, politics, and governance for many years. Globalization is dealt with quite well in the writings of Joseph Stiglitz, the Nobel Prize–winning economist (2007, 2010). The globalization model envisions an increasingly extensive and intensive trading and interchange system as the prime force for greater overall economic growth and "progress" among the peoples, places, provinces, and nations of the world. A key element is the thrust of huge growing cities and urban regions as the great engines of development. Witness Toronto, Southern Ontario, Northern Ontario, Canada, and beyond.

In unfettered free-market fashion, global cities compete intensely with one another for technical and other advantages leading, theoretically, to ever-higher growth and stature internationally. Global city ideology has been embraced by numerous scholars, educators, business people, members of the media, planners, and decision makers as a guiding vision for Toronto and Ontario. We did not envision discussing the global city and globalization in any detail when initially planning this book. However, in 2008, toward the end of our work on the first complete draft, the great global recession struck. Toronto was hit hard, as were other georegions, in accordance with their ties to the global city, the globalization network, and their own particular vulnerabilities. This turn of events led us to look beyond global-city and globalization ideology in thinking about the future planning of Toronto and Ontario. The results, which are indicative rather than comprehensive, are presented in the final chapter, "Retrospect and Prospect."

So this book arises from growing concern about the ideological basis for the planning of Ontario. The origins of this concern lie at least as far back as the 1970s. About that time the now well-known view of Toronto as the great economic engine of the province began to dominate planning policy and practice. In the last decade the idea of Toronto as the driver of Ontario's socio-economic future has become so strong that it overshadows other planning perspectives and approaches and has resulted in a kind of flat-earth view of Ontario. Toronto and adjacent areas are seen as the prime concern of planning; surrounding areas are destined to be absorbed into its development envelope. They are imperfectly recognized and tend to be described and treated very generally in relation to the growing Toronto conurbation as rural, wildland, or open space. They are destined to become part of the great city or serve as a source or market for resources, goods, and services.

From the standpoint of globalization theory, Toronto is seen as competing with major urban growth centres around the world. Its robust growth is deemed necessary for the health of the rest of the province. This kind of thinking reinforces the tendency to see only Toronto in detail. Other parts of the province are viewed as a development field for this growth engine.

Such thinking became quite pronounced by the late 1990s. One basic expression of it is the massive urban sprawl extending from Toronto far into surrounding regions: north to Barrie and Lake Simcoe; west to Hamilton, the Fruit Belt, and Kitchener; and east to Port Hope and Oshawa. The reach of the city goes beyond newly built-up areas to include the green belt and other "open space" programs designed to limit urban growth and meet recreational and other needs of the more than four million people living in the metropolis. Extensions of provincial and city power reduce the authority of the local governments whose lands and waters are absorbed into the metropolitan envelope. These extensions also push investment toward meeting Toronto's needs rather more than those of people living in outlying areas.

The distinctive character of these outlying lands – their social, economic, and natural history, landscapes and demography – are insufficiently appreciated as Toronto marches, with legislative and other support from the province, into what are frequently perceived as undeveloped open spaces. Yet these lands have a distinctive historic and geographic character of their own. Their heritage involves a trajectory that generally is not well considered in the expansion of Toronto's global growth engine.

One reason for this Toronto-based planning ideology has been the rise since the 1980s of neo-liberal thinking and the growing focus on the market and wealth as determinants and symbols of success. Another has been the continuing emphasis on centralized planning and the associated lack of civic understanding and dialogue in evaluating what has been happening. Underlying this has been the absorption of the schools into the model. History, geography, environmental studies, and other contextually enlightening fields of study have been curtailed in favour of greater stress placed on "productive" disciplines such as science, technology, computing, and business, notably by the Harris government in the late 1990s.

One big result – and this is the major reason for this book – is widespread loss of citizen capacity to recognize and understand the complexity and individual needs of the diverse landscapes, land uses, and communities making up the great province of Ontario. Many, if not most, citizens view the intricate history and geography of Ontario as an essentially flat surface – or blank slate – surrounding the city and awaiting its arrival to gain character. Younger people taught in the truncated curriculum of Ontario's schools in the late 1990s and early 2000s are especially affected by this situation. They need writings like those in this book as a remedy.

The foregoing discussion provokes consideration of a more informed, diverse, and participatory path beyond the Toronto-centred growth model. As one basic step in this direction, this book aims to portray Ontario in terms of its natural and cultural diversity and richness. We aim to provide the broader understanding that is neglected in the flat-earth model and to give the reader the information needed for a more comprehensive and participatory dialogue about the planning of all of Ontario, especially among younger residents of the province.

To do this, we have approached Ontario in terms of the regional details that underlie the rural, the wild, and open space. We work in terms of georegions, a broad concept that extends beyond the bioregions, ecoregions, or other ideas that have been prominent on the regional stage in the last two or three decades, when the regional concept was largely set aside by geographers and other scholars for more specialized, systematic studies, such as studies of urban transport, waste generation and disposal, pollution, migration, economic development, and governance. This neglect and its consequences have provided another motivation for this book. Some geographers have, of course, continued to work usefully in a regional framework, although its application varies, as can be seen, for example, in the historical geographer Cole Harris's *Reluctant Land* (2008) or Robert Bone's broad *Regional Geography of Canada* (2005). Other scholars have pursued regional studies in diverse ways and at various scales in Ontario (Campbell 2005; Chapman 1966; Clarke 2001; Heidenreich 1971; Hodge 2001; Kelly 1974; Wood 2000; Yeates 2001. As will be apparent shortly, however, the focus in *Beyond the Global City* is on a broad and more systematic regional approach embracing the diverse interactions among the natural and human dimensions of a place called Ontario. One major reason for a systematic approach is to make it easier to compare the different georegions of the province; another reason is encourage authors to consider all major natural and human influences at work in their regions.

The regional approach in disciplines such as biology and ecology has strengthened in recent decades. Major advances in ecological theory and method were made in the late 1970s and 1980s, and some were closer to science than others, for example, island biogeography, biodiversity (Wilson 1988), and landscape ecology (Foreman and Godron 1986). These changes in theory interacted with one another so that it became possible to summarize quantitatively the species and landscape diversity of a region and to relate the conservation of this diversity to the size and other characteristics of the region in which it was located. Other notable influences on diversity were the type and range of human activities in an area and the degree to which they affected the natural system, for example, by fragmenting and isolating plant communities (Noss 1983). Without connections or corridors to source areas, species are susceptible to decline and extinction, potentially reducing the diversity, resilience, and sustainability of the system of which they are a part. From these perspectives, then, human activities such as agriculture, transport, and lumbering can be seen as threats to regional biodiversity.

Other changes in bioregional or ecological thought in the 1980s and later were not as conducive to thinking of human activities as threats. A leading example is bioregionalism (McGinnis 1999), a principal aim of which is to identify historical, economic, and social activities that are more or less in harmony with the natural character of the region and to plan the type and scale of human development along these lines. This is a relatively powerful line of thought that developed strongly, for example, in the rainforest bioregion in coastal Oregon, Washington, and British Columbia. It led to the development of socio-ecological concepts such as "home place" and to aligning the kind and intensity of development to sustain the character of a place (Rowe 1990; Hiss 1990).

Among the features that led to considerable criticism of bioregional thinking was its somewhat deterministic flavour. Some historical activities were seen as ecologically appropriate to the region; others were not. A second concern was that the approach was problematic in areas already undergoing major urban, industrial, or other changes that were out of line with and damaging to the natural and cultural character of the region. Such situations were especially problematic when the intensity, scale, and significance of such development was highly valued among influential people inside and outside the region. One example of such a situation is Toronto.

In the 1980s and 1990s planning also underwent changes supportive of regional theory and practice. Some emphasis on top-down corporate planning endured, but much attention shifted to grassroots participatory, co-operative, and civic approaches at the local and regional levels. Emerging planning theory and practice stressed the interactive and the adaptive and involved more initiatives that crossed local, provincial, national, and international scales. These initiatives tended to be linked, or integrated, at the regional level.

Recently, concern about the region has attained greater prominence in the humanities. One hundred and fifty poets, writers, geographers, musicologists, historians, and others met in Lincoln, Nebraska, in November 2003 to explore the theme of regionalism in the humanities. According to the editors of the proceedings of this meeting, most participants shared the general perception that, while once viewed as a reaction against the forces of globalization and modernism, the region has become an expression of the role of place and space in efforts to understand "ourselves and what it means to be human." The distinguishing feature of regionalism is its focus on "the space lived in, inhabited, made home, or travelled through." Regionalism arises from basic human interactions with nature: "the land, climate, flora and fauna, and the physical environment" (Mahoney and Katz 2008).

With this rich set of perspectives on regional thinking in mind – and it is amplified upon in chapter 7, "Peterborough: A Region in Transition," and chapter 9, "The Kingston Georegion" – we decided to take a dynamic view of the region in this book. A deliberate effort is made to organize and encourage a broad yet systematic and generally integrative approach to the regions of Ontario that moves beyond the ecoregional or bioregional thinking dominant when our project began. We decided to use the term georegion to indicate a region thought of not only in terms of its geologic, hydrologic, biotic, and land-use characteristics, but also to a large degree in terms of its demographic, ethnic, institutional, and other human dimensions.

In practice, however, although a georegion may initially be thought of in this broad way, certain of its characteristics generally receive more emphasis than others because of the author's interests or his views of their importance. Each author's choice of foci for analyzing, describing, and interpreting the character of and challenges to a georegion reflects his or her conclusions about its "sense of place." Read collectively, the georegional chapters in *Beyond the Global City* generally portray the diversity facing decision makers in Ontario.

What we focus on is the core region, the central or definitive part of a region, whose distinctive natural, economic, social, land-use, and other qualities have remained consistent or coherent through the years while its boundaries have expanded, contracted, or fluctuated in interactions with surrounding regions. We do not attempt to map, or describe in detail, the nature of these fluctuations, although a reading of *Beyond the Global City* certainly reveals examples of them, as in Toronto, Peterborough, Kingston, and the Upper St Lawrence and Ottawa regions. Again, our focus is on the cores of these regions, their persistence through much of Ontario's recent history, and their indications of the province's diversity.

All of this presents a major opportunity for geographers, planners, ecologists, historians, political scientists, and other scholars. One precondition for finding ways of understanding and planning for the growing Toronto conurbation is to understand the regional character of surrounding areas and find ways to link them together more sustainably. In this respect society needs to build on the concept of the bioregion by more fully understanding the economic, social, and institutional characteristics of regions, as well as their interactions with natural systems and one another. With such georegional understanding we should be in a position to plan more effectively for the tapestry that is Ontario.

DEVELOPMENT OF THE BOOK

Beyond the Global City began in a workshop by the Canadian Association of Geographers (CAG) at the annual meeting of the Learned

Societies of Canada in London in 2004. Researchers from geography and related disciplines were invited to present papers on challenges to some selected regions of Ontario in the light of the prevailing Toronto-centred approach to planning for the province. Subsequently, workshop participants developed a framework of eleven georegions and worked with the editor to identify knowledgeable authors for development of a book on a georegional approach to Ontario. Such a study was thought to be useful not only in Ontario but as a model for similar work in other provinces of Canada. This possibility is still seen as one of the potential benefits of *Beyond the Global City*.

To implement a broad, yet focused, approach we adopted the ABC method as the organizational framework for the various georegional chapters and, ultimately, for the book as a whole. The ABC method encourages an author to consider initially the abiotic, or geologic, hydrologic, and other physical influences; the biotic, or plant, animal, and other biological influences; and the cultural, or land-use, economic, ethnic, social, governance, or other human influences before selecting those that the historical and other evidence indicate are most significant in determining the overall character and planning issues facing a georegion. The use of the ABC method in this book is intended to be broad but not encyclopedic, serving as a screen that a knowledgeable person can use systematically in determining the features and challenges that are considered significant for understanding and planning for a region.

A deep appreciation of the intricacies of the ABC method is not essential to understanding its use in *Beyond the Global City*. For those interested, details are available in landscape guides prepared for Old Town Toronto, the Grand River Watershed, the Ridgetown-Rondeau area, the Upper St Lawrence Valley, and the Thames Watershed in Ontario (Nelson, Porter, Lemieux, et al. 2003; Nelson, Porter, Farassoglou, et al. 2003; Nelson, Veale, et al. 2004; Nelson, Fournier, et al. 2005; Troughton and Quinlan 2009). A manual on preparing landscape guides, which describes the ABC method in some detail (Nelson and Preston 2005), has been produced, and a book titled *Places: Linking Nature, Culture and Planning*, which describes the evolution of the ABC method over approximately the last three decades, has also been published (Nelson and Lawrence 2009).

It is important to emphasize that *Beyond the Global City* is built on an inductive approach to Ontario's georegions and their potential role in understanding and planning more fully for Ontario. In other words the approach is bottom-up. The book begins with two historical chapters that analyze and describe the interweaving of abiotic, biotic, and cultural factors in producing the diverse georegions of Ontario. These historic chapters are followed by separate chapters on each georegion that describe the evolving characteristics of individual georegions such as the Great Lakes, Toronto, Huron, or the Northwest and Northeast from approximately the early twentieth century to the present day, highlighting significant current planning issues but generally not addressing them in any detail, a task for another book. The eleven georegional chapters are followed by two cross-cutting chapters describing and interpreting the natural and cultural features and challenges of Ontario from a provincial perspective.

Aside from asking them to think in terms of the ABC method, no guidance was given to authors on what theory or guiding concepts they should use in preparing their chapters. They selected the factors and theories they considered significant based on their experience and learning. Many theoretical or conceptual perspectives are consequently found in these chapters. They include some that the reader may find useful in reflecting on other chapters or on the book as a whole. Examples are considerations of regional theory in chapter 7; slow food, smart growth, and other urban theory in chapter 9; and ecosystem or cultural heritage theory in chapters 14 and 15.

The final chapter, "Retrospect and Prospect," was prepared after the completion of the georegional and cross-cutting chapters. It briefly summarizes the purposes of the book and major findings in the chapters and then assesses their implications for some of the major concepts or theories used to justify the global city approach in planning for the development of Toronto and Ontario in the last several decades. Among the concepts discussed are neo-liberal thinking, trickle-down theory, the regional state, and creative destruction, all of which are found wanting as a basis for understanding and planning effectively for all of Ontario.

While the last chapter was being written, Ontario and other parts of the world suffered from some of the consequences of adherence to these ideas, entering the great recession of 2008–10. This development downturn is still unfolding and is discussed in the light of critiques by Stiglitz (2007, 2010), the Nobel Prize-winning economist, and others about globalization and the global city. Their ideas and the findings in *Beyond the Global City* offer useful food for thought in planning for the future in Ontario.

To summarize, the chapters in *Beyond the Global City* are written by authors in their own voice. While all authors attempt to deal with the natural (abiotic and biotic) and the cultural histories and interactions to some degree, their emphases and styles compare and contrast, as do the extent to which they deal with planning issues, implications, and recommendations. In doing so, they tell the story of each of Ontario's georegions in somewhat different ways that, to some degree, reflect differences in perception and preference. Yet collectively they paint a vivid multi-faceted picture of the diversity of Ontario, as well as of the particular challenges and opportunities facing each region and Ontario as a whole. The hope is that these writings will provide government officials, business people, professionals, students in schools and universities, and citizens generally with some of the food for thought that is needed to plan more sensitively, effectively, and equitably for the complexity that is Ontario.

GUIDANCE FOR THE READER

The reader can approach this book in several ways. It can be read from end to end in the spirit of cumulative learning and as a basis for understanding the final three interpretive chapters, 14, 15, and 16. It can also be read to address or build on the reader's background and preferences, for example, through the ecological concepts in chapters 3 and 14 or the regional or cultural concepts in chapters 7, 9, and 15. The busy reader can also go directly to the final chapter, which summarizes and interprets the chapters in this book and relates the findings to concepts and approaches that have recently guided understanding and planning for the global city, such as trickle-down theory or other ideas. The choice is yours.

REFERENCES

Bone, Robert M. 2005. *The Regional Geography of Canada*. Don Mills, ON: Oxford University Press.

Campbell, Claire Elizabeth. 2005. *Shaped by the Wind: Nature and History in Georgian Bay*. Vancouver and Toronto: UBC Press.

Chapman, L.J., and D.F Putnam. 1966. *The Physiography of Southern Ontario.* 2d ed. Toronto: University of Toronto Press.

Clarke, John. 2001. *Land, Power and Economics on the Frontier of Upper Canada.* Montreal and Kingston: McGill-Queen's University Press.

Foreman, Richard T., and Michel Godron. 1986. *Landscape Ecology.* New York: John Wiley and Sons.

Harris, Cole. 2008. *The Reluctant Land: Society, Space, and Environment in Canada before Confederation.* Vancouver: UBC Press.

Heidenreich, Conrad. 1971. *Huronia: A History and Geography of the Huron Indians, 1600–1650.* Toronto: McClelland and Stewart.

Hiss, Tony. 1990. *The Experience of Place.* Vintage Books. New York: Random House.

Hodge, Gerald, and I.M. Robinson. 2001. *Planning Canadian Regions.* Vancouver: UBC Press.

Kelly, Kenneth. 1974. Damaged and Efficient Landscape in Rural Southern Ontario, 1880–1900." *Ontario History* 66:1–14.

Mahoney, Timothy R., and Wendy J. Katz, eds. 2008. *Regionalism and the Humanities.* Lincoln: University of Nebraska Press.

McGinnis, M.V., ed. 1999. *Bioregionalism.* London and New York: Routledge.

Myers, Rollo, Heather Thomson, and Gordon Nelson. 2001. *Old Town Toronto: A Heritage Landscape Guide.* Waterloo, ON: Heritage Resource Centre, University of Waterloo.

Nelson, J.G, J. Porter, C. Lemieux, C. Farassoglou, S. Gardiner, C. Guthrie, and K. Van Osch. 2003. *Ridgetown and Rondeau Area: A Heritage Landscape Guide.* Waterloo, ON: Heritage Resource Centre, University of Waterloo.

Nelson, J.G., J. Porter, C. Farassoglou, S.Gardiner, C. Guthrie, C. Beck, and C. Lemieux. 2003. *The Grand River Watershed: A Heritage Landscape Guide.* Waterloo, ON: Heritage Resource Centre, University of Waterloo.

Nelson, Gordon, Barbe Veale, Beth Dempster, et al. 2004. *Towards a Grand Sense of Place.* Waterloo, ON: An Environments Publication in association with the University of Waterloo, Waterloo.

Nelson, Gordon, Susan Fournier, Christopher Lemieux, Eric Tucs, Natalie Korobaylo, Cortar Farassoglou, and Lucy Sportza. 2005. *The Great River.* Waterloo, ON: Environments Publication, University of Waterloo.

Nelson, J. Gordon, and Susan Preston. 2005. *Towards a Grand Sense of Place: Preparing Heritage Landscape Guides. A Manual for Urban and*

Rural Communities in Ontario. Waterloo, ON: Environments Publication. University of Waterloo.

Nelson, J. Gordon, and Patrick Lawrence. 2009. *Places: Linking Nature, Culture and Planning*. Calgary, AB: University of Calgary Press.

Noss, R.F. 1983. A Regional Landscape Approach to Maintain Biodiversity. *Bioscience* 33:700–6.

Rowe, Stan. 1990. *Home Place: Essays in Ecology*. Edmonton, AB: Newest.

Stiglitz, Joseph E. 2007. *Making Globalization Work*. New York-London: W.W. Norton.

– 2010. *Freefall: America, Free Markets, and the Sinking World Economy*. New York-London: W.W.Norton.

Troughton, M.J., and Cathy Quinlan. 2009. *The Thames River Watershed: A Heritage Landscape Guide*. London: Carolinian Canada Coalition and the Thames Canadian Heritage River Committee.

Wilson, E.O. 1988. *The Diversity of Life*. New York: W.W Norton Company.

Wood, D.J. 2000. *Making Ontario*. Montreal and Kingston: McGill-Queen's University Press.

Yeates, Maurice, 2001. *Main Street: Canada*. 2d ed.

BEYOND THE GLOBAL CITY

1

The Origin of Ontario's Georegions: Pre-European Times to the 1850s

GORDON NELSON AND MICHAEL TROUGHTON

This chapter and the next set the stage for the rest of the book by presenting the historical evidence for the settlement of Ontario and the development of the major georegions in some detail. They also describe the effects of First Nations and Euro-American immigrants on the environment and on one another. The immigrants arrived primarily from Europe and the thirteen American colonies to form an ethnic and cultural mosaic that is strongly reflected in ensuing chapters on Ontario's eleven georegions. An awareness of settlement history and its effects on environmental, economic, land-use, and other dimensions also provides useful background for chapters 14 and 15, on the natural and cultural heritage of Ontario, as well as for the concluding chapter, "Retrospect and Prospect."

SETTING THE SCENE

First Nations have been at work changing the face of Ontario for thousands of years. In northern Ontario, beyond Georgian Bay (figure 1.1), collecting, hunting, and fishing persisted over the millenia among small groups of nomadic native people, causing relatively subtle and uncertain effects on forests and ecosystems. In contrast, the adoption of farming by growing numbers of shifting native cultivators resulted in land clearing and some regrowth of forests and savannahs on more tillable soils in much of southern Ontario (Heidenreich 1971).

European invasion in the early seventeenth century set off a series of great changes. The first was the introduction of the fur trade, which had far-reaching effects on animals, plants, and the native people themselves. The second was early-nineteenth-century European colonization, when tens of thousands of settlers exploited forests, prairies, and water for agriculture, lumbering, fishing, and mining. The third series of changes, described in chapter 2, came with the mid-nineteenth-century construction of the railroads, which promoted industrialization, urbanization, mass migration, and farming, logging, fishing, and other activities in remote parts of Ontario. The fourth began about 1900 with the wide-ranging introduction of the telegraph, the telephone, electrification, the automobile, and other new ways of accelerating communication among the millions of people increasingly concentrated in urban areas, notably the growing metropolis of Toronto and the Golden Horseshoe around the western end of Lake Ontario.

The pre-1900 changes in the human role and in human numbers transformed the rocks, the landforms, hydrology, plant and animal life, and the people of Ontario. The province evolved from a predominantly wild environment into a diverse array of communities, neighbourhoods, land uses, natural areas, and landscape patterns. These landscape qualities, these differences in the wild, rural, and urban scene, formed a changing tapestry across Ontario. As with a Persian carpet, the patterns can be seen in finer and finer detail, but it is the broad pattern that makes the rug.

Chapters 1 and 2 focus on understanding the large patterns and how they have evolved over the years into the broad economic, social, structural, institutional, environmental, and land-use differences that we call the georegions of Ontario. Today we can discern eleven major georegions in the province: Toronto, Ottawa, Kingston and the Upper St Lawrence, Peterborough, Southwestern Ontario or Carolinian Canada, Huron Country, Georgian Bay and Muskoka, Northeastern Ontario, Northwestern Ontario, and the Great Lakes (figure 1.2). A recent addition is the Niagara Escarpment Georegion. Most georegions began to appear quite early. By the end of the colonization era, about 1850, the general shape of many of them was apparent. The ensuing industrial era worked to elaborate these patterns into the twentieth century. In the post-World War II development era, the tapestry is being shaken and even rewoven, mainly by the sprawling Toronto conurbation. The core areas of most of the historic georegions are more or less intact but the georegions are

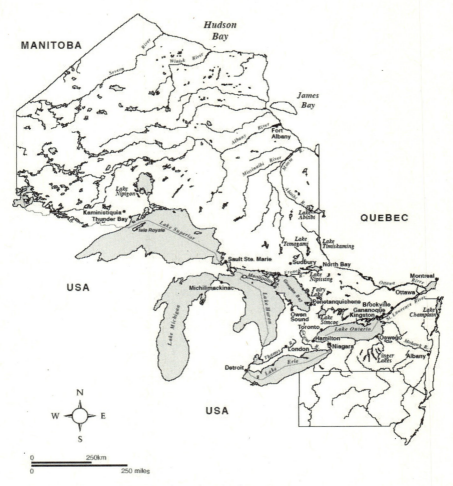

Figure 1.1 Location map. By the authors, with permission

being reshaped, and in some cases their integrity is at risk. Their historic contribution to the richness of the provincial tapestry may fade from the scene. More thought is needed in weaving the new into the old.

THE FIRST NATIONS ERA AND ITS LANDSCAPES AND SOCIETIES

The ancestral roots of Ontario's georegions extend deep into its ancient history. The earliest native people recognized in the province are the Paleo-Indians of ten thousand years ago. Overhunting

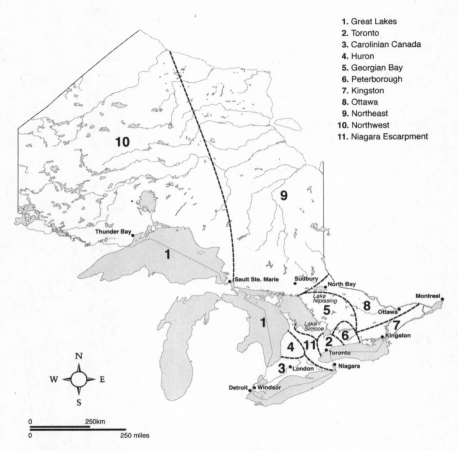

Figure 1.2 Georegions of Ontario. By the authors, with permission

by these people elsewhere in North America is thought to have contributed to the elimination of large Pleistocene or Ice Age fauna such as the mammoth and the mastodon. Yet no archaeological sites are known to link human worked-stone hunting tools with mammoths or other extinct Ice Age animals in Ontario. In later Archaic and early Woodland times, a few thousand to five hundred years before the birth of Christ, or what is now called the beginning of the Common Era (CE), the predecessors of Ontario's northern Algonquin people hunted caribou, deer, elk, bear, and other fauna, as well as fishing and collecting plants. The effects of such long-continued activities on ecosystems are difficult to ascertain.

First Nations' lifestyles and landscapes varied across the province in late Woodland times, just before European arrival (Wright 1972; Ellis and Ferris 1990). On the ancient acidic, granitic rocks of the Canadian Shield a few thousand people collected, hunted, and fished in a forested landscape of long winters. Another approximately forty thousand people cultivated warmer and more tillable soils in forests on the lowland Paleozoic rocks of southern Ontario. These folk followed a shifting village pattern, moving approximately every fifteen years as declining fertility exceeded the natural capacity for soil renewal. Surrounding uncultivated forests were used by the natives for hunting, fishing, gathering, and trade, and they used fire to drive game and attract deer and other animals to green growth in the spring. Their collecting changed vegetation patterns by encouraging the spread of preferred species such as sugar maple, mast oak, or medicinal plants.

There were similar effects in the forests of the Canadian Shield. Cultivation was virtually absent, but the long-term impacts of gathering and hunting were similar. Fire, the selection of preferred plants, and other modifying mechanisms were at work here just as they were amid the forested tracts of southern Ontario. Seasonal settlements were established at favourable fishing sites, notably in the spring, with the spawning runs of salmon and other anadromous fish. Spring was also a time for social activity after months hunting in small groups in the cold forest.

Overall, then, two broad landscape types were found in Ontario in pre-European times. The first was the mixed agricultural and forested landscape of the lowland south, the product of centuries of cultivation as well as hunting, gathering, fishing, and trade. The second was the forested system of the rugged acidic Shield, the homeland of ancient gathering, hunting, fishing, and trading groups that slowly modified the landscape through long-continued deliberate burning for hunting and exploitive use. Man-made burns were also mixed with lightning fires. These northern forest effects are more subtle and less well understood; the people were much more entwined in the rhythms of the wild system.

Within these two broad landscape types, various indigenous tribes or groups were identified by Caucasian historians and archaeologists. They were seen as differing in language, kin affinities, and territorial and other factors by Caucasians, although perhaps not as much by the native people themselves. A major division lay between the people of Algonquian language affiliation in the north and Iroquoian language affiliation in the south. The southern Iroquoian people, with their more concentrated village settlements and ways of life, were most affected by the European invasion, notably by the widespread epidemics of smallpox, measles, and other diseases to which they had not previously been exposed.

Native cultivation of crops began in southern Ontario about twenty-five hundred years ago (Ellis and Ferris 1990). The clearing and disruptive effects were concentrated where people of broad Iroquoian affiliation found soil and other conditions favourable to agriculture while also pursuing hunting and trade. The favoured areas included Huronia, on the sandy soils between Lake Simcoe and Georgian Bay; part of the Lake Ontario shore, for example, in the Toronto-Hamilton-Grimsby area; the eastern Upper St Lawrence Valley; and numerous sites in the deciduous or Carolinian forests and savannahs of the southwest, such as the valleys of the Grand and the Thames Rivers. Here, in the years around the first coming of the European fur trade, Huron, Petun, Neutral, and other folk eventually came into major conflict with their linguistically affiliated brethren, the Seneca and other Iroquois of New York. The Seneca and their brethren triumphed, and the agricultural peoples of Ontario disintegrated and moved away. Much formerly cultivated land became trapping and hunting ground for the southern Iroquois. And the cleared land afforested or succeeded to savannah, becoming the "primeval wilderness" perceived by early European explorers, traders, and settlers in eighteenth- and nineteenth-century Ontario.

THE FUR TRADE: OPENING ROUTEWAYS FOR COLONIZATION

The launching of the fur trade in the early seventeenth century led to competitive wars among native people for control of commerce with Europeans and also to the destructive effects of Old World diseases. For example, when Jacques Cartier came up the St Lawrence to Isle de Montreal in the 1530s, he found native villages growing corn and other crops around the junction of the Ottawa and Upper St Lawrence Rivers. But when Samuel de Champlain came only seventy years later, little or no trace was found of these people. They may have been displaced by other native groups, but more likely they succumbed to European diseases.

The Origin of Ontario's Georegions

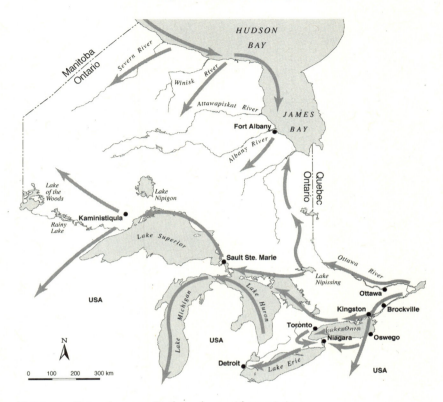

Figure 1.3 Fur trade routes. By the authors, with permission

Routeways

The fur trade moved into what became Ontario along a number of different routes (figure 1.3). Trappers went up the Ottawa Valley, west along the French River and Lake Nipissing, and into Georgian Bay and the Upper Great Lakes within a few years of Champlain's first voyage to Montreal in 1608 (Barry 1995). Priests accompanied these moves and set up missions at Sainte Marie among the Hurons and other sites south of the present Georgian Bay towns of Penetanguishene and Midland. By the late 1600s traders and missionaries had traced the Georgian Bay and Huron coasts and passed into Lake Erie. Others voyaged around Manitoulin Island into Lakes Huron, Michigan, and Superior.

The second major route into the province was west along the St Lawrence River into Lake Ontario. This route developed slowly because of the great rapids and because of resistance by the Iroquois of New York. Henry Hudson, as a Dutch employee, moved up what became the Hudson River in 1608, at the same time that Champlain was on the St Lawrence. By the mid-1600s the Dutch had a major trading post at Albany, south of Lake Champlain. While the Dutch appear not to have moved into Lake Ontario, they did attract native people from this area to Albany. In response the French moved more aggressively south into what is now New York State and inland. Missionaries moved into the Finger Lakes country, and posts were established on the south shore of Lake Ontario, for example, near present-day Oswego.

Traders also moved along the north shore of Lake Ontario. The great traveller La Salle built Fort Frontenac near present-day Kingston in 1673 (Osborne and Swainson 1988). After a few years of profitable trade, he used Fort Frontenac to launch exploration toward the great river of the west, the Mississippi. In 1678 he sent men to build a post at Niagara and the following year at Detroit (Marchand 2005). From the St Lawrence the French also moved along the north shore of Lake Ontario to the mouth of the Humber River and present-day Toronto. Traders then moved north and inland by canoe and portage to the Holland Marsh, Lake Simcoe, and Georgian Bay and the way to the west.

From the Ottawa, Upper St Lawrence, and Humber River routes many streams and trails were followed throughout central and southern Ontario. The legendary Étienne Brûlé travelled from the Mission among the Hurons to the Great Lakes and beyond in the early 1600s. His route may have included travel along the Humber, on to the head of Lake Ontario, at present Hamilton, across to the Grand River, and downstream to Lake Erie. This route was used increasingly after 1700 to avoid the great falls at Niagara.

By the early 1700s Detroit had become the launching pad for another branch route into southern Ontario. Detroit was by then a major entrepôt between the lower and the upper Great Lakes and the site of a growing fur trade and agricultural settlement, which developed on both sides of the Detroit River, as well as up the Thames and along the marshy northwest Lake Erie coast toward Point Pelee.

Opening the North and the West

On their passages up the Ottawa River into the French River and Georgian Bay, the French encountered the Nipissings and other Algonquin speakers, including the Temagami, an inland group, said to be resident on the "North Sea," or Lake Temagami. These people were exchanging meat and skins originating as far north as the James Bay lowlands for corn and cornmeal from Huronia on the south side of Georgian Bay. The Nipissings were middlemen in this ancient trade between the hunting and fishing peoples of the north and the predominantly agricultural peoples of the south. Such indigenous trade networks extended over great distances in North America before European arrival, and, as with the Temagami traffic, these ancient avenues were incorporated into the fur trade (Hodgins and Benidickson, 13–15).

By the 1640s traditional nature exchange systems and the emerging fur trade came under heavy pressure from assaults by the Seneca and the southern Iroquois. They attacked the French in the St Lawrence area and by 1650 destroyed Huronia and other indigenous agricultural areas in southern Ontario. In the 1650s and 1660s the Iroquois followed the ancient river routes into northern Ontario, reaching eastern James Bay, where they continued raiding into the 1670s (Hodgins and Benidickson, 15). In the early 1700s, however, they were beaten by the combined attacks of the French and the Algonquin peoples and driven into their Finger Lakes homeland of central New York. Trade and communication then strengthened between the French and the northern Algonquin peoples. The extent of French exploration, which was enormous, is poorly understood by Ontarians today.

The English entered the scene in the sixteenth century, building posts at the mouths of the Rupert, Moose, and Albany Rivers on Hudson and James Bays and establishing the Hudson's Bay Company. A struggle for control of the trade ensued throughout northern Ontario, as well as further west in the Great Plains, the far Northwest and the Mississippi Basin. In the Seven Years' War (1756–63), the English triumphed and the French Montreal trade was taken over by "pedlars": men from Canada, Scotland, New England, and New York. Their great freight canoes began annual journeys up the Ottawa into Georgian Bay and on to Kaministiguia (Thunderbay), the great entrepôt at the Lakehead, the launching point for trails to the west and north (Hodgins and Benidickson, 23–5). Intense competition once again reared its head. The Hudson's Bay Company moved into northern Ontario and set up posts such as Brunswick House on the upper Missinaibi River from 1781 to 1791. The pedlars retaliated with posts at sites such as Lake Mistinikon. The wave of competition intensified after 1787, when the pedlars united to form the Northwest Company. By about 1813 the Hudson's Bay and Northwest Companies were overextended. In 1821 the two companies amalgamated into a new Hudson's Bay Company.

The intense competition between principal trading organizations such as the Hudson's Bay and the Northwest Companies ruined much of southern and northern Ontario for fur. Sir George Simpson, the governor of the amalgamated Hudson's Bay Company, introduced conservation policies; for example, quotas were set for exhausted districts, and some were set aside from trapping to "recruit" again. But the policy was not successful, because other trappers were moving into all parts of Ontario and competing for fur, leaving few animals to restore populations. In the 1820s and 1830s the American Fur Company of New York had trappers in the Georgian Bay area and northern Ontario.

The fur trade era reached its peak in the decades before 1820. By 1770 the trade followed established routes up the Ottawa into Georgian Bay and the upper Great Lakes. Arms of this northern trade extended via Temiskaming, the Albany, and other river systems to the James Bay shore and also west along the Huron and Superior coasts to the Thunder Bay area, Quetico, and Lake of the Woods. Tentacles reached south into Lake Michigan. Here the Ottawa and northern traffic encountered the second major fur trade routeway from Montreal into Lake Ontario, Lake Erie, Detroit, Lakes Huron and Michigan, and the Mississippi watershed.

Numerous seasonal and temporary posts were located around the northern and southern routeways. Only a few major posts were established at long-lived sites such as Fort Frontenac (Kingston), Niagara, Detroit, and Kaministiguia (Thunder Bay). These major posts had fluctuating populations of a few score to over a hundred people, depending on the season and the trade. The major posts attracted missionaries to serve the Indians and settlers who grew corn and other native crops, as well as introducing wheat, pigs, and other European plants and animals to the fur country. Native people also brought corn, wild meat, and country food to the posts, along with furs. In all cases the major posts were near large water bodies, which facilitated transport and provided large quantities of fresh and salted fish for use by residents and visiting voyageurs, especially in winter.

EFFECTS

Major effects of the fur trade were the decimation of the beaver and other animals of northern and southern Ontario and the introduction to the people of new crops and metals, of religion and economic and social ideas and arrangements. In return the Europeans received new crops and ideas, such as the Iroquois Confederacy system of governance. The traders also introduced new diseases and incited internecine and native-Caucasian wars, which greatly reduced First Nations populations. They also displaced native groups on a wholesale basis, indirectly by force of their advance and directly by setting the stage for treaties that conveyed land to Caucasians.

The effects were not uniform across Ontario. The greatest impacts were felt in southern and central Ontario, where native land use, and the native economy and culture were transformed and to a large extent displaced or disappeared. In the north, while disease and other effects were substantial, native people and their institutions survived in a more coherent form, although the economy, culture, technology and way of life were altered to be amenable to the trade that continues to be pursued by native people into the twenty-first century.

The fragile nature of the fur trade construction and technology, notably its wooden posts and other buildings, left little trace on the land except in archaeological remains and exhumed ruins like those at Fort Frontenac or attempts at reconstruction like those at Fort William. However, numerous fur trade routes were incorporated into contemporary routeways and settlements. Fur trade corridors such as the Upper St Lawrence, the Humber-Simcoe-Georgian Bay route, the French River, and the Albany are today marked by Heritage River designations or by canoe routes and recreational trails. And some of the major posts have become cities and towns, including Fort Frontenac (Kingston), Niagara (Niagara on the Lake), Detroit, the Sault, and Thunder Bay. Yet the evidence of the fur trade is predominately literary, the historic fabric has largely worn away.

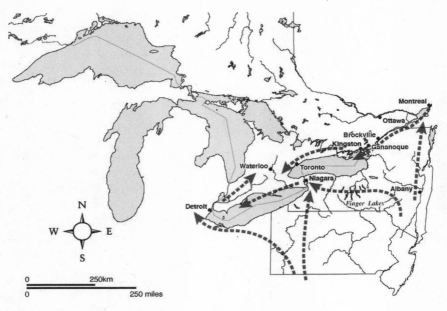

Figure 1.4 Schematic of Loyalist routes to Canada. By the authors, with permission

ERA OF COLONIZATION AND THE EMERGENCE OF SOUTHERN GEOREGIONS: THE LOYALISTS AND THEIR BRETHREN

The hammer truly fell on the fur trade with the American Revolution of 1776–83. Before and after the 1783 Treaty of Paris, which confirmed the establishment of the United States, thousands of migrants from the Thirteen Colonies moved to Canada as Loyalists to the English Crown. One major route led from New England and New York along Lake George, Lake Champlain, and the Richelieu River to Sorel Island and the Montreal area (figure 1.4). The migrants were English, Scottish, Irish, and German as well as Yankee settlers who had opposed the now independent Colonists. Many of these Loyalists moved as military units into block settlements along the Upper St Lawrence River west of the Ottawa (Nelson et al. 2005). They established river settlements such as Prescott, Brockville, and Gananoque. Others went to the Fort Frontenac area at the head of the St Lawrence where the Cataraqui River flows into Lake Ontario.

The easternmost Upper St Lawrence settlers tended overwhelmingly to be Scots, who were dominant in what became Glengarry

County near the Ottawa River. They were soon supplemented by countrymen leaving their homeland because of the land clearances undertaken to make way for sheep and wool to supply the new towns of the Industrial Revolution. Settling west of the Scots on the Upper St Lawrence were the Palatine Germans who had migrated into north central New York in the mid- to late eighteenth century in search of economic opportunity and religious freedom. They were soon joined by more Germans from mainland Europe, who worked to build villages and towns such as Morrisburg. Further west still, around Brockville and Gananoque, the settlers tended to be of English or old Yankee stock from Vermont or Massachusetts. The same was generally true of people settling near Fort Frontenac, or Kingston.

On their arrival these Euro-American migrants encountered people of native, Metis, and French origins. These folk had been traditional users of this land for generations; the First Nations for thousands of years, the French for close to two hundred years. Some of the French – or French Canadians – had secured title to land through grants from the French Colonial regime. For some, these seigneurial rights were respected; for others, like the French coureurs de bois, Metis and native voyageurs, trappers and labourers in the fur trade, traditional rights were more or less swept away. The host of names that they had established for islands, stopping places, and other sites along the St Lawrence or other territories well known to them were also replaced by English names.

A second major Loyalist migration route ran north from Pennsylvania or west along the Mohawk River in New York to the Niagara River and Lakes Ontario and Erie. Many crossed near the Falls and moved along the sandy western Lake Ontario shore to settle at places such as Niagara-on-the-Lake, St Catharines, and Dundas on Hamilton Bay. Many landed on the north Lake Erie shore at embayments and creek mouths where villages such as Port Dover and Port Ryerse grew up. From the Lake Ontario shore some crossed the Niagara Escarpment, spilling into the Grand River Valley near present-day Caledonia and Cayuga. Among the main migrants were Loyalist First Nations – the Mohawks and the Iroquois of New York. These Revolutionary War allies of the British were rewarded with a large land grant extending about six kilometres on either side of the Grand River from Lake Erie well inland, even to its headwaters three hundred kilometres upstream, although the northern boundary is still contentious today (Nelson et al. 2005).

Another major group that crossed mainly at Niagara were the Plain Folk, people with simple living styles and strong religious convictions. These included the Quakers and the Mennonites. The Quakers tended to disperse more than the Mennonites, who moved in groups into the Markham region north of Toronto and into the Grand Valley near present-day Galt, Preston, Kitchener, and Waterloo. Unlike other settlers, who were inclined to hug the shore and avenues of travel, commerce, and communication, the Mennonites settled inland where they could live a plain, devoted life and avoid being drawn into the conflicts and distractions of the day.

A third line of migration crossed at Detroit and moved up the Thames River and along the Lake Huron and Lake Erie shores, following fur trade routes (Hamill 1971). Earlier, the French and their colonial government had supported a major military and fur trade post and agricultural settlement at Detroit. Farmers were encouraged to settle on both the north and the south, later the Canadian side of the Detroit River, between Lake St Clair and Lake Erie. A French engineer prepared a map in 1754 that shows the strip of farm settlement along the Detroit River. By the late 1780s, excluding First Nations' Reserves, 150 farm lots lay along the river front from La Rivière aux Dindes to Lake St Clair (Lajeunesse 1960, ix–lxv). These farms were laid out in typical rang fashion in relatively narrow parallel strips that gave all farmers access to the river for water, transport, fishing, hunting, and trapping. The people lived in log cabins, and orchards of apples, cherries, and pears were to be found on nearly every farm.

Little settlement by either the French or the English took place between the 1760s and the American Revolutionary War in 1783. In 1792 John Graves Simcoe, the new governor of Upper Canada – the colony carved out of old Quebec to provide a home for the Loyalists from the United States – created a series of counties, including Essex, which contained much of the earlier French settlement. Loyalists began to cross the Detroit River in the late 1780s, many of them in military units like those in the Upper St Lawrence region. Delaware native people and Moravian missionaries moved up the Thames following displacement and conflict with American settlers pushing inland from the eastern seaboard. Their settlement near Thamesville was destroyed by American troops in the 1812–14 War between Britain and the United States, and a new site was established south of the Thames, where a reserve remains today. Numerous indigenous people were displaced by land surrenders pushed by the Euro-Americans after 1800.

Loyalist land grants were of two types. The first was the large grant of hundreds to thousands of acres to officers and high officials. Soldiers and less prominent settlers usually were granted 150 to 200 acres, or they were allotted or sold land located in the large blocks given to the elite. This block system is evident in the large grants to Thomas Talbot, who distributed land to migrants from a Lake Erie outpost near Port Stanley in the early nineteenth century.

SETTLEMENT AFTER THE 1812–14 WAR AND THE NAPOLEONIC WARS

The next major wave of settlement came after the 1812–14 War and the Napoleonic Wars in Europe before 1815. The British government had sent thousands of English, Irish, Scottish, and other troops to fight the Revolution in North America. Many of these veterans were disbanded and settled in Upper Canada. They were joined by thousands of other troops of various origins who had fought with the British against the French in Spain, France, and mainland Europe. These troops were dispersed to different parts of the colony, including more remote locations in the bush. Some, for example, went to a block settlement at Perth, north of the Rideau as well as to the original Loyalist settlements along the St Lawrence, some went to the unopened Peterborough area, and others settled in still-forested land on the morainal, or higher, ground along the inland, or upper, watersheds northwest of the Lake Erie shore (Wood 2000; White 1985; Nelson, Veale, Dempster et al. 2004).

The country north of the Thames River and west of the Niagara Escarpment was at this time a place apart. It lay outside the main north-south and east-west flow of central and southern Ontario. The area was settled differently than most of Ontario, through private rather than public means. The main agency was the Canada Company (Lee 2004), which was set up in the early 1820s and received its charter in August 1826. It was established to sell land to settlers as a business and to finance land settlement at a time when the normal colonizing agency, the government, was in poor shape financially owing to compensation claims from the War of 1812 and the ongoing expenses of the growing new colony of Upper Canada. Two tracts of land, much of it recently secured from the native people, were involved in the Company's work. The first was a block around Guelph (Matheson and Anderson 2001). The second was a huge tract of about one million acres located mainly east and south of Goderich. This block encompassed much of what is now Huron County, as well as north Middlesex and other counties. The first active commissioner of the Company was the Scottish novelist John Galt, who induced many of his countrymen to take up Canada Company land.

Galt and Tiger Dunlop, the "Warden of the Forest," set up the original blocks and got settlement under way with planned villages at Guelph and Goderich. These two settlements were joined by a forest-cut route known as the Huron Road. The road, which for many years was little more than a narrow and difficult path through the bush, was an example of the difficulties that plagued the early decades of the Company and led to a succession of commissioners before it became a reasonably efficient and effective business. The Company was created without any real understanding of the preparatory and continuing costs of building large new settlements. Migrants normally secured seed, implements, and other support from government, but the company initially did not provide this. There was also the question of who would pay for what many viewed as public services: schools, medical facilities, bridges, and churches (Lee 2004, 78).

The Huron Tract extended from near London on the Thames north toward Georgian Bay and the Bruce Peninsula. The northern borderlands were not actually part of the Canada Campany Tract, but logging and settlement expanded into this area from the Company holdings. By the early 1850s the tract included the beginnings of most of the places that would become the villages and towns of today. Bigger towns such as Goderich, Stratford, St Mary's, Exeter, Mitchell, Seaforth, and Clinton were all on the map.

THE IRISH MIGRATION

The arrival of tens of thousands of Irish immigrants marked the third great wave of settlement. These people were often victims of the infestation and destruction of the potato crop, a mainstay in the diet of the island. Successive crop failures in the 1840s and news of economic opportunity overseas combined to drive immigration, which was encouraged by Old Country governments. After weeks or months of travel by sea, trail, carriage, and small boat, the immigrants reached Quebec, Montreal, the southeastern triangle of Upper Canada, and St Lawrence towns such as Cornwall, Johnstown, Prescott, Brockville, and Kingston. From there they moved north toward the Rideau and Perth, northwest to Peterborough and west to Toronto and beyond.

Conflict between Britain and the United States in 1812–14 led to American invasion of the Thames, the Niagara region, and the Upper St Lawrence (Hamill 1951). In the aftermath the British government encouraged settlement, communication, and trade away from the Great Lakes and the US border in order to make Upper Canada less vulnerable to attack by Americans. One of the results was the construction of the Rideau Canal in 1822–31 (Legget 1955). The canal extended through a series of locks along the Rideau and Cataraqui Rivers, crossing the rugged, forested Canadian Shield between the lumber site at Bytown (Ottawa) and the head of Lake Ontario at Kingston. The decade-long project, which was financed by the British, required thousands of troops, engineers, and labourers to complete; many of the troops and labourers were Irish. They settled along the canal at places such as Merrickville and Smiths Falls and also moved into the Shield country between Brockville and Kingston. Although this acidic land had been judged as unsuitable for settlement by surveyors at the time of the late-eighteenth-century Loyalist migrations, in the 1830s and 1840s the Irish cleared, logged, and carved this land into fields, pastures, and woodlots that many of their descendants still inhabit today.

Large numbers of the Irish settled in the Peterborough area, as well as in Toronto and Hamilton and at the head of Lake Ontario. They moved inland and westward onto higher forested ground not taken up by early Loyalist, Mennonite, and other settlers. They moved beyond and north of Waterloo, Guelph, and Fergus onto the unsettled banks of the upper Grand River Valley and central Ontario. This was logging country: timber was floated downriver to the lower Grand and the growing nearby Lake Ontario towns of Hamilton and Toronto. Great quantities were used for homes, construction, plank or corduroy roads, harbour development, boat building, and fuel. Raw wood was the common fuel of houses and businesses, but much was heated into charcoal, a more efficient source of energy for the iron smelters and other new industries.

Enormous amounts of wood were cut and reduced in small furnaces to make charcoal. For example, thousands of acres of forest were cut to fuel the 1820s–30s iron smelters at Normandale near

Lake Erie and at Furnaceville northeast of Kingston. By about 1900 only about 10 percent of the forested land remained in southern Ontario. The practice grew of logging and lumbering in the cold season, when it was much easier to pull the logs to streams for rafting to market after the ice breakup. This kind of logging eventually was carried out to varying degrees all over Ontario. The Irish engaged extensively in logging, for example, in the upper Grand River Valley.

Among the most famous loggers in the assault on the Canadian Forest were the French and the Scots. The French were cutting along the Upper St Lawrence and the Ottawa in the 1600s, moving logs in huge rafts to the growing towns of Montreal and Quebec City. They were joined in the early 1800s by Scottish migrants from the United States and Europe. The two groups intermingled and competed in places like Glengarry County, where both were also mainstays in the fur trade. Men such as Alexander Mackenzie and Simon Fraser rose to prominence in the late 1700s going west as explorers and traders along the Saskatchewan, the Mackenzie, and the Fraser Rivers of the north and west (Nelson et al. 2005).

COLONIZATION PATTERNS, 1850

The historical geographer David Wood (2000) has summarized in vivid maps and text the 1780–1850 era of Caucasian agricultural colonization of Ontario. Figures 1.5 and 1.6 are his portrayals of urban nuclei and passable roads in the mid-1840s and of the settlement area or ecumene and urban places in the late 1840s. The first map shows conditions in the mid-1840s. At this time settlement was largely confined to the borderlands of the Upper St Lawrence, Lake Ontario, Lake Erie, the Thames and Grand Rivers, and central Lake Huron. The King's Highway (number 2) ran west from the Quebec border, along Lake Ontario to the Grand Valley, the Thames, and on toward the US border. Most of the large colonial villages and towns were located along this route: Cornwall, Brockville, Kingston, Cobourg, Toronto, Hamilton, Brantford, London, Chatham, and Windsor. A few roads led north and south to emerging areas of settlement such as Ottawa, Peterborough, Penetanguishene, Niagara, Owen Sound, and Goderich. From Hamilton the predecessor to today's Highway 6 ran southwest to Port Dover and Long Point.

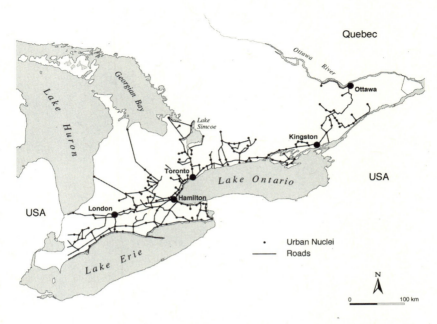

Figure 1.5 Urban nuclei and passable roads. From Wood (2000), with permission

This road intersected the route west, now Highway 3, to the Talbot settlement and Port Stanley, St Thomas, Blenheim, and Windsor.

These settlement patterns are confirmed and extended by Wood's map "Ecumene and Urban Places" for the late 1840s. Here we see the rapid westward extension of clearing and settlement up the Ottawa Valley and north of the St Lawrence, as well as northwest from the King's Highway toward Georgian Bay, the Huron Coast, and the Bruce Peninsula. In the years before 1850 the bush or wilderness was pushed back extensively. A wildland, agricultural, and urban system was emerging with a varied and tiered hierarchy from woodland to field, village, town, and city. Some of Ontario's georegions were also emerging, for example, Kingston and the upper St Lawrence, Peterborough, Toronto, the southwest, or Carolinian, and the Huron.

THE CANALS

The "Ecumene and Urban Places" map records the development of an important new form of transport, the canals. Since they provided

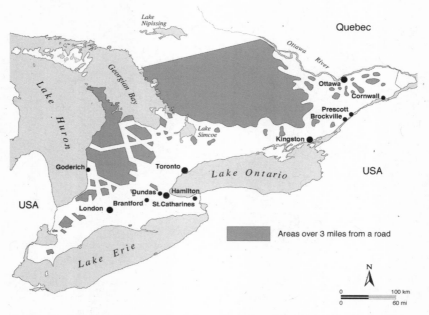

Figure 1.6 Ecumene and urban places, late 1840s. From Wood (2000), with permission

for greater and faster movement of goods and people than the poorly developed roads, they became magnets for new settlement. The 1820s, 1830s, and 1840s constituted the canal era in North America and other parts of the world. The state of New York led the way in its struggle with Montreal for control of wheat and other traffic from the Great Lakes and the west, to the east coast and Europe. The state supported the construction of the Erie Canal along the Mohawk and Hudson River corridor between Buffalo, Albany, and New York City. Ontario, then Upper Canada, answered in 1824 with the construction of the Welland Canal between Lakes Erie and Ontario to make Toronto and Montreal more competitive for the western traffic. The Rideau Canal was also built at this time to carry Montreal-Great Lakes traffic around the Upper St Lawrence rapids and away from the threat of American attack (Leggett 1955). The Grand River Canal was built in the 1830s between Lake Erie and Brantford, with a feeder line to the Welland Canal, to give inland southwestern Ontario better access to the lake trade. Of these historic canals only the Welland remains commercially active today.

The Rideau is now a heritage and recreational waterway. The Grand River Canal was abandoned long ago, although decaying traces can still be seen along the river today near Dunnville, Cayuga, and Brantford. The canals were to be the freight ways of the era of agricultural colonization. Their brief heyday ended with competition from the new railways of the 1850s.

SETTLEMENT MODELS AND THE SHORT DISTANCE SOCIETY

Before going on to the railways and the industrialization of Ontario, we need an image of what the landscapes were like toward the end of the colonization era in the mid-nineteenth century. The work of the geographer and planner Tony Fuller (1997) can help us build this picture. Fuller has suggested that four different societies and landscape patterns succeeded one another in Ontario since initial European invasion and the beginning of the fur trade. These are the Short Distance, the Industrial, the Open, and the Arena Societies. Each of these societies and associated land uses and landscapes developed at different times following technical or other changes. We are mainly interested here in Fuller's Short Distance and Industrial Societies and landscapes. The Short Distance landscape was typical of central and southern Ontario about 1850. The evolution of the Industrial landscape is discussed in chapter 2.

The Short Distance Society was a product of the era of European colonization. The landscape was predominantly agricultural with growing farms and fields amid dwindling forests and wildlands. The focus was on producing a wide range of crops and some livestock for subsistence, barter, and whatever cash sales were possible. Because transport was normally by foot or horse, goods were moved slowly and not in great bulk. Markets were overwhelmingly local in small, evenly spaced villages that served farmers and settlers within a radius of about twenty miles. Life was focussed on neighbours and what could be reached in a day.

A vivid description of this pattern has been given by Ruth Mckenzie (1967) in her account of land settlement along the Upper St Lawrence near Brockville and Gananoque in what became Leeds and Grenville Counties. As she describes it, "Every village and hamlet ... was a small industrial centre. Most of the industries were small scale. Each community had at least one blacksmith shop, tannery and saddlery, shoemaker, tailor, cabinet maker and mason, and cooper or barrel maker. A chandler made candles to light houses and buildings in the evenings ... Often there was a brick yard and a cheese factory. Through such industries the community largely met its own needs."

THE SHORT DISTANCE SOCIETY AND THE EMERGING GEOREGIONS

The extent and intensity of the Short Distance Society tended to differ among the emerging georegions of the province. It was relatively well developed in the incipient georegions of Kingston and the Upper St Lawrence, the Peterborough region, and southwestern Ontario, or the Carolinian zone, beginning at Hamilton and extending through the Grand, Thames, and other country to Chatham and the US border.

The Short Distance Society and landscape were less apparent elsewhere. In the 1840s Bytown and the Ottawa Valley were still heavily forested with an economy focussed on seasonal logging and transport up and down the river or inland along the Rideau Canal, the wood often intended for long-distance markets in Montreal, Europe, and the United States. In the Georgian Bay area villages were scattered along the coast and engaged in the fur trade, lumbering, and fishing; the products were shipped by steamer to US ports on the Great Lakes. In the adjacent eastern lake country of Muskoka, the forest still reigned supreme, with ephemeral or incipient logging, mineral exploration, and other activities scattered amid vast wildlands. In western and northwestern Ontario toward the Huron coast and the Bruce Peninsula, the Short Distance Society and its agricultural landscape was extensive in the Huron County area; logging moved into the bush to the north in what became Grey and Bruce Counties. Further north across Georgian Bay the great forest remained the home of the First Nations and the fur trade. Towns such as Sault St Marie and Fort William and Port Arthur (Lakehead) were emerging because of their long role in the fur trade. Many small fishing settlements were scattered along the Upper Great Lakes coasts, and exploration for minerals was under way. But the Northeast and Northwest Georegions were not yet on the map.

All these emerging georegions consisted of people of diverse origins who nevertheless settled in varying concentrations in accordance with the settlement process. Native people were dispersed to the north and west or scattered in diminishing reserves and remaining land holdings in the south. The Scots and Irish were most evident in the St Lawrence Valley, the Peterborough area, and a band north of Lake Erie. Loyalists of English, German, and other origins were diffused through the south, engulfing the early French.

REFERENCES

Barry, James. 1995. *Georgian Bay: The Sixth Great Lake*. Erin, ON: Boston Mills Press.

Dear, M.J., J. Drake, and L.G. Reeds, eds. 1987. *Steel City: Hamilton and Region*. Toronto: University of Toronto Press.

Ellis, C.J., and Neal Ferris. 1990. *The Archaeology of Southern Ontario to AD 1650*. Toronto: Society Inc.

Fuller, Tony. 1997. "Changing Agricultural, Economic, and Social Patterns in the Ontario Countryside." *Environments* 24(3):5–11.

Guillet, Edwin C. 1969 (1933). *Pioneer Travel in Upper Canada*. Toronto, ON: University of Toronto Press.

Hamill, Fred Coyne. 1951. *The Valley of the Lower Thames*. Toronto, ON: University of Toronto Press.

Heidenreich, Conrad. 1971. *Huronia*. Toronto, ON: McClelland and Stewart.

Hodgins, Bruce, and Jamie Benidickson. 1989. *The Temagami Experience*. Toronto, ON: University of Toronto Press.

Lajeunesse, Ernest J., ed. 1960. *The Windsor Region: Canada's Southernmost Frontier, a Collection of Documents*. The Champlain Society for the Government of Ontario. Toronto, ON: University of Toronto Press.

Lee, Robert C. 2004. *The Canada Company and the Huron Tract 1826–1853*. Toronto, ON: Natural Heritage Books.

Leggett, Robert. 1955. *Rideau Waterway*. Toronto, ON: University of Toronto Press.

MacKay, Donald. 1998. *The Lumberjacks*. Toronto, ON: Heritage/Natural History.

Marchand, Philip. 2005. *Ghost Empire: How the French Almost Conquered North America*. Toronto, ON: McClelland and Stewart.

Matheson, Dawn, and Rosemary Anderson, eds. 2001. *Guelph: Perspectives on a Century of Change, 1900–2000*. Guelph, ON: Guelph Historical Society, Ampersand Printing.

McKenzie, Ruth. 1967. *Leeds and Grenville: The First Two Hundred Years*. Toronto, ON: McClelland and Stewart.

Nelson, J.G., with Susan Fournier, Christopher Lemieux, Eric Tucs, Natalie Korobaylo, Costas Farassoglou, and Lucy Sportza. 2005. *The Great River: A Heritage Landscape Guide to the Upper Saint Lawrence Corridor.* An Environments Publication Waterloo, ON: University of Waterloo.

Nelson, J.G., J. Porter, C. Lemieux, C. Farassoglou, S. Gardiner, C. Guthrie, and K. Van Osch. 2003. *Ridgetown and Rondeau Area: A Heritage Landscape Guide.* Heritage Resources Centre, Waterloo, ON: University of Waterloo.

Nelson, J.G., Barb Veale, Beth Dempster, et al. 2004. *Towards a Grand Sense of Place.* An Environments Publication, Waterloo, ON: University of Waterloo.

Osborne, Brian S., and Donald Swainson. 1988. *Kingston: Building on the Past.* Westport: Butternut Press.

Troughton, Michael, and Cathy Quinlan. 2009. *A Heritage Landscape Guide to the Thames River Watershed.* Department of Geography. London, ON: University of Western Ontario.

White, Randall. 1985. *Ontario, 1610–1988.* Toronto, ON: Dundurn Press.

Wood, David, J. 2000. *Making Ontario: Agricultural Colonization and Landscape Re-Creation before the Railway.* Montreal and Kingston: McGill-Queen's University Press.

Wright, James V.C. 1972. *The Prehistory of Ontario.* Ottawa: Natural Museums of Canada.

2

The Industrial Era and the Further Development of Ontario's Georegions: The Railroads, 1850s–1900s

GORDON NELSON AND MICHAEL TROUGHTON

SUMMARY

A picture of the Short Distance Society and the Industrial Society was presented in chapter 1. The Industrial Society took off with the rapid construction of railroads beginning in the 1850s. This is not to say that some of the villages and towns had not undertaken small-scale manufacturing as part of the Short Distance Society. Because some of the settlements of the colonization era were positioned so that they had advantages over the others, industry became more concentrated and advanced in them. For example, upstream from Kingston, Gananoque had a good harbour on the St Lawrence and was located at the mouth of the Gananoque River with access to inland forests and good sources of water power on the lower river. It also had some early innovative entrepreneurs and by the 1840s was known as "the Birmingham" of Ontario, manufacturing much-needed bolts, screws, nails, shovels, and wheel heads. Peterborough occupied a key position in east central Ontario astride major streams and lakes with access to power, transport, lumber, and productive land. The village became a town with grain-milling operations, furniture industries, and waterway transport facilities for its region. Kingston was located at a transport node, had access to water power and the Great Lakes, and developed grain, shipbuilding, and transshipment facilities, as well as serving, from a very early date, as an administrative and military centre. And we have not cited other places such as Toronto, Hamilton, London, and Collingwood.

However, the building of railways quickly favoured the villages, towns, and agricultural areas along their routes and disadvantaged those not so well positioned. In addition to their capacity to ship goods in greater bulk more quickly, the railroads were easier to schedule regularly, since mud, snow, and ice did not prevent them from operating all year, as was the case with roads and canals. The advantages of a rail location led to great competition for lines in the mid- to late-nineteenth century. Some of the lines were regional, national, or international in scale; others were local links between neighbouring towns.

THE RAILWAYS

Only the major lines, which opened up new or large markets and had a fundamental impact on land use, the economy, society, and the natural environment, can be considered here. Four such lines brought trains and steam power to central and western Ontario in the 1850s. By 1855 the Great Western Railway linked Windsor and Detroit with Hamilton, Toronto, and Niagara. The Northern Railway, which joined Toronto with Collingwood and Georgian Bay in the following year, carried loads of Western and Northern grain and lumber to Toronto for processing and shipping to other parts of Canada, the United States, and Europe. By 1859 the Grand Trunk Railway ran between Montreal, Toronto, and Sarnia. And by 1858 the Buffalo and Lake Huron Railway linked the salt and fish port of Goderich on Lake Huron with Fort Erie and access to US markets (figure 2.1). The links that all these railways provided to increasing agriculture, fishing, mining, and other production in the west were strengthened by completion of the Sault Ste Marie Canal between Lakes Huron and Superior in 1855, enabling transport by steam boats between Toronto and other southern centres, the Lakehead, and the Far West.

As a rail hub for these major lines Toronto grew rapidly (Careless 1984; Lemon 1985; West 1979). By the 1850s its population was in excess of thirty thousand. Other favoured urban areas also grew well beyond the Short Distance model. Hamilton, a junction for several of the major new lines, attained a population of greater than fourteen thousand by this time. Kingston was the largest urban area in Upper Canada until the 1840s. Less favoured by rail connections

The Industrial Era

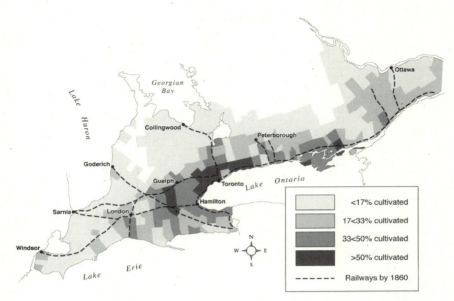

Figure 2.1 Railways and the extent of transformation, ca 1852. From Wood (2000), with permission

and access to northern and western resources than Toronto or Hamilton, Kingston then housed 11,700 people, placing it third after Toronto and Hamilton. London's inland location made access to lake trade difficult, but it was on the Thames waterway, in the agricultural heartland. With growing access to the new rail lines, in the 1850s it reached a population in excess of seven thousand. Bytown, later Ottawa, at the head of the Rideau Canal and a node for lumbering and the Ottawa River trade, had a population close to eight thousand people (White 1985, 108–9).

EXPANSION UP THE OTTAWA VALLEY AND INTO NORTHERN ONTARIO

Bytown was in a position to benefit from the construction of a series of major new rail lines (figure 2.2). The Ottawa Valley was the site of small-scale agricultural settlement in the eighteenth century by the French and, after the Revolutionary War, by Loyalists and migrants from Europe. A colony of New Englanders grew up at the

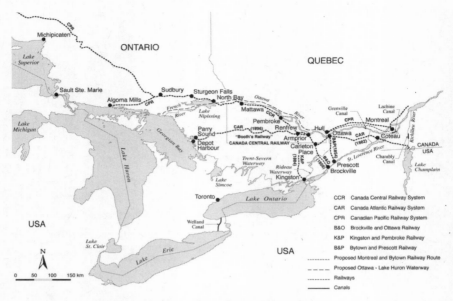

Figure 2.2 Canals and railways. By the authors, with permission

mouth of the Gatineau River opposite Bytown in 1800. Potatoes, wheat, and other crops were planted, and a tannery and grist and sawmills were soon built. Sawmills and logging became a major source of export revenue.

Rafts of logs were floated down the Ottawa to Montreal and became part of a booming trade in lumber with Britain during the Napoleonic Wars of the early 1800s. The long-term depletion of England's forests and the French blockade of Scandinavian forests forced Britain to rely more and more heavily on Canadian lumber. By 1811 five hundred ships had sailed from Quebec City carrying seventy-five thousand loads of pine and white oak and twenty-three thousand masts for the British fleet (Mackay 1998, 17). By the 1820s British North America was supplying 75 percent of England's lumber, much of it from the Ottawa Valley and the Upper St Lawrence region. And the reach of the logging industry continued to grow into eastern Ontario and the Algonquin Park and Peterborough areas. The products of the industry became more diverse, progressing from felled trees, through square-cut timber to planks, fine-cut timber, and eventually, later in the nineteenth century, pulp and paper.

THE PETERBOROUGH GEOREGION

The timber trade reached inland to the Peterborough area in the 1830s. The streams and drainage of the Trent and other rivers and of the Kawartha Lakes were dammed and modified for shipping and logging operations. In the 1820s–40s this led to the building of the Trent-Severn system of canals and locks linking Lake Ontario and Georgian Bay through Peterborough. The town and surrounding villages such as Lakefield on the Otanabee, Bobcaygeon on Sturgeon Lake, Buckhorn on Buckhorn Lake, and settlements on Rice Lake were tied together in an emerging regional lumbering economy. This lumbering era peaked in the 1850s and 1860s (Adams and Taylor, 106).

A rail network grew up around Peterborough in the ensuing years, and the town developed into a railway communication centre. The passage from an agricultural settlement and a lumbering hub to a manufacturing economy resulted in forty-six industrial enterprises, including a woollen mill, agricultural-implements production, factories, and sawmills. By the 1890s, as a result of development efforts by council and citizen groups, Peterborough became the site of two major industries serving national and international markets: the Edison Electric Light Company, later General Electric, and the American Cereal Company, later Quaker Oats. Both these firms were attracted by the hydropower of the region, and Quaker Oats also drew on the diverse local production of cereal crops. However, industrial growth struggled thereafter, since improving rail and road connections to the big city brought development in Peterborough and the surrounding region increasingly into competition with Toronto. One positive change for the Peterborough region was a rising involvement in recreation and tourism, much of it originating from the Toronto area.

By the 1840s the loggers had moved up the Ottawa and French Rivers and were at work on red pine in the Lake Temiskaming area. Logging and eventually pulp and paper production stimulated growth at Bytown and lower river settlements such as Hawkesbury. English-speaking places like Bytown, Hawkesbury, and Hull began to attract Quebec workers and became more and more French-speaking. The influx of the French Canadians and their language influenced the selection of Bytown as capital of a united Upper and

Lower Canada in 1841. Aside from its inland position away from the perceived threat at the US border, Bytown was in an intermediate position between the frequently contending French and English speakers of Upper and Lower Canada. The valley and surrounding region were seen as a buffer between the French and the English, a place where they could intermingle economically and socially, working out their differences and common interests (Greening 1964; Taylor 1986). And after the Confederation of Canada in 1867, Bytown became Ottawa, the capital of the new Dominion.

THE OTTAWA GEOREGION

The growth of the Ottawa Valley and northern Ontario's lumber and pulp and paper industry was greatly aided by the construction of the Canada Central and related rail lines beginning in the 1860s (figure 2.2). Routes were built from Montreal to Ottawa and Pembroke and extended south to Brockville and the Upper St Lawrence by 1870. In 1880 the Canadian Pacific Transcontinental Railway Company (CPR) was formed and gained control of the Canada Central. In 1882 the old Canada Central line was extended to Lake Nipissing, where it connected with the new CPR line running westward through northern Ontario to Port Arthur at the head of Lake Superior and on to Winnipeg. By 1885 the CPR had completed its line across the prairies and the Rockies to the West Coast. And the expansion of the network continued when the lumber baron J.R. Booth brought together a line from Ottawa south to the US border and Lake Champlain (figure 2.2). Another line ran from Arnprior and Renfrew north of Ottawa into the forests of the upper Ottawa watershed, and another led from Renfrew through what was to become Algonquin Park to the port of Parry Sound on Georgian Bay. A fleet of steamboats then travelled from the Bay to lake ports such as Chicago, Duluth, and Port Arthur. Timber, pulp, and other wood products could now be shipped in large quantities in virtually all directions to and from the Ottawa Valley, the Great Lakes country, New England, New York, the Midwest and Montreal and Europe.

In 1867 Ottawa still relied on the lumber industry while embracing its new role as capital of the new Dominion of Canada. Ottawa's strategy was to become an extra-regional and a national transport hub through rail lines to Brockville and the Great Lakes, Montreal, and Quebec. But this role was soon usurped by Montreal, which became the node of the great Canadian Pacific transcontinental railroad route. The lines built by J.R. Booth to New York, Parry Sound, and the North slowed the loss of the extra regional and national transport role for a time, but Booth eventually sold out to the Grand Trunk, which was centred in Toronto. This city eclipsed Ottawa in the growth of banking and financial services and was well on its way to a prime role in development of southern and, increasingly, of northern Ontario by the 1880s.

Ottawa became a service and marketing hub for the surrounding Ottawa Georegion and the site of the capital of the new Dominion (Brandt 1946). By the end of the nineteenth century, Sparks Street, in Upper Town, was recognized as the Broadway of Ottawa, with its street lights, Victorian "skyscrapers" and tram cars. The "Civic Gothic" and Second Empire features of the parliament buildings stimulated similar architecture in churches as well as in the nearby Market Annex.

Plans were put forward in the early 1900s to meet the everyday needs of the town and the scenic and symbolic necessities of the capital of Canada. By 1910 a proposal had been advanced for a comprehensive plan to create "a city worthy of the capital." It involved straightening out the railway and industrial uses of the city and the Parliament Hill area and advocated a great Circle of Parks, now the Greenbelt, and other attributes commensurate with Wilfred Laurier's 1893 image of a "Washington of the North" (Taylor 1986, 146–8). This national image was overlaid on a working city landscape and emerging georegion reaching north beyond Renfrew, east into Hull, the Gatineau, and Quebec, south to Hawkesbury and the Rideau, and west beyond Perth toward the Muskoka Lakes and Georgian Bay.

LOGGING, MINING, GREAT LAKES TRADE, AND THE OPENING OF NORTHERN ONTARIO

Toward the end of the nineteenth century the CPR built a branch line to Temiskaming in the boreal forest of northeastern Ontario. Logging grew apace, and by the turn of the twentieth century thousands of men were logging in the bush in the winter season. The clay soils of this area attracted attention and comprehensive surveys were made of agricultural, forestry, mining, and water power prospects by the Ontario government. Interest in agricultural colonization of the rugged area between the Ottawa Valley, Georgian Bay, and the Temiskaming district led to the building of a rail line from North Bay to Lake Temiskaming and plans to extend it to James Bay, as well as schemes for roads to open the land for farming.

French Canadians were interested in creating a French-speaking and Roman Catholic society in this region, in part to balance the migration and loss of their brethren to growing industries in New England and New York. Townships were surveyed on the eastern side of Lake Temiskaming and settlers were brought in who established farms that marketed vegetables, milk, hay, and other products to lumber camps. In spite of the advantages of the new North Bay line, the agricultural settlements over time did not prove to be as successful as was initially hoped, because of the shifting markets for lumber and other products, the northern location, and the cold climate of the area.

Mining was another story. Prospectors and companies began to work the shores of Georgian Bay, Lake Huron, and Lake Superior in the 1840s. Although some mining of copper for tools had been undertaken by First Nations in the North "since time immemorial," the two earliest Canadian mining operations were for copper in 1846 at Bruce Mines about 30 miles east of the Sault and about the same time at Silver Islands near the head of Lake Superior. Major mining efforts did not begin, however, until the construction of the railways later in the nineteenth century. Silver, copper, and other minerals were discovered along the route of the North Bay–Temiskaming line and in the vicinity of Temiskaming Lake itself. The line met the lake and went north to Cobalt, leading to successful mines and growing towns. Other mining operations began in the early 1880s with the construction of the CPR. Copper was discovered at Sudbury in 1883, but nickel soon became the most valued mineral because of its importance as a hardening alloy for steel. By the early 1890s Sudbury was the main source of nickel globally.

THE NORTHEASTERN AND THE NORTHWESTERN GEOREGIONS

With the support of the Ontario government mineral deposits were discovered and new mining operations grew rapidly at numerous places in northeastern Ontario. These included large operations at Timmins and Kirkland Lake for gold, silver, copper, and other minerals (Greening 1964). In northwestern Ontario gold strikes came into production in the Lake of the Woods, Kenora, and Rainy River districts about 1890 (Bray and Epp, n.d.). In 1899 iron ore was found near Michipicoten. The Algoma rail line was built to take the ore to the Sault, where finished steel was produced by the Algoma Steel Corporation beginning in 1902, and the expansion of the mining industry continued into the twentieth century.

A big part of the story of development in the North has been instability, or boom and bust. Deposits were mined for several years, attracting miners, housing, and other development. Then the ores declined, or markets lagged, the mines closed, and the enterprise moved on. However, because of their locations and other advantages some places emerged as long-lasting settlements. Often they were the focus for lumbering or mining or a combination of these with manufacturing, power generation, and transport (Bray and Epp n.d.).

Examples of long-lived communities include Sault Sainte Marie, Sudbury, Thunder Bay, and North Bay. But smaller places have persisted as well, for instance, Manitowaning, Gore Bay, and Little Current on Manitoulin Island. In the early years of the nineteenth century, as settlement progressed in southern and central Ontario, questions arose about the future of First Nations people who were being displaced by Caucasian invasions. For a time serious consideration was given to setting aside Manitoulin Island as an Indian Reserve where native people could continue their lifestyles. This idea gave way to its impracticality and pressure for island land and resources from Caucasian settlers. Some lumbering and pulp and paper production developed on the island, along with recreation and tourism, especially for visitors from Michigan and elsewhere. Manitowaning, Gore Bay, and Little Current became nodes for much of this activity and main settlements for First Nations and newcomers.

Sault Sainte Marie began to grow after the building of its early-nineteenth-century canal. It received strong impetus from the iron discoveries and development of the iron and steel industry led by the remarkable American entrepreneur Hector Clerque. An array of manufacturing enterprises developed along with iron and steel manufacturing. Many European immigrants came to work in the steel and manufacturing industries, including Italians, who constitute a strong element in the population today.

Thunder Bay grew out of the two Lake Superior communities of Port Arthur and Fort William, the site of the old fur trade entrepôt. These settlements became major transshipment points for Western grain after the CPR opened the West in the 1880s. Large grain elevators were built, and the economy of the twin cities expanded with the aid of a huge lumbering mill, a ship repair and construction facility, railway operations, and flour milling. North Bay grew early on as a transportation and communications node and expanded as lumbering, mining, and other development spread farther north. Lumbering, pulp and paper, and other processing increased along with its role as a service centre for an array of northeastern towns and also for hunting, fishing, and tourism, which started well before the turn of the twentieth century.

With the expansion of the CPR in the early 1880s many company towns like Sudbury developed across the country (Wallace and Thomson 1993, 14–15), but as the line moved away to the west, Sudbury declined. Prospecting finally resulted in the creation of the Canadian Copper Company of Cleveland, Ohio, in 1886. The ore boom was on. Sudbury became a "French town" with a new Roman Catholic Church, schools, banking, merchandising, and other facilities, and a population of 2,400 people by 1891.

In the 1890s the value of nickel as a hardening alloy for armour plate and steel was recognized internationally, and Sudbury embarked on its voyage as a global nickel-producing district. The town also became the main business centre for a huge lumbering district stretching from fifty miles north of the settlement to the Georgian Bay shore and east-west along the CPR line for two hundred miles (Wallace and Thomson, 51). Over 5,000 lumbermen worked in logging camps in this broad region in 1900. By 1911 Sudbury's population was about 4,200, with nearby Coppercliff at 3,000, and the Sudbury basin as a whole at 12,500 (Wallace and Thomson, 59). People of French and English ancestry dominated at about 36 percent and 53 percent, respectively, of the total, with Germans, Italians, and Jewish at about 2 percent each.

Sudbury's role as a regional centre was boosted when the Ontario government made the area a judicial district. Lumbering expanded into an approximately one-hundred-mile timber limit near Killarney on the Georgian Bay north shore. Farming grew with the clearing of the forests, the rising population, and the development of fertile land, notably in the nearby Blezard Valley, and mining expanded greatly in the years leading to World War I. Road and rail corridors were improved locally and also to Toronto and farther north. By about 1910 Sudbury had consolidated its position as the regional centre of the nickel belt and was assuming this role in the greater Northeastern Ontario georegion (Wallace and Thomson 1993, 79).

Over time, with the unpredictable give and take of lumbering, mining, and other development, Northern Ontario has tended to organize into two broad regions: the Northwest and the Northeast. The Northwest has supported mining, lumbering, and pulp and paper and has been important for steel and other manufacturing and for transcontinental rail and lake trade. The focus has been on the twin cities of Port Arthur and Fort William, now amalgamated into Thunder Bay. French speakers form a smaller part of the population, which is, however, ethnically varied. This region has attracted Finns, Ukrainians, the English, and other nationalities to its lumbering, pulp-and-paper, mining, and, more recently, recreational economy (Bray and Epp n.d.).

The Northeast has had greater success with mining, although lumbering and pulp and paper have been important. It is a fuzzy region at its boundaries, extending for some observers as far south as North Bay and also encompassing the lowlands along James and Hudson Bay. Some also see Manitoulin Island as part of the Northeast, but it can also be seen as part of the Great Lakes. Foci for the Northeast are more numerous: Sudbury and also more northerly mining centres such as Kapuskasing and Timmins. Sault Sainte Marie falls in the boundary zone between the Northwest and the Northeast, and given its position at the entrance to Lake Superior, it can be seen as the threshold of the Northwest. The characteristics of Northwestern and Northeastern Ontario were by the turn of the twentieth century different enough environmentally, economically, socially, and culturally to transcend a gross image of one Northern Ontario.

GEORGIAN BAY AND THE MUSKOKA LAKES GEOREGION

The railways also opened the Georgian Bay and adjoining Muskoka Lakes region to more intensive exploitation and industrialization (Barry 1995; White 1985; Campbell 2005) (figure 2.1). A line was built from Toronto to Collingwood and the Bay in the 1850s and another along its east shore to North Bay and Temiskaming at about the same time. In the 1880s J.R. Booth built his line from the Ottawa Valley to Parry Sound, where it tied into the emerging steamer transport industry of Georgian Bay and the Great Lakes (figure 2.2). Georgian Bay had been part of the early exploration and missionary efforts of Champlain, Brûlé, and other French traders from the early 1600s, and it had provided a routeway for the fur trade throughout the seventeenth, eighteenth, and nineteenth centuries. An inland road to Georgian Bay from Toronto and the south was built by 1814 as part of the British response to the 1812–14 conflict with the United States. The road led to Penetanguishene, where the British planned a ship-building and naval-operations port. At the end of the war this plan died, but Penetanguishene remained as the only settlement on Georgian Bay.

The tract of land around Penetanguishene was obtained from the Ojibwa in the late 1700s, and then in 1818 a treaty was signed for a larger parcel between Penetanguishene and Lake Simcoe. Another big block of 1,592,000 acres was secured from the native people afterwards. This heavily wooded tract along the southern and southwest shores of Georgian Bay was added to the treaties for new lands in the Bruce Peninsula and the future Grey County in 1836 and 1857. The gift and annual payment ceremonies for these lands were moved to Manitoulin Island with the intent of getting all the native people to concentrate there. Officials were moved to the island, and a settlement was built that became the village of Manitowaning. Reserves were created for indigenous people on the Bruce and later at places like Christian Islands in the waters of southeastern Georgian Bay. Small fishing and trading settlements developed, often ephemerally, in numerous places around the shores of Georgian Bay, as well as Lake Huron. Fish and maple sugar were shipped by schooner to the growing US cities of Detroit and Chicago (Barry 1995, 51–4).

By the mid-nineteenth century numerous small clearings had been opened in the forest between Collingwood, in the southeast part of the Bay, and Sydenham, the village to the west at the base of the Bruce Peninsula, that later became the major fishing and shipping port of Owen Sound. In these clearings a few hundred people survived on fishing and hunting and, with time, a little farming. The wild lands inland and to the southwest, between the Bay and the advancing fringe of agricultural settlement and logging from southern Ontario, were known as the Queen's Bush. They were under settlement from Toronto, the upper Grand, the Thames, and the Canada Company lands in the Huron County country. These lands were settled by folk of many origins: Scots, Irish, English, and Germans, as well as Blacks, many of whom were slaves who had fled from the

United States and the confinement and conflicts that resulted in the American Civil War of 1860–65.

The history of Black folk in Ontario is as old as the European invasions of North America in the 1600s (Prince 2004). Blacks were among the slaves then held by the French, the English and other Europeans, and some free Blacks were also part of their societies. Slaves were to be found in the early settlements of Montreal and Quebec and much more widely in the Thirteen Colonies and the South. In the 1700s Black people from various parts of Africa numbered in the tens of thousands on Southern plantations and in domestic, farm, and other capacities in the Northern colonies. Many of them tried over the years to escape and become free, and their efforts increased after the American Revolutionary War ended in 1783. More and more Blacks began to try to reach the northern British Colonies in succeeding years and especially after slavery was declared illegal throughout the British Empire in 1833. The loss of Black labour prompted Southern plantation and other owners to put increasing pressure on the US federal government to expedite the recapture of their "lost property." The result was the new Fugitive Slaves Act of 1850, which required all US residents to help in recapturing escaped slaves and returning them to their owners.

In the course of this growing pressure to confine Blacks to their owner's lands and wishes, thousands escaped north after the 1830s with the growing assistance of the famous Underground Railroad. Black escapees and free men, with abolitionists, church folk, and others, worked to move escapees from safe houses or stations along various routes to places such as Rochester and Buffalo, New York, Sandusky and Cleveland, Ohio, and Niagara, where they passed by schooner, steamer, raft, or swimming to the Niagara Peninsula, Hamilton, Toronto, and other places mainly in southwestern Ontario (Brown-Kubisch 2004). Some more northerly settlements such as Oro on Lake Simcoe and Owen Sound also were host to significant numbers of Black escapees. In the 1840s, 1850s, and 1860s colonization schemes established concentrated Black settlements at places such as Sandwich, near Windsor; Wilberforce, near Lucan and the town of London; and Buxton, near Chatham. North Buxton is now a National Historic site. Although many Blacks returned to the United States after the Civil War ended in 1865, many stayed to establish a lasting presence in Ontario, notably in Carolinian

Canada, although signs of their history remain to the north in places such as Owen Sound on Georgian Bay.

In 1856 the Bruce Peninsula was officially opened for settlement, but once the forest had been cut, the soils were found to be unproductive, a thin mantle over dolomite bedrock largely swept clean by ancient glaciers. Settlement began on the east and north shores of Georgian Bay in the 1860s, notably at a town site purchased by the entrepreneurial Beatty family of Thorold. This site became Parry Sound, a fishing, lumbering and trading centre. Another lumbering centre was established at Byng Inlet on the Magnetawan River north of Parry Sound. A road was built from Parry Sound inland to the Muskoka Lakes and the new village of Bracebridge, which developed overland connections to the south. The inland Muskoka Lakes and Shield country was subsequently settled from two principal directions: the Bay and the growing settlements around Toronto, Georgian Bay, and Muskoka.

THE TORONTO GEOREGION

The development of roads and railroads and steamship lines out of Toronto made it a marketing node for parts of northern and southern Ontario. Most of the grist mills, foundries, tanneries, and distilleries were relatively small in the 1840s, and the population was concentrated in a narrow band between Front Street and the harbour, the Don River, and Garrison Creek, just west of present-day Bay Street in downtown Toronto (see Sportza, chapter 4 of this volume; Brown 1997; Careless 1984). With the railroads and other transportation improvements, as well as technical advances in manufacturing, large enterprises began to grow.

In the 1850s and 1860s the shore of Toronto Bay was filled to make room for the railroads, grain mills for Western wheat, and distilleries such as Gooderham's at the mouth of the Don (Myers et al. 2001). The central city became a major wholesale and warehouse district and the banking and financial centre for Ontario. Between 1851 and 1881, as firms began to move from Brantford, Hamilton, and other towns to take advantage of the transportation, labour force, and financial services, Toronto's population tripled.

Between 1887 and 1912 the city was launched into massive growth leading to a population of about four hundred thousand on an area of approximately eighty-two square kilometres, which formed the early dimensions of the expanding georegion of today (Roots et al. 1999). One direction of expansion was toward the Muskoka Lakes, Georgian Bay, and farther north. In 1868 a system of free land grants was introduced into Muskoka and other northern areas to promote settlement. The Muskoka Lakes and areas further east were explored from Georgian Bay and the Severn waterway, as well as the Ottawa area. Although people moved from Toronto and other centres in response to the free land, their efforts were largely fruitless because of thin soils and the glaciated, acidic bedrock of the Canadian Shield.

Lumbering and extractive industries such as tanning proved to be more successful. With the further development of roads and rail lines, sawmills and other enterprises expanded at Bracebridge as well as at new communities such as Huntsville, at the west end of the Fairy and Peninsula Lakes and not far from the eastern shore of Georgian Bay (Nelson et al. 1996). These lakes became a focus for development in the surrounding lands and waters. Travellers began to visit the two lakes and others nearby as an aftermath to searches for minerals or lumber. They came from Toronto and centres in southern Ontario, as well as Ohio and other states in the northern United States. Steamships soon crossed the lakes, and inns were built for fishermen and outdoor tourists.

MUSKOKA AND GEORGIAN BAY

Logging and tourism around the Fairy and Peninsula Lakes increased after the Northern Railway reached Huntsville in 1886. Industries associated with lumbering were established at Huntsville, where a leather tannery opened in 1891 and a hardwood products manufacturing plant opened in 1900. A canal was built between the Fairy and Peninsula Lakes, which boosted the logging industry and tourism. By the 1890s the first cottages were being built on the shores of the Fairy and Peninsula Lakes, and big lodges and inns were opened to steamboat tourists. Roads were built to connect with other Muskoka Lakes such as the Lake of Bays, and large inns developed there as well. By the 1890s the growth of logging was slowing as tourism grew in economic importance, and the Muskoka Lakes and nearby areas such as Algonquin received increasing recognition as vacation destinations. After the 1890s cottage growth increased substantially, and the Muskokas and the northern lake country were on their way to their current high status as an international recreation and tourism region. In 1885 much of the Algonquin Country to the east of the Muskokas was set aside by the Ontario government as a forest reserve, and in 1892 most of this became Algonquin Park, where logging, tourism, and nature conservation contest for dominance into the twenty-first century.

Historically, then, strong links were formed between Toronto, the interior Muskoka Lakes country, and the waters and borderlands of Georgian Bay. The interior and the borderlands shared many natural, economic, and social attributes: glaciated bedrock and thin soils, much of it on the Canadian Shield; extensive pine and other forests; some mining; intensive lumbering; relatively small isolated communities; and the wild scenery and outdoor opportunities needed to build a growing recreational and tourism industry. Artists such as A.Y. Jackson, Frank Carmichael, Arthur Lismer, Tom Thomson, and other members of the Group of Seven were later attracted by and memorialized the pink rocks and tall pines around the Bay (Murray 2006).

FISHING, SHIPPING, AND THE GREAT LAKES GEOREGION

The deep expansive waters of Georgian Bay, which were not shared by the interior, formed the basis of the nineteenth- and early twentieth-century shipping and fishing regime that extended throughout the Great Lakes (figure 2.3). Long-distance shipping and fishing gradually formed the core of a distinctive Great Lakes region, a place connected to, but apart from, the surrounding shores. It was along these shores that navigation and fishing began in pre-European days and continued into the early decades of the Caucasian invasion. Fur traders travelled in their great canoes along the lake shores, and fishing was an important source of food for posts along the lakes. Fishing among settlers began early along the coasts as well as along inland tributaries, for example, fishing for anadromous salmon for subsistence and for barter or sale in local markets. The toll on salmon and other river fish was heavy, and the damage was exacerbated by dam construction, sawdust, and other pollution, which eventually eliminated the species from their breeding grounds and from Lake Ontario. Commercial fisheries developed slowly to meet rising demand in growing settlements such as York, Hamilton, and Detroit. The fish were plentiful and the catches often huge, so that

Figure 2.3 Political divisions of the Great Lakes. By the authors, with permission

fishing was being regulated by the city council in the Toronto Islands by the 1830s (Boque 2000, 28–9).

THE GREAT LAKES GEOREGION

The great expansion of fishing and shipping began in the 1820s and 1830s and from that time on the Great Lakes emerged as a georegion with a distinctive environmental, resource, economic, and institutional system. The driver for this was the successive introduction of a series of technological changes. The main product for export to greater than local markets was salt fish. Its transport was improved gradually after the 1830s by the introduction of steamboats and later by construction of the Erie and Welland Canals, among others. The canals connected sources of supply and demand all around the Great Lakes and extended trade to hitherto very distant markets such as New York City. With steamboats of larger capacity and more secure storage, entrepreneurs at places such as Detroit began to take clothing and other goods to settlements around the lakes in return for fish for sale in the United States.

Two other major technological changes in the 1850s and 1860s led the fishing and shipping industries to higher levels of endeavour. The widespread construction of the railroads resulted in a more advanced transportation network that was linked to growing steamer traffic throughout the Great Lakes. The two drove one another, leading to new initiatives such as large-scale passenger service, Great Lakes cruises and tourism, and combined rail and steamer traffic by companies such as the CPR.

Freezing techniques were introduced in the 1860s, and frozen fish was shipped in growing volumes by rail from lake ports throughout the United States and Canada. The returns were lucrative, and by 1872 Chicago, Detroit, Toledo, Buffalo, Toronto, and other towns had developed into major fish-processing and shipping centres. Much of the fish came from incursions by US operators into Canadian waters, a source of international friction into the twentieth century. In Buffalo and Chicago large dealerships with extensive connections throughout the Great Lakes were commonplace, and these two cities annually marketed 6.4 million and 7.5 million pounds of fish, respectively. Their dominance arose from their first-rate steamship and railroad network, their excellent processing and freezing facilities, and their marketing, economic, and technical expertise. For similar reasons Toronto dominated the Canadian fishing, shipping, and lakes industry. In 1892 ammonia freezing was introduced, and the entrepreneurs expanded their fishing system into the far corners of the lakes aided by freezers that held fish until markets were more favourable for sale.

Changes in the fishery were only part of the overall industrialization of the Great Lakes in the mid- to late-nineteenth century. With the introduction of propeller-driven steamers about 1860, larger and safer passenger craft carried more and more business people and tourists throughout the Great Lakes, whose ports developed the inns and other facilities required to serve this clientele. Larger, faster craft also drove the rising exploitation of copper and other minerals from around the Great Lakes. Iron ore was discovered in good quality and quantity on the US side of Lake Superior, and bulk cargoes began to flow to growing steel industries at Chicago, Cleveland, Toronto, and Hamilton.

Hamilton had an early start as a fishing, shipping, and manufacturing centre and as a gateway between the lakes and the interior. In the 1830s and 1840s the city was beginning the metal-work activities that are its hallmark today. Local entrepreneurs imported metal parts from southwestern sources such as the Normandale forge on Lake Erie, which produced iron from local bog ores in the Long

Point area from 1820 to 1860. In the mid-1850s the Great Western Railway was built from Buffalo through Hamilton and westward to London and points west. It bypassed competing towns such as Dundas and Ancaster and established Hamilton as a threshold to and transport focus for London and the southwest, where much of the doubling of Ontario's wheat production occurred between 1851 and 1857. The Great Western Railway also put Hamilton in a position to compete with Toronto for traffic to Georgian Bay and east through Buffalo to New York and Boston. New companies were established, such as the Ontario Rolling Mills, the Hamilton and Bridge Company, and the Greening Iron Works. Hamilton's population grew from twenty-five thousand in 1871 to fifty thousand in 1891, and development continued apace beyond the turn of the twentieth century with the Steel Company of Canada, a global competitor, being assembled from predecessor companies about 1907.

At this stage Hamilton was a principal entryway and trade centre for the southwest, or the Carolinian Canada Georegion, and was expanding into international markets with great visions for the future (Weaver 1982). At the same time Toronto was en route to preeminence with growing financial and investment interests. Although Hamilton fell behind in the competition for Georgian Bay and the northern trade, it did, however, engage more and more in the growing Great Lakes traffic. Coal came from Oswego and Rochester, New York. Iron ores came in from the upper Great Lakes. Limestone for the fluxing of steel arrived from hinterland sites such as St Marys in west-central Ontario. The expansion of canal systems at the Sault and Welland and in the Upper St Lawrence Valley and central New York turned the Great Lakes into an international waterway linking Lake Superior, Chicago, Cleveland, Hamilton, Toronto, Kingston, and other lake ports to markets along the eastern seaboard of the United States, as well as to Europe and elsewhere overseas.

During the 1840s Kingston prospered as a port and as a transshipment and forwarding centre. Trade came and went via the Rideau Canal and the Upper St Lawrence. Trans-lake shipment also took place with the United States, and local trade was conducted with communities along the Upper St Lawrence and throughout eastern Ontario. However, Kingston's port trade slowed, along with that of Brockville and the other Upper St Lawrence ports, between 1850 and 1880, in part because of competition from the railroads and because larger ships and the better harbours favoured direct shipping from Montreal and the St Lawrence to Toronto and Hamilton. Both these towns had rich agricultural and economically productive hinterlands, whereas Kingston backed onto thin soil and the acidic Canadian Shield. Roads and railroads were built into this area, and for a time the lumber trade fed Kingston, but it and the mineral industry did not have lasting power for the Kingston economy.

The efforts to develop manufacturing were a long-continued struggle. By 1861 the city included a steam-powered flour mill, two woolen mills, two shipyards, two tool factories, and other more specialized manufacturing enterprises. But Kingston did not have the natural advantages to secure any of the major new enterprises of the industrial era, such as the great steel mills of Hamilton or the banking and financial services of Toronto. It did reach out for such enterprises, for example, publishing, and finally achieved some success through the efforts of some astute businessmen. John Monton went from success in brewing and the lumber business to manufacturing locomotives for the new and rapidly expanding railroad system. By 1858 his foundry was Kingston's leading employer, and despite financial difficulties it continued to produce well into the twentieth century (Osborne and Swainson 1998).

James Richardson began as a tailor, worked in the grain trade, and ended up in the forwarding business. He built and ran his own fleet and transported grain down the St Lawrence and across Lake Ontario to the United States. He bought land on the lakefront and built and managed an ever larger set of grain elevators; he got into the mining game, developing a forwarding business in coal, iron, and other minerals; and he went west, following the CPR, and began to forward grain to the east and overseas. His firm established an office in Winnipeg and kept the executive office in Kingston until 1923. Richardson and his entrepreneurial colleagues took Kingston and the Upper St Lawrence area from its early military and transshipment mode into manufacturing, finance, and the new industrial age (Osborne and Swainson 1998).

In these years another distinctive characteristic of the Great Lakes became increasingly apparent. They constituted a binational region, shared by Ontario and the US Great Lakes states, as well as by Canada and United States. The conflicts inherent in this situation were part of the development of the fishing and shipping industries from the mid-1850s and extended into related difficulties such as the sharing of trans-boundary watersheds and their uses, as well as the construction of dams, stream diversions, pollution, and other overlapping challenges. They led to the formation in 1907 of the Canada–US Joint Commission as a means of resolving these difficulties.

THE SOUTHWEST, OR CAROLINIAN, GEOREGION

Towards the end of the nineteenth century much of Ontario was passing out of the Industrial Era into that of the Open Society. Included here was the southwestern Ontario deciduous forest known widely today as the Carolinian Zone. It had grown to become the agricultural core as well as an important manufacturing georegion in Ontario. The eastern threshold of this georegion, Hamilton, was heavily industrialized and increasingly open to international trade. Kitchener, other settlements in the Grand River basin, and London were being affected in similar ways in the southwestern hinterland.

Known earlier as Berlin, Kitchener, like nearby Waterloo, initially developed inland and away from other settlements, mainly, it appears, by design. The location was in very swampy terrain west of, but fairly close to, the Grand River. The first large group of settlers to come into the Kitchener-Waterloo area were the Mennonites about 1800. They migrated from Pennsylvania to escape worries about their religious freedoms under the new US government and also to find inexpensive land on which to grow their community. After some uncertainty about ownership they purchased what became known as the German Company Block of about sixty thousand acres. The layout was quite different from the Anglo-Saxon areas round about. The lots were surveyed into equally sized farms with no road allowances, no Crown reserves, clergy reserves, or Loyalist land grants typical of most of Upper Canada. The Mennonite migration from Pennsylvania was steady into the 1820s, when their dominance was paralleled by the migration of many German speakers from Europe. They contrasted with the Mennonites in their commitment to crafts, trade, and industry, rather than agriculture. Following the arrival of Germans from the homeland, the Mennonite hamlet of Ebystown was renamed Berlin. With the combination of fine farmers and skilled craftsmen and entrepreneurs, Berlin grew in about two decades to become the seat of the new Waterloo County. The settlement was home to administrative buildings, banks, many trades, steam-powered manufacturing, and a wide range of artisans and craftsmen (English and McLaughlin 1983, 8–10).

Berlin was quite distinct from nearby villages such as Galt, only about fifteen miles away on the Grand, with water power for manufacturing and a situation favourable to transport, trade, and commerce. These advantages were magnified by the construction of railways, including a branch line from Hamilton to Galt. Yet Berlin continued to grow, steadily aided by the increasing availability and use of steam power. The Grand Trunk Railway arrived in the mid-1850s, linking Berlin to Hamilton as well as the expanding markets in western Ontario. Steam power provided energy for the development of a range of manufacturing, including furniture, food processing, button making, and tanning based mainly on resources from the nearby southwest.

By the 1870s Berlin had about twenty-seven industrial firms with approximately seven hundred employees (English and McLaughlin 1983, 27–9). The atmosphere was strongly German. Growth flowered into the twentieth century when the town was shaken by World War I, and its overtly Germanic character led to a name change to Kitchener. In many ways Kitchener and Waterloo underline a characteristic of southwestern Ontario – its ethnic diversity. In the Grand River Valley alone, the Germans find their place among the Scots of Fergus and Guelph, the Six Nations people near Brantford, the Irish in the river's headwaters and the Loyalist and American folk of the lower river. And to the north of the growing town of Waterloo, a Black settlement of several hundred people farmed for a few decades around the time of the US Civil War of 1860–65, but they eventually left, leaving little but gravestones to mark their former presence in a predominantly Mennonite area today.

Another major centre in southwestern Ontario was London. This early settlement was on the Governor's Road, the major link between Toronto, Hamilton, and Chatham on the lower Thames (Troughton and Quinlan 2009; Hamill 1951). Its site at the junction of two branches of the Thames also provided entry upstream toward St Marys and the communities in the Grand watershed, as well as westward into the lower Thames River Valley. Grist and sawmills were built in London to process the growing amount of timber and grain being produced in the region. Two breweries, eventually Carling and Labatt, and a tannery developed close to the Forks of the Thames. Early bridges serviced the north-south Talbot Road and east-west Dundas Street, which became London's main commercial avenue. With the growth of railroads, the town expanded to the north, east and south and turned its back on the Thames, which presented a constant threat for floods. The Thames also became less desirable because of industrial and domestic pollution.

In 1853 the Great Western Railroad linked London and Chatham with Hamilton, Niagara Falls, Buffalo, Detroit, and other centres in the United States. In 1886 the Grand Trunk Railway was built from Toronto to Stratford and St Marys, en route to Sarnia. A branch line was built to London from St Marys in 1858. The railways arrived about the time that London achieved city status, and a population of about ten thousand people reinforced its role as a major regional marketing, manufacturing, and distribution centre. A branch line was built from London to Port Stanley on Lake Erie in 1858, facilitating imports of US coal, local grain, and passenger traffic. In 1882 the Grand Trunk Railway took over the Great Western and established locomotive shops in London and Stratford. The peak of the railroad era was reached in 1890–1914, when the Thames corridor was served by two national systems, the Grand Trunk and the Canadian Pacific, which had extended a line into the area in 1884. London was also the focus of numerous smaller lines, including the London to Port Stanley on Lake Erie bridge lines to smaller communities in nearby southwestern Ontario. Electrified inter-urban and street railways were introduced in this period linking London to St Thomas, Chatham, and the north Erie shore at Erieau near Rondeau Provincial Park, first established as a joint federal-provincial endeavor in 1884. These lines were especially valued for recreational travel and put London at the centre of a transport and communication network linked to Hamilton and Toronto and serving much of the Southwest Georegion.

THE OPEN SOCIETY

The Open Society is typified first and foremost by an increasing separation of individuals and organizations from local circumstances. In the Colonization Era and the early part of the Industrial Era strong ties were forged among people, their economies, technologies, social and institutional systems, and their places of domicile. Later in the Industrial Era, with the introduction of the railway, steam power, and improved transport, various forces worked to separate individuals, organizations, and their economic, social, and other activities from home locations. These forces grew from the 1880s onward with the introduction of the telegraph, the telephone, and electricity, which greatly increased the speed and range of industrial growth and communication by the turn of the twentieth century. These inventions spanned large expanses of land, as well as oceans, and made individual, economic, social, and institutional connections much more efficient and effective. However, they were not necessarily more equitable because, like the railways, their distribution system tended to favour some places over others. The telephone, for example, did not reach many rural areas in Ontario until the 1940s.

These late-nineteenth-century improvements in communication made it possible for many people to move into the Open Society, where they could live in one area and work in another, carry their social activities beyond the local and live on a regional, national, and international scale. These changes were accelerated by other technological advances of the very early twentieth century. The automobile was introduced commercially about 1910 with enormous subsequent effects on roads, land use, and economic and social systems generally.

The airplane became an operational reality in World War I and moved into commercial production and widespread use in airline communication for mail and passenger service in the 1920s. The effects of both the car and the plane were gradual, but they worked steadily to place people and their activities into the Open Society and ultimately into the Arena Society in which we now find ourselves.

In the Arena Society we are increasingly free to pursue economic, social, organizational, and political interests without strong ties to place. We are now more and more visible to the scrutiny of others, who are also caught up in this great game. We are also undergoing waves of immigration from non-traditional places in Africa and Asia that are changing economic, social, and environmental characteristics in yet incompletely understood ways. While these "liberating" forces are at work, other impulses and influences are pulling us back to "place," particularly concerns about home, the environment, and human or cultural heritage. Many people are concerned about diminishing kin connections, separation from place and environment, loss of architectural, engineering, and built heritage, and valued landscapes and lifestyles of the past.

To summarize, Ontario has a rich, diverse, and multi-faceted ethnic, social, land-use, and landscape history that is strongly reflected in its different georegions. Beginning with ten thousand years of First Nations history, the Caucasian invasion and development in Ontario passed, by about 1900, through the Fur Trade, Colonization, and Industrial Eras. Each of these eras was marked by distinctive economies, technologies, land uses, landscapes and societies. As in earlier First Nations times, the fur trade was associated with a predominately wild landscape containing numerous ephemeral but a few large, longer-lasting settlements. Fur trapping and trading transformed indigenous people, their tool kits, economies, societies, and belief systems. The trade also caused the decline of the wildlife upon which the natives depended.

The Era of Colonization focused on the settlement of tens of thousands of migrants, the development of farming, villages and small towns, and an agricultural economy and society. These changes were accompanied by rapidly diminishing wildlands, mainly in southern and central Ontario. The north remained remote, the scene of the declining fur trade and increasingly acculturated native people.

By the 1850s a number of Ontario's georegions had emerged, particularly in the south, including the Kingston and the Upper St Lawrence Georegion, with its military activity, transshipment, and small-scale manufacturing; the Peterborough Georegion, an agricultural settlement and the hub of a regional lumbering economy; rapidly growing Toronto, a fishing, manufacturing, and financial centre already sprawling into the surrounding countryside and assuming a central role among Ontario's georegions; the agricultural and manufacturing Southwest, with a growing number of interconnected manufacturing and marketing centres; and the Huron County Georegion, a predominantly agricultural area of villages and small towns.

The Industrial Era was driven by the railway and associated technical, economic, social, and institutional changes. In southern Ontario these changes promoted industrial growth and trade among advantaged, growing urban centres and stagnation and decline among less favoured villages and agricultural areas. Overall, however, the railroad and industrialization worked to firm up the georegions of the Colonization Era in southern Ontario.

The railways also opened up central and northern Ontario to the large-scale lumbering and mining that underlay the development of the Ottawa, the Georgian Bay–Muskoka Lakes and the Northeastern and Northwestern Ontario Georegions. The two northern georegions were distinguished by differing ethnic, religious, and economic patterns amid resource-based economies and diminishing forest ecosystems. In many parts of central and northern Ontario growing numbers of travellers and recreationists were also shaping the framework of an outdoor recreation and tourism industry. In the late nineteenth century the telegraph, telephone, electricity, and other technical changes accelerated industrial growth and the speed and range of communications. The result was the Open Society, in which people, their work, and their social and economic activities were increasingly separated from the home place. These changes also produced the georegional framework that is addressed in detail in this book.

In succeeding chapters knowledgeable authors will tell more detailed stories about the subsequent evolution of Ontario's georegions. Their futures and their contributions to their residents and to the diversity of the province are being challenged by a number of forces, notably the rapid growth of the Toronto megalopolis and recent attempts to manage this by provincial measures such as the creation of an extensive greenbelt around the city, with associated controls on land use and land governance. This development was preceded by the identification and creation of a provincial regulatory system for the most recent of Ontario's ecoregions, the Niagara Escarpment. What are the implications of these changes and associated forces such as globalization for the future of Ontario's georegions and the province as a whole? Can better ways be found of melding the growth of the megalopolis with the historic georegions and heritage mosaic that is Ontario?

REFERENCES

Adams, Peter, and Colin Taylor, eds. n.d. *Peterborough and the Kawarthas*. Peterborough, ON: Heritage Publications, Gould Graphic Services.

Barry, James. 1995. *Georgian Bay The Sixth Great Lake*. Erin, ON: Boston Mills Press.

Boque, Margaret Beattie. 2000. *Fishing the Great Lakes: An Environmental History, 1783–1933*. Madison, WI: University of Wisconsin Press.

Brandt, Lucian. 1946. *Ottawa Old and New*. Ottawa: Ottawa Historical Information Institute.

Bray, Matt, and Ernie Epp, eds. n.d. *A Vast and Magnificent Land: An Illustrated History of Northern Ontario*. Lakehead University and Laurentian University with Ontario Ministry of Northern Affairs.

Brown-Kubisch, Linda. 2004. *The Queens Bush Settlement: Black Pioneers, 1839–1865*. Toronto, ON: Natural Heritage Books.

Brown, Ron. 1997. *Toronto's Lost Villages*. Toronto, ON: Polar Bear Press.

Campbell, Claire Elizabeth. 2005. *Shaped by the West Wind: Nature and History in Georgian Bay*. Vancouver, BC: UBC Press.

Careless, J.M.S. 1984. *Toronto to 1918: An Illustrated History*. James Lorimer and Company and National Museum of Man, Toronto, ON: National Museums of Canada.

Dear, M.J., J. Drake, and L.G. Reeds, eds. 1987. *Steel City: Hamilton and Region*. Toronto: University of Toronto Press.

English, John, and Kenneth McLaughlin. 1983. *Kitchener: An Illustrated History*. Waterloo, ON: Wilfrid Laurier University Press.

Greening, W.E. 1964. *The Ottawa*. Toronto, ON: McClelland and Stewart.

Hamill, Fred Coyne. 1951. *The Valley of the Lower Thames*. Toronto, ON: University of Toronto Press.

Lemon, Jim. 1985. *Toronto since 1918: An Illustrated History*. James Lorimer and Company and National Museum of Man. Toronto, ON: National Museums of Canada.

Lower, A.R.M. 1938. *The Assault on the Canadian Forest*. Toronto, ON: Ryerson Press.

MacKay, Donald. 1998. *The Lumberjacks*. Toronto, ON: Natural Heritage/Natural History.

Murray, Joan. 2006. *Rocks, Carmichael, Franklin, Arthur Lismer, and the Group of Seven*. Toronto, ON: McArthur and Company.

Myers, Rollo, Heather Thomson, and J. Gordon Nelson. 2001. *Old Toronto: A Heritage Landscape Guide*. Waterloo, ON: Heritage Resources Centre, University of Waterloo.

Nelson J.G, A.J. Skibiki, K. Wilcox, and P. Lawrence. 1996. *Assessing Environment and Development: Fairy and Peninsula Lakes*. Waterloo, ON: Heritage Resources Centre, University of Waterloo.

Osborne, Brian S., and Donald Swainson. 1988. *Kingston: Building on the Past*. Westport, ON: Butternut Press.

Prince, Bryan. 2004. *I Came as a Stranger: The Underground Railroad*. Toronto, ON: Tundra Books.

Roots, Betty L., Donald A. Chant, and Conrad E. Herdenreich, eds. 1999. *Special Places: The Changing Ecosystems of the Toronto Region*. Vancouver, BC: University of British Columbia.

Taylor, John H. 1986. *Ottawa: An Illustrated History*. Toronto, ON: James Lorimer and Company and Canadian Museum of Civilization, National Museums of Canada.

Troughton, Michael, and Cathy Quinlan. 2009. *A Thames Watershed Landscape Guide: A Heritage Landscape Guide to the Thames River Watershed*. London, ON: Department of Geography, University of Western Ontario, mimeo.

Wallace, C.M., and Ashley Thomson, eds. 1993. *Sudbury: Rail Town to Regional Capital*. Toronto, ON, and Oxford: Dundurn Press.

Weaver, John C. 1982. *Hamilton: An Illustrated History*. Toronto, ON: James Lorimer and Company and National Museum of Man, National Museums of Canada.

White, Randall. 1985. *Ontario 1610–1985*. Toronto, ON, and London: Dundurn Press.

West, Bruce. 1979. *Toronto*. Toronto, ON: Doubleday and Company.

Wood, David, J. 2000. *Making Ontario: Agricultural Colonization and Landscape Re-Creation before the Railway*. Montreal and Kingston: McGill-Queen's University Press.

3

Sweetwater Seas and Shores: Waters and Coasts of the Great Lakes Georegion

PATRICK L. LAWRENCE

SUMMARY

With over twenty-six thousand square kilometres of water surface and ten thousand kilometres of shoreline, the lakes and watersheds of the Great Lakes comprise a significant regional land and waterscape within the province of Ontario. The essence of this region is the presence of the water bodies of the Great Lakes and their importance in the abiotic, biotic, and cultural aspects of human history and land use in Ontario. The Great Lakes serve as essential natural ecosystems with important roles in hydrology, drainage, and wildlife habitat. The waters and coasts of the Great Lakes have also been important in terms of human settlement and resource use from First Nations to European times, based on an abundance of food, forests, productive soils, a moderate continental climate, and access to a large supply of fresh water. The history of human uses has led to severe degradation of many of these resources, as well as significant environmental impacts including overfishing, loss of forests and wetlands, pollution and water quality concerns, shoreline flooding and erosion hazards, diversion of water supplies, and impacts on native species. Recent increased pressures of land and resource uses and the awareness of environmental issues associated with the lakes have resulted in several serious emerging issues, including climate change, water withdrawals, fundamental chemical and biological changes to the aquatic ecosystem, and continued contamination of the lakes from a large number of pollutants.

INTRODUCTION

The Great Lakes – Superior, Michigan, Huron, Erie, and Ontario – are an important part of the physical and cultural heritage of Ontario. Spanning more than twelve hundred kilometres from west to east, these vast inland freshwater seas have provided water and natural resources for food, manufacturing, water supply, transportation, power, recreation, and a host of other uses. The water of the lakes and the many resources of the Great Lakes basin have played a major role in the history and development of Ontario. For the early European explorers and settlers, the lakes and their tributaries provided the means for penetrating the continent, extracting valued resources, and exporting local products. Currently, the Great Lakes basin is home to more than one-quarter of the population of Canada. Some of the world's largest concentrations of industrial capacity are located in the Great Lakes region. Nearly one-quarter of total Canadian agricultural production is located in the basin (Environment Canada/USEPA 2005).

The geographic size of the Great Lakes water system is difficult to appreciate, even for those who live within the basin. The lakes contain about 23,000 km^3 (5,500 cu. mi.) of water, covering a total area of 244,000 km^2 (94,000 sq. mi.). They provide the largest surface fresh-water system on earth, with approximately 18 percent of the world supply (Ashworth 1986). In spite of their large size, the Great Lakes are sensitive to the effects of a wide range of air and water pollutants, whose sources include the runoff of sediments and chemicals from agricultural lands, waste water from cities, discharges from industrial areas, and leachate from landfill sites (Colborn 1990). The large surface area of the lakes also makes them vulnerable to direct atmospheric pollutants that fall with rain or snow and as dust on lake surfaces.

Outflows from the Great Lakes are relatively small (less than 1 percent per year) in comparison with the total volume of water held in the basin. Pollutants that enter the lakes from direct discharge, via the tributaries, from land use, and from atmospheric sources are retained for long periods in the system. Many pollutants can also remain in the lakes because of re-suspension and accumulation with subsequent cycling through biological food chains (Dempsey 2004; Environment Canada/USEPA 2005).

As a result of the large size of the watershed, physical characteristics such as climate, soils, and topography vary greatly across the basin. To the north, the climate is cold, and the terrain is dominated by the granite bedrock of the Canadian Shield consisting of Precambrian rocks under a generally thin layer of acidic soils and coniferous forests. In the southern areas of the basin, the climate is much warmer. The soils are deeper with layers or mixtures of clays, silts, sands, gravels, and boulders deposited as glacial drift or as glacial lake and river sediments. Since the lands are usually fertile and can be readily drained for agriculture, the original deciduous forests have given way to agriculture and sprawling urban development.

Although part of a single system, each lake is different. Table 3.1 provides some of the key characteristics of each of the Great Lakes. In volume, Lake Superior is the largest and also the deepest and coldest of the five. It has a water volume retention time of 191 years. Lake Michigan, the second largest, is the only Great Lake entirely within the United States. Lake Huron, which includes Georgian Bay, is the third largest of the lakes by volume. Many Canadians and Americans own cottages on the shallow, sandy beaches of Huron and along the rocky shores of Georgian Bay. Lake Ontario, although slightly smaller in area, is much deeper than its upstream neighbour, Lake Erie, with an average depth of eighty-six metres and a retention time of about six years. Major urban industrial centres, such as Hamilton and Toronto, are located on the Lake Ontario shoreline. Lake Erie is the smallest of the lakes in volume and is exposed to the greatest effects from urbanization and agriculture. The lake receives runoff from the agricultural area of southwestern Ontario and parts of Ohio, Indiana, and Michigan. Seventeen metropolitan areas with populations over fifty thousand are located within the Lake Erie basin. Although the area of the lake is about twenty-six thousand square kilometres, the average depth is only about 19 metres. It is the shallowest of the five lakes and therefore warms rapidly in the spring and summer and frequently freezes over in winter. It also has the shortest retention time of the lakes, 2.6 years. The western basin, comprising about one-fifth of the lake, is very shallow with an average depth of 7.4 metres and a maximum depth of 19 metres (US EPA 1995).

HISTORY AND CULTURE

The Great Lakes basin has long been a home to humans. In early Woodland times, thousands of years ago, natives were hunting, fishing, and gathering along the shores and on the waters of the lakes.

Table 3.1
Great Lakes Physical Features and Population

	Superior	Michigan	Huron	Erie	Ontario	Totals
Elevation (feet)	600	577	577	569	243	
(metres)	183	176	176	173	74	
Length (miles)	350	307	206	241	193	
(kilometres)	563	494	332	388	311	
Breadth (miles)	160	118	183	57	53	
(kilometres)	257	190	245	92	85	
Average depth (feet)	483	279	195	62	283	
(metres)	147	85	59	19	86	
Maximum depth (feet)	1,332	925	750	210	802	
(metres)	406	282	229	64	244	
Volume (cu. miles)	2,900	1,180	850	116	393	5,439
(km³)	12,100	4,920	3,540	484	1,640	22,684
Water area (sq. mi.)	31,700	22,300	23,000	9,910	7,340	94,250
(km²)	82,100	57,800	59,600	25,700	18,960	244,160
Land Drainage area						
(sq. miles)	49,300	45,600	51,700	30,140	24,720	201,460
(km²)	127,700	118,000	134,100	78,000	64,030	521,830
Total area (sq. mi.)	81,000	67,900	74,700	40,050	32,060	295,710
(km²)	209,800	175,800	193,700	103,700	82,990	765,990
Shoreline length (miles)	2,726	1,638	3,827	871	712	10,210
(kilometres)	4,385	2,633	6,157	1,402	1,146	17,017d
Retention time (years)	191	99	22	2.6	6	
Population: US (1990)	425,548	10,057,026	1,502,687	10,017,530	2,704,284	24,707,075
Canada (1991)	181,573		1,191,467	1,664,639	5,446,611	8,484,290
Totals	607,121	10,057,026	2,694,154	11,682,169	8,150,895	33,191,365
Outlet	St Marys River	Straits of Mackinac	St Clair River	Niagara River/ Welland Canal	St Lawrence River	

Source: By the author with permission, data from Coordinating Committee on Great Lakes Basic Hydraulic and Hydrologic Data, Coordinated Great Lakes Physical Data, 1992; Extension Bulletins E-1866-70, Michigan Sea Grant College Program, Cooperative Extension Service, Michigan State University, East Lansing, Michigan, 1985.

The waters of the lakes provided a rich and often plentiful supply of food stocks, which led to population expansion and the emergence of many native Indian tribes in the basin. Early forms of agriculture were developed with the burning and clearing of lands and planting of corn and other crops. Seasonal settlements were often located along the Great Lakes shore to take advantage of the fish and wildlife found there. For a more detailed discussion of the role of the First Nations in Ontario see chapters 1 and 2 of this volume.

The modern history of the Great Lakes region, from discovery and settlement by European immigrants to the present day, can be seen as an intensifying use of a vast natural resource. The first European arrivals had a modest impact on the system, which was limited to the exploitation of fur-bearing animals and fish. However, with the subsequent immigration of European populations into the region, logging, farming, and fishing become common and often resulted in serious ecological changes. Trees were clear-cut from the forests, drainage patterns disrupted, soils altered, and fish populations extensively harvested (Weller 1990).

As settlement and exploitation increased, many significant features and processes of the natural ecosystems of the lakes were drastically changed. Logging removed protective shades from streams and left them blocked with debris. When the land was plowed for farming the exposed soils washed away more readily, destroying streams and river-mouth habitats (Colborn 1990). With industrialization, untreated waste water and solid-waste debris increased, resulting in the degrading of many rivers draining into the Great Lakes (Ashworth 1986). The urbanization that accompanied industrial development added to the effects on water quality, creating conditions such as bacterial changes, increasing storm water, excessive nutrient loading, and the introduction of contaminants.

As industrialization progressed and as agriculture intensified after the turn of the twentieth century, new chemical substances came into use, such as PCBs (polychlorinated biphenyls) and DDT (dichlorodiphenyl-trichloroethane). Non-organic fertilizers were used to enrich soils to enhance production. The combination of synthetic fertilizers, nutrient-rich organic pollutants, such as untreated human wastes from cities, and phosphate detergents caused an acceleration of biological production leading to eutrophic conditions in the lakes (Colborn 1990; Caldwall 1988; Regier et al. 1990).

Toxic contaminants pose a threat not only to aquatic and wildlife species but to human health as well, since humans are at the top of many food chains. Some toxic substances biologically accumulate or are magnified as they move through the food chain. Consequently, top predators such as lake trout and fish-eating birds can receive extremely high exposures to them. Aquatic and wildlife species have been studied intensively, and adverse effects such as cross-bills and egg-shell thinning in birds, and tumors in fish are well documented (USEPA 1995). There is less certainty about the risk to human health of long-term exposure to low levels of toxic pollutants in the lakes, but there is no disagreement that the risk to human health will increase if toxic contaminants continue to accumulate in the Great Lakes ecosystem (IJC 2000, 2006; Fuller and Shear 1996).

THE PHYSICAL ENVIRONMENT

The foundation for the present Great Lakes basin was set about three billion years ago, during the Precambrian Era (Hough 1958; Prest 1970). The Michigan Basin was a major area of deposition of sediments, from the Cambrian period (about 600 million years ago) to the Jurassic period (200 million years ago). During these periods forces of erosion were at work removing areas of the Precambrian mountains and carrying the sediments to the basin. Numerous times the waters alternately spread over and retreated from the land, causing environmental conditions leading to the formation and deposition of differing bedrock formations (figure 3.1) (Sly and Thomas, 1974).

With the cooling of the global climate, large continental glaciers advanced from northern North America, and the region experienced the arrival of the ice sheets that advanced and retreated several times over a five-million-year period (Hough 1958; Karrow and Calkin 1985). During the stages of glaciation, giant sheets of ice flowed across the land, eroding highlands and carving out massive valleys. Where they encountered more resistant bedrock in the north, generally only the overlying layers were often removed. Further to the south, the softer sandstones and shales were more affected and greatly modified by the erosional forces of ice and water. Approximately two million years ago, the final advance of the ice sheets started southward from various centres of ice accumulation in northern Canada and subsequently covered the northern half of North

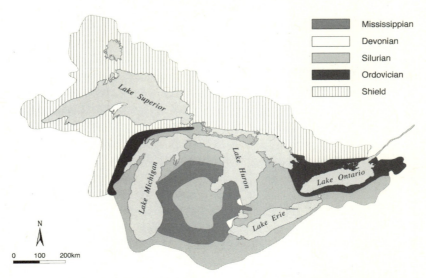

Figure 3.1   Geology of the Great Lakes Region. From www.kwic.com/~longpointbio/Reserve/Publications/FOLIO/chap2/chpt2.htm, with permission

America as far south as the Missouri and Ohio Rivers (Karrow and Calkin 1985).

The current land and water surfaces of the Great Lakes region are largely the result of the erosion and deposition of glacial materials during the last two million years in the Pleistocene epoch. Several stages of continental glaciation affected the region, but the most important was the Late Wisconsin stage (100,000 to 10,000 years ago). Since the retreat of the glaciers, water levels of the Great Lakes have continued to undergo dramatic fluctuations, in the magnitude of more than one hundred metres. These extremes were caused by changing climate conditions, crustal rebound within the basin, and the natural opening and closing of outlet channels greatly affecting drainage conditions (Karrow and Calkin 1985).

Presently, four of the five Great Lakes are at different elevations, leading like a series of steps down to the St Lawrence River and the Atlantic Ocean. The five individual lakes are connected to each other through channels that form into one drainage system for the Great Lakes. Water continually flows from the headwaters of the Lake Superior basin through the remainder of the system, draining out from Lake Ontario (figure 3.2).

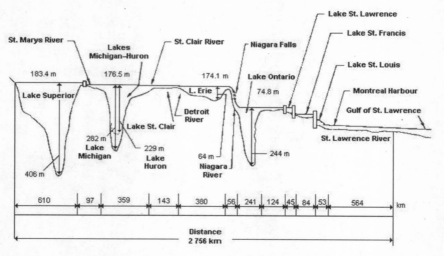

Figure 3.2 Great Lakes drainage. With permission of Natural Resources Canada (2011), courtesy of *Atlas of Canada*.

Water levels are part of the ebb and flow of the natural hydrology of the Great Lakes. The difference between the amount of water coming into each lake and the amount going out is the determining factor in whether water levels will rise, fall, or remain stable. When several months of above-average precipitation occur with cooler, cloudy conditions that cause less evaporation, the water levels of the lake will gradually rise. During prolonged periods of lower-than-average precipitation and warmer temperatures the result is the lowering of water levels (Bruce 1984). The modern range of Great Lakes water levels is in the order of one metre from high to low levels relative to the long-term mean. The Great Lakes have also been subject to several diversions, channel modifications, and water-control structures on Lake Superior and Lake Ontario. The recent decline of Great Lakes' water levels since 1997, resulting in lows not seen since the mid-1960s, is believed to be due mostly to increased evaporation during the warmer-than-usual temperatures of the past several years, a series of mild winters, and a below-average winter snow pack in the Lake Superior basin (NOAA GLERL 2006).

THE BIOLOGICAL ENVIRONMENT

The aquatic ecosystem of the lakes provides for a very rich and diverse habitat for many plants and animals and the supporting natural systems that sustain them. The Great Lakes ecosystem includes such diverse elements as northern evergreen forests, deciduous forests, tall grass and lake plain prairies, sandy barrens, alvars, dunes, and coastal wetlands. Over thirty of the basin's biological communities and over one hundred species are globally rare or found only in the Great Lakes basin (Nature Conservancy 2006). The terrestrial and aquatic natural biodiversity and complex ecosystems are important characteristics of the Great Lakes and essential to understanding their processes and human uses.

When the first Europeans arrived in the basin nearly four hundred years ago, it was a lush, thickly vegetated area. Vast timber stands consisting of oaks, maples, and other hardwoods dominated the southern areas. The forest and grasslands supported a wide variety of life, such as moose in the wetlands and coniferous woods and deer in the grasslands and brush forests of the south (Flader 1983). Habitat within the Great Lakes basin has been significantly altered, especially during the last 150 years. Nearly all the existing forests have been cut at least once, and the forest and prairie soils suited to agriculture have been plowed or intensively grazed. The construction of dams and urbanization has created vast changes in the plant and animal populations.

Streams have been changed not only by direct physical disturbance but by sedimentation and alterations in runoff rates owing to changing land use and by increases in temperature caused by the removal of shading vegetation. In the mid-1800s, commercial logging became an important industry in the region. The earliest loggers harvested the easy-to-cut and abundant white pine, which was much in demand for shipbuilding and construction. The trees were hundreds of years old and could not be replaced quickly. Hardwoods, such as maple and oak, were then cut to make furniture, barrels, and specialty products. Unfortunately, the region was often not reforested during this time of intense logging (Flader 1983).

It is estimated that as many as 180 species of fish were indigenous to Great Lakes waters. Those inhabiting the nearshore areas included smallmouth and largemouth bass, muskellunge, northern pike, and channel catfish. In the open water were lake herring, blue pike, lake whitefish, walleye, sauger, freshwater drum, lake trout, and white bass (Bogue 2000). Because of the differences in the characteristics of the lakes, the species composition varied for each of the Great Lakes. The warm and shallow water of Lake Erie create the most productive aquatic ecosystem, while deep, cool

Superior is the least productive in terms of the biological conditions currently present.

The Great Lakes can be characterized by their biological productivity, that is, the amount of living material supported within them, primarily in the form of algae. The least productive lakes are called oligotrophic, those with intermediate productivity are mesotrophic, and the most productive are eutrophic. The variables that determine productivity are temperature, light, depth and volume, and the amount of nutrients received from the environment. Except in shallow bays and shoreline marshes, the Great Lakes were oligotrophic before European settlement and industrialization (Ashworth 1986). These conditions have changed over the past two hundred years of human activities within the basin. Temperatures of many river and streams have been increased by removal of vegetative shade cover and thermal pollution. The amount of nutrients and organic material entering the lakes has increased with intensified urbanization and agriculture and the resulting inputs of materials, including, phosphorous (USEPA 2000). The overall water quality of the Great Lakes ranges from excellent to marginal, with more severe problems occurring in the lower Great Lakes of Erie and Ontario.

The depletion of oxygen through the decomposition of organic material is referred to as biochemical oxygen demand (BOD). Increases in BOD, which can have serious effects on aquatic ecosystems and the species that occupy them, are generated from two different sources. In tributaries and harbours, high BOD is often caused by materials contained in the discharges from wastewater treatment plants. The other principal source is decaying algae. In large embayments and open lake areas such as the central basin of Lake Erie, algal BOD is the primary problem. Changes in species of algae, bottom-dwelling organisms (or benthos), and fish are biological indicators of oxygen depletion. Turbidity in the water as well as an increase in chlorophyll also accompany accelerated algal growth and indicate increased eutrophication (USEPA 1995).

Many toxic substances tend to bioaccumulate as they pass up the food chain in the aquatic ecosystem. While the concentrations in water of chemicals such as PCBs may be so low that they are almost undetectable, biomagnification through the food chain can increase levels in predator fish such as large trout and salmon by a million times. Further biomagnification occurs in birds and other animals that eat fish (USEPA 2000). The bioaccumulative toxic substances can affect aquatic organisms in the lakes and birds and animals that eat them. Public health and environmental agencies in the Great Lakes basin warn against human consumption of certain fish (Fuller and Shear 1996; IJC 2006).

Currently, the Great Lakes are home to more than 130 non-native invasive species, including round goby, sea lamprey, and zebra mussels, that result in associated costs of $5 billion per year (www.HeathlyLakes.org). Zebra mussels (Dreissena polymorpha), which are small, fingernail-sized mussels native to the Caspian Sea region in Asia are believed to have been transported to the Great Lakes in ballast water from a transoceanic vessel. According to a 1995 Ohio Sea Grant study, large water users on the Great Lakes spend an annual average of $350,000 to $400,000 per user just to clear zebra mussels from their intake pipes (GLC 2007). The mussels are also affecting the tourism industry, since their sharp-shell remnants clutter beaches and are encrusting historically significant shipwrecks throughout the Great Lakes.

ECONOMICS AND RECREATION

The wealth of natural resources has long made the Great Lakes region a heartland of the Canadian industrial economy. The lakes became major highways of trade and were exploited for their fish. The fertile land that had provided the original wealth of furs and food yielded lumber, then wheat, then other agricultural products. Bulk goods such as iron ore and coal were shipped through Great Lakes ports, and manufacturing grew (Weller 1990). Canals led to broader commodity export opportunities, allowing farmers to expand their operations beyond a subsistence level. Wheat and corn were the first commodities to be packed in barrels and shipped abroad. Grist mills – one of the Great Lakes' first industries – were established on rivers and streams that drained into the lakes.

Railroads replaced the canals after mid-century, establishing important transportation links between the Great Lakes and both seacoasts. With the completion of the St Lawrence Seaway system in 1959 modern ocean-going vessels were able to enter directly into the Great Lakes. However, in recent decades commercial shipping on the Great lakes has not further expanded further because of intense economic competition from the other modes of mass transportation – trucking and railroads. Currently, the three main commodities shipped on the Great Lakes are iron ore, coal, and grain. Recently the shipment of iron ore on lake freighters has declined because many steel mills in the region have shut down or reduced production.

As a result of clear-cutting without proper rehabilitation of the forest, bare soils suffered serve erosion when large volumes of sediments drained into streams and rivers and lakes. The first paper mill was built on the Welland Canal in the 1860s. Eventually the basin became the world's leading producer of pulp and paper products which continues today along the shorelines of Lake Superior. The pulp and paper industry (along with chloralkali production) contributed to the mercury pollution problem on the Great Lakes until the early 1970s, when mercury was banned from use in the industry (Weller 1990).

Commercial fishing began about 1820, and the largest harvests were recorded in the late nineteenth century when annual levels approached sixty-seven thousand tonnes (Bogue 2000). However, by the 1880s some preferred species in Lake Erie had declined. Catches increased with more efficient fishing equipment, but the Great Lakes commercial fishery peaked by the late 1950s. Since then, average annual catches have been around fifty thousand tones. The value of the commercial fishery has also declined drastically because the more valuable, larger fish have given way to small and relatively low-value species. Over-fishing, pollution, shoreline and stream habitat destruction, and the accidental and deliberate introduction of exotic species such as the sea lamprey all played a part in the decline of the commercial fishery in the basin (GLFC 2003).

The Great Lakes form the largest system of surface freshwater on earth and contain roughly 20 percent of the world's supply, totalling about six quadrillion gallons of water (Annin 2006). Approximately fifteen million people in Canada rely on the Great Lakes as a source of drinking water. The Great Lakes Commission has conducted a preliminary examination of water-use data (1987–93) in the Great Lakes basin. Water uses comprise two categories: consumptive uses and removals. Close to 90 percent of withdrawals are taken from the lakes themselves, with the remaining 10 percent coming from tributary streams and groundwater sources. An estimated 5 percent of the water withdrawn from the Great Lakes is consumed and is therefore lost to the basin. Consumptive use in the Great Lakes basin was estimated to be 116 cubic metres per second and withdrawals about

2,493 cubic metres per second. If current trends continue, total water use in the Canadian portion of the basin is expected to increase by close to 20 percent between 1996 and 2021 (Annin 2006).

The economic activity in the Great Lakes basin exceeds $200 billion a year. There are notable concentrations of steel, pulp/paper, and manufacturing facilities. The international shipping trade annually transports fifty million tonnes of cargo. The main commodities shipped through the basin are grains, iron ore, coal, coke, and petroleum products. Almost 50 percent of this cargo travels to and from oversea ports, especially in Europe, the Middle East, and Africa (USEPA 1995). As the world's longest deep-draft inland waterway, the system extends from Duluth, Minnesota, on Lake Superior, to the Gulf of St Lawrence on the Atlantic Ocean, a distance of more than 2,340 miles. This shortcut to the continent's interior was made possible with the construction of a ship canal and lock system opened in 1855 at Sault Ste Marie, Michigan; the development of the Welland Canal since 1829; and the completion of the St. Lawrence Seaway in 1959.

Recreation is also an important part of the Great Lakes economy. The annual value of the commercial and sport fishery is estimated at over $4.5 billion dollars (GLFC 2003). Since early in the industrial age, the waterways, shorelines, and woodlands of the Great Lakes region have attracted leisure-time activities. Although boating and fishing were once important commercial activities, they are now primarily leisure pursuits. Recreation in the Great Lakes became an important economic and social activity with the age of travel in the nineteenth century. A thriving pleasure-boat industry based on the newly constructed canals developed, bringing people into the region in conjunction with rail and road travel. Niagara Falls attracted travellers from considerable distances and was one of the first stimulants to the growth of a leisure-related economy (Ashworth 1986).

With industrial growth, greater personal disposable income, and shorter workweeks, more people began to spend more of their leisure time outside the cities, often on the shore lands and waters of the Great Lakes. An extensive system of parks, wilderness areas, and conservation areas was established by the federal and provincial governments to protect valuable local resources and provide outdoor recreational opportunities. However, there was also great interest in private development of the Great Lakes shoreline for seasonal residential uses that resulted in about 20 percent of the Canadian shoreline of the lower Great Lakes (Huron, Erie, and Ontario) being privately owned and not accessible by the public (USEPA 1995).

The increasingly intensive recreational development of the Great Lakes has had mixed effects. Extensive development of cottage areas, summer-home sites, beaches, and marinas has resulted in the loss of wetland, dune, and forest areas. Shoreline alteration by developers and individual property owners has changed the shoreline erosion and deposition process, often to the detriment of important beach and wetland systems that depend on these processes. The development of areas susceptible to shoreline flooding and erosion has caused considerable public concern, resulting in considerable political pressure to control lake levels in order to prevent changes that are part of natural weather patterns and processes. Pollution from recreational sites and boats has also caused water-quality degradation in many locations on the Great Lakes.

THE GREAT LAKES COAST

Although the focus of many studies and interests regarding the Great Lakes is on the water bodies themselves, the coastal zones, or lakeshores, are also very important to consider in regard to human activities and impacts within the basin. The ten thousand kilometres of Great Lakes shoreline, or coast, represents not only a geologically, chemically, and biologically dynamic environment, it also is the site of some of the most intense economic, social, and political pressures within the Great Lakes basin. Throughout the Great Lakes in Ontario, the estuarine and coastal environments are characterized by intensive development, urbanization and industrialization, areas of high biological productivity and natural diversity, and sites of high recreational value. The coastal zone of the lakes is the focus for multiple, and often competing, land and resource uses. In many locations intense land-use pressures exist between interests in further human development and the concern for the conservation of natural areas and processes.

The Great Lakes shoreline is under continual stress. In the lower lakes region of Lake Erie and Lake Ontario, little remains undeveloped (figure 3.3). Urban, rural, residential (including cottages), and industrial land uses occupy much of the US portions of coastal zone of the lower Great lakes. Most lakefront properties are in private

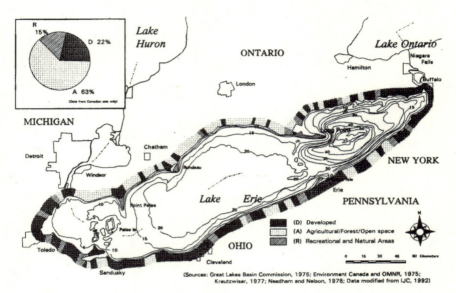

Figure 3.3 Lake Erie shoreline uses. From Lawrence and Nelson (1999), with permission

ownership and thus under limited control by public authorities wishing to protect the coast. Shoreline erosion impacts have been significant, resulting in high economic costs to society and individuals owing to intensive development and the loss of vegetative cover and other natural protection in many locations along the Great Lakes shoreline. The loss of natural habitats, especially forests and wetlands, has been very extensive on Lakes Erie and Ontario. Large parts of the inland area of this portion of the Great Lakes basin are also taken up by intensive agricultural uses, including cultivated crops such as corn, wheat, and soybeans, and livestock farming.

Human activities and development have had their most immediate and direct impact along the coasts of the Great Lakes and its tributary rivers and streams. Land use in coastal areas of the Great Lakes is changing in response to the region's evolving economy and industrial restructuring as well as the relentless forces of urban sprawl. With over thirty-three million people living in the basin the pressures from urban, residential, and industrial land uses and associated activities are significant. The aesthetic and recreational attraction of the shores is also spurring renewed public appreciation and use of this asset. Several national parks and other forms of protected areas have been established along the Great Lakes in both Canada and the United States. Long Point, on the north shore of Lake Erie, has been designated as a World Biosphere Reserve, and three US National Wildlife Refuges exist within the western basin of Lake Erie.

Human activities in the coastal zone have become a leading stressor of the Great Lakes Basin ecosystem. The coastal areas are environmentally sensitive because they represent the natural interface between the land and the lake and support special communities of aquatic and terrestrial flora and fauna. The Great Lakes coast has been identified as an important location for wetlands, fish habitat, shorebirds, and waterfowl. The loss of wetlands and unique habitat has been significant along the Great Lakes coast, especially for Lake Erie and Lake Huron. Development has also disrupted the natural process of beach creation and replenishment by interfering with the supply and distribution of sand (Lawrence and Nelson 1994).

Damages to properties, roads, and other infrastructure along the coast owing to flooding and erosion are also of concern, particularly during periods of high lake levels (Lawrence and Nelson 1994). Damages to private property from flooding and erosion lead to public pressure on governments to further regulate lake levels through diversion and control structures on outlet channels. The demand for public access to the lakes for recreation has grown steadily in recent years and can be expected to continue. Currently, the greatest growth is in the development of marinas for recreational boating.

It has been estimated that approximately 25 percent of the Great Lakes shoreline is prone to erosion and that 10 percent is prone to flooding. The result of these natural hazards is severe inland flooding of beaches, dunes, and wetland areas and erosion of beach and bluff sediments. Property damages from these natural hazards have increased sharply over the last four decades (Nelson et al. 1975; Lawrence and Nelson 1999). Flooding and erosion damages to public and private properties along the Great Lakes shoreline in the early 1970s and again in 1985/86 were valued in the millions of dollars (Kreutzwiser 1987).

The increasing costs suggest that inappropriate land use continues and that government policies and programs have been ineffective in reducing losses from Great Lakes shoreline flooding and erosion. Planning and management mechanisms within the Great Lakes basin, through the use of emergency responses and land-use setbacks, have been insufficient in resolving the rising costs of damages to ecosystems and property (Lawrence 1995).

POLICY AND PLANNING

As we have seen, the Great Lakes basin extends into both the United States and Canada, and all the lakes except one (Lake Michigan) are bordered by both countries. Therefore, state-by-state or province-by-province policy does not work as effectively as a regional approach. Governments of the Great Lakes region have implemented several agreements, programs, policies, and agencies intended to protect the Great Lakes ecosystem. The 1909 Boundary Waters Treaty was created to provide the principles and mechanisms necessary to help prevent and resolve disputes concerning water quantity and water quality along the boundary between Canada and the United States. The International Joint Commission (IJC) was created as a result of the treaty and serves as an independent binational organization responsible for carrying out the requirements of the Boundary Waters Treaty (Lee et al. 1982; Rabe and Zimmerman 1995; IJC 2006).

In 1972, Canada and the United States signed the Great Lakes Water Quality Agreement in recognition of the urgent need to improve water quality in the Great Lakes. In 1978, the agreement was amended to include toxic contamination, which was seen as a serious emerging concern within the basin. The agreement also evolved into an ecosystem approach, recognizing the importance and interconnectedness of all components of the environment of the Great Lakes: water, air, and land (IJC 1978). It includes a number of objectives and guidelines to achieve its goals, such as the elimination of toxic chemicals in Great Lakes waters, financial assistance to construct waste water treatment facilities, and the development of best management practices to control most sources of pollutants. Many binational programs have been developed out of the agreement, such as the Great Lakes Binational Toxics Strategy and the State of the Lakes Ecosystem Conferences (SOLEC) (Fuller and Shear 1996; IJC 2006).

The economic well-being of the Great Lakes region is closely tied to the health of the ecosystem. The challenge of Great Lakes environmental protection and natural resource management is to balance the use of the resources of this unique ecosystem with their conservation. In 1987, management aspects of the ecosystem approach were further defined in revisions to the Great Lakes Water Quality Agreement, which called for management plans to restore fourteen beneficial uses within the lakes. They include unimpaired use of the ecosystem by all living components, including humans (Mackenzie 1996; Edwards and Regier 1990). The agreement called for Remedial Action Plans (RAPs) to be prepared for geographic Areas of Concern (AOCs), where local use impairments existed at forty-three sites distributed widely through the lakes. It also called for Lakewide Management Plans to be prepared for critical pollutants that affect whole lakes or large portions of them. The purpose of these management plans is to clearly identify the key steps needed to restore and protect the lakes (Hall et al. 2006; IJC 2006).

The millions of people who live in the Great Lakes basin and their governments face an immense challenge for the future. The management needed to address Great Lakes issues will require greater public awareness, political will, and innovative actions and cooperation. The Great Lakes are within two sovereign nations, a Canadian province, eight American states and hundreds of local, regional, and special-purpose governing bodies with jurisdiction for management of some aspect of the basin or the lakes. Cooperation is essential because problems such as water quality, invasive species, water consumption and diversions, climate change, lake levels, and shoreline management do not respect political boundaries. Efforts at regional collaboration have been developed, including the Great Lakes Regional Collaboration (2005), Great Lakes Charter Annex (2005), the Healing Waters campaign (NWF 2006), and the renewal of the Canada/Ontario Agreement (2007).

In Canada, the British North America Act assigns the authority for navigable waters and international waters to the federal government, while pollution control and the management of natural resources are primarily provincial responsibilities. Consequently, the initiative to establish water quality objectives under the Great Lakes Water Quality Agreement within Canada has been a federal-provincial joint partnership, and the implementation has been primarily a provincial responsibility of the Ontario Ministry of Environment and Energy. Environment Canada is the lead agency for the Canadian federal government roles in regard to the Great Lakes (www.on.ec.gc.ca/greatlakes/). The federal Canada Water Act provides for federal-provincial agreements setting out responsibilities for both levels of government. The Canada/Ontario Agreement (COA) provides for joint work on activities required by the Great Lakes Water Quality Agreement. In 1988, the Canadian Environmental Protection Act (CEPA), which provides a framework for controlling toxic substances, was developed. Under this act, for example, dioxins and furans could be virtually eliminated from pulp-and-paper mill discharges.

The Department of Fisheries and Oceans is a major contributor of scientific and research support to Canada's Great Lakes program. Other federal departments directly involved include Health Canada, Agriculture and Agrifood Canada, Transport Canada and the Department of Government Services. The major responsibility for water quality at the provincial level rests with the Ontario Ministry of Environment and Energy (MOEE). The MOEE is responsible for establishing individual control orders for each industrial discharger. It also provides funding for sewage treatment. The Ontario Ministry of Natural Resources provides leadership for public lands, shoreline management, and fisheries, forestry, and wildlife management.

Public consultations that include residents, private organizations, industry, and government are considered to be an essential part of the decision-making process for managing the resources of the Great Lakes ecosystem. Residents of the basin have been empowered to participate in the problem-solving process, promote healthy sustainable environments, and reduce their personal exposure to Great Lakes contaminants. Great Lakes United serves as an international coalition of non-government member organizations with interests in the environment and Great Lakes issues (www.glu.org).

THE FUTURE

The Great Lakes continue to be plagued by a complex set of environmental and human-use issues. As population increases, so do water demand, energy use, land-use changes, increased urbanization, pollutant concerns, the introduction of non-native exotic species, and threats from water diversions (Fields 2005). Although the Great Lakes have been the focus of considerable management and planning efforts in recent decades in both Canada and the United States, many concerns about human impacts on the ecosystem remain. It is also believed that the Great Lakes region will be greatly affected by global climate change, leading to alterations to winter climate, reduced water inputs, increased variation in climate conditions, and more extreme events. Current climate change models predict a drop in average annual water levels of between 0.5 and 1 metre over the next thirty years (Sousounis and Bisanz 2000).

As much of the heavily populated and developed regions of the western and southwestern United States increasingly suffer severe drought conditions, continued political and public pressure will be placed on the Great Lakes as a source of water. As host to the largest supply of freshwater in North America, this resource is seen as important to sustaining the growing population and industrial base of the region in both countries. The economic and political influence of the Great Lakes region has also been on the decline, resulting in fewer ways to combat this pressure.

In recent years significant efforts have been developed within the region to increase the coordination of policies related to many basin-wide issues requiring the cooperation of federal, state or provincial, and local government agencies and organizations within the Great Lakes. In 2008, the Great Lakes states all ratified the Great Lakes-St Lawrence Water Resources Compact, which includes a ban on any new large-scale water diversions to areas outside the basin. The International Joint Commission is currently reviewing the water levels and outflow regulations for both Lake Ontario and Lake Superior. Ongoing discussions over the renewal of the US Canada Great Lakes Water Quality Agreement reflect both the urgency of regional cooperation, but also the challenges faced by such an approach. Many serious environmental issues remain, and there is renewed concern for more complex and emerging issues that will pose significant scientific and management challenges to our collective need to respond and address.

The governments of Canada and the United States have completed the Great Lakes-St Lawrence Seaway Study (GLSLS), which is intended to evaluate the infrastructure needs of the seaway system, specifically the engineering, economic, and environmental implications of those needs as they pertain to commercial navigation. The study assesses long-term maintenance and capital requirements to ensure the continuing viability of the system as a safe, efficient, reliable, and sustainable component of North America's transportation infrastructure. Since the 1960s there has been a lengthy history of focused research efforts to understand and address many serious Great Lakes issues. Areas of extensive study have included water chemistry, aquatic biology, fisheries, shoreline flooding and erosion hazards, and climate change impacts. However, unfortunately many issues have been unevenly addressed, because of a lack of consistent funding sources and dedicated efforts to support such work by universities and government-supported programs. Within Environment Canada, the Department of Fisheries and Oceans, the Ontario Ministry of Natural Resources, and the Ontario Ministry of Environment, staff cutbacks and reduced financial support have made continued monitoring and research and the development of new research efforts very difficult.

The Great Lakes, as a binational resource shared between Canada and the United States, also serve as an international border and thus are becoming a focus of a debate on security issues that will have wide-ranging impacts on cross-border shipping, transportation, travel, trade, and addressing environmental issues. As transboundary negotiations continue on issues such as border policies, economic policies, water resources as a commodity, and potential lake-bed energy extractions, the future of the Great Lakes will increasingly be tied to the cultural, economic, and political landscape currently emerging between and within the two nations.

The history of human activities, resource use, and environmental degradation is well known and documented for the Great Lakes. There are many great success stories to reflect on, such as the return of Lake Erie from its declared "dead" status in the 1960s and 1970s. The Great Lakes have also been highlighted as excellent examples of the implementation of an ecosystem approach for addressing environmental issues and for the spirit of international cooperation between two sovereign nations sharing such a large water body. The future will require renewed and continued efforts in both these areas and by all public and private parties, government and the citizenry, in order to ensure ongoing human and ecological health for the next generations.

The Great Lakes will continue to be a vital cultural, social, economic, and political region within the province of Ontario. The historical context and future prospects will require several important planning and management efforts to ensure that the Great Lakes remain an important and vital georegion in Ontario:

- Continued public and political efforts within the province are necessary to maintain a strong voice in regard to the many transboundary issues that face the Great Lakes with special concern for

water diversions, non-native exotic species, water quality, and climate change impacts.
- Cooperation between the Canadian federal government and the Ontario provincial government needs to be maintained and strengthened in decision making, binational relationships with the United States, ongoing and future environmental research and monitoring, and economic and trade issues.
- Efforts are needed to revitalize the Ontario Great Lakes Shoreline Management Program and to consider the linkages to already existing federal programs and initiatives in coastal and ocean management that all too often do not fully acknowledge the Great Lakes coast.
- Various management organizations are now playing a key role in Great Lakes issues and merit greater support because of their role at grass roots levels.
- Stronger efforts are required in regard to rapid population growth, housing development, and the associated urbanization occurring along much of the Great Lakes shoreline, especially along Lake Erie and Lake Ontario, and within the rural landscape areas of the region.
- The important natural and economic values of the Great Lakes need to be carefully evaluated in considering large-scale development, including urban expansion and industrialization, in order to more fully understand and appreciate the fundamental impacts these changes are causing to this georegion and its people.
- Finally, the province of Ontario should take under advisement the creation of one central-planning authority to deal with Great Lakes issues that are now under the responsibility of numerous provincial acts, agencies, and programs that often lack cohesiveness and coordination. The Ohio Lake Erie Commission and Great Lakes Commission are examples of such entities currently functioning in this regard in the United States.

REFERENCES

Annin, P. 2006. *The Great Lakes Water Wars*. Washington, DC: Island Press.

Ashworth, W. 1986. *The Late, Great Lakes: An Environmental History*. Toronto, ON: Collins Press.

Bogue, M.B. 2000. *Fishing the Great Lakes: An Environmental History, 1783–1933*. Madison, WI: University of Wisconsin Press.

Bruce, J.P. 1984. "Great Lakes Levels and Flows: Past and Future." *Journal of Great Lakes Research* 10(1):126–34.

Caldwell, L.K. (1988). *Perspectives on Ecosystem Management for the Great Lakes*. Albany, NY: State University of New York Press.

Colborn, T.E. 1990. *Great Lakes, Great Legacy?* Ottawa: Institute for Research on Public Policy.

Council of Great Lakes Governors. 2005. *The Great Lakes Charter Annex: A Supplementary Agreement the Great Lakes Charter.*

Dempsey, D. 2004. *On the Brink: The Great Lakes in the Twenty-first Century*. East Lansing, MI: Michigan State University Press.

Donahue, M.J. 1987. *Institutional Arrangements for Great Lakes Management: Past Practises and Future Alternatives*. Ann Arbor, MI: Michigan Sea Grant College Program.

Dworsky, L.B. 1986. "The Great Lakes: 1955–1985." *Natural Resources Journal* 26:291–336.

Edwards, C.J., and H.A. Regier. 1990. "An Ecosystem Approach to the Integrity of the Great Lakes in Turbulent Times." In *Proceedings of a 1988 Workshop Supported by the Great Lakes Fishery Commission and the Science Advisory Board of the International Joint Commission. Great Lakes Fishery Commission Special Publication 90-4.*

Environment Canada and US Environmental Protection Agency. 2005. *Our Great Lakes*. Ottawa, Canada. http://binational.net/ourgreatlakes/ourgreatlakes.pdf

– 2003. *State of the Great Lakes 2003*. Ottawa, Canada.

Fields, S. 2005. "Great Lakes: Resource at Risk." *Environews* 113(3): A164–A173.

Flader, S.L. 1983. *The Great Lakes Forest: An Environmental and Social History*. Minneapolis, MN: University of Minnesota Press.

Fuller, K., and H. Shear. 1996. *State of the Lakes Ecosystem Conference '96: Integration Paper*. Prepared for the SOLEC Steering Committee. Burlington, ON: US Environmental Protection Agency Great Lakes Program Office, Chicago, Illinois and Environment Canada.

Great Lakes Commission. 2007. *Great Lakes Aquatic Invasions*. Ann Arbor, MI.

Great Lakes Fishery Commission. 2003. *Protecting and Restoring the Great Lakes Fishery*. Ann Arbor, MI.

Great Lakes Regional Collaboration. 2005. *Great Lakes Regional Collaboration Strategy*. http://www.glrc.us/

Hall, J.D., K. O'Connor, and J. Ranieri. 2006. "Progress toward Delisting a Great Lakes Area of Concern: The Role of Integrating Research and Monitoring in the Hamilton Harbour Remedial Action Plan." *Environmental Monitoring and Assessment* 113(1–3): 227–43.

Hough, J.L. 1958. *Geology of the Great Lakes*. Chicago, IL: University of Illinois Press.

International Joint Commission. 1978. *The Ecosystem Approach: Scope and Implications of an Ecosystem Approach to Transboundary Problems in the Great Lakes Basin*. Windsor, ON: Great Lakes Research Advisory Board.

International Joint Commission. 2000. *Protection of the Waters of the Great Lakes*. Ottawa, Canada.

– 2006. *Thirteenth Biennial Report on Great Lakes Water Quality*. Ottawa, Canada.

Karrow, P.F., and P.E. Calkin. 1985. *Quaternary Evolution of the Great Lakes*. Geological Association of Canada. Special Paper 30. Ottawa, Ontario.

Kreutzwiser, R.D. 1987. "Managing the Great Lakes Shoreline Hazard." *Journal of Soil and Water Conservation*. 42(3): 150–2.

Lawrence, P.L. 2004. "Managing the Great Lakes as a Transboundary Coastal Ecosystem." *Petermanns Geographische Mitteilungen* 1:20–9.

– 1995. "Development of Great Lakes Shoreline Management Plans by Ontario Conservation Authorities." *Ocean & Coastal Management* 26(3): 205–23.

Lawrence, P.L., and J.G. Nelson. 1999. Great Lakes and Lake Erie Floods: A Life Cycle and Civics Perspective. *Environments* 27(1): 1–22.

– 1994. "Flooding and Erosion Hazards on the Great Lakes Shoreline: A Human Ecological Approach to Planning and Management." *Journal of Environmental Planning and Management* 37(3): 289–303.

Lee, B.J., H.A. Regier, and D.J. Rapport. 1982. "Ten Ecosystem Approaches to the Planning and Management of the Great Lakes." *Journal of Great Lakes Research* 8(3): 505–19.

Mackenzie, S.H. 1996. *Integrated Resource Planning and Management: The Ecosystem Approach in the Great Lakes Basin*. Washington, DC: Island Press.

National Wildlife Federation. 2006. *Healing Our Waters: An Agenda for Great Lakes Restoration*. http://www.healingourwaters.org/.

Nature Conservancy. 2006. *Towards a New Conservation Vision for the Great Lakes*. The Nature Conservancy. Chicago, IL: Great Lakes Program.

Nelson, J.G., J.G. Battin, R.A. Beatty, and R.D. Kreutzwiser. 1975. "The Fall 1972 Lake Erie Floods and Their Significance to Resources Management." *Canadian Geographer* 19(1): 35–58.

NOAA Great Lakes Environmental Research Lab. 2006. *Great Lakes Water Levels*. Ann Arbor, MI. http://www.glerl.noaa.gov/pubs/brochures/wlevels/wlevels.pdf

Prest, V.K. 1970. "Quaternary Geology of Canada." In *Geology and Economic Minerals of Canada.* Economic Geology Report No. 1. Geological Survey of Canada.

Rabe, B.G., and J.B. Zimmerman. 1995. "Regime Emergence in the Great Lakes Basin." *International Environmental Affairs* 7(4): 346–63.

Regier, H.A., G.R. Francis, A.P. Grima, A.L. Hamilton, and S. Lerner. 1990. "Integrity and Surprise in the Great Lakes Basin Ecosystem: Implications for Policy." In *An Ecosystem Approach to the Integrity of the Great Lakes in Turbulent Times,* edited by C.J. Edwards and H.A. Regier. Proceedings of a 1988 Workshop Supported by the Great Lakes Fishery Commission and the Science Advisory Board of the International Joint Commission. Great Lakes Fishery Commission Special Publication 90-4, 17–36.

Sly, P.G., and R.L. Thomas. 1974. "Review of Geological Research as It Relates to an Understanding of Great Lakes Limnology." *Journal of Fisheries Research Board of Canada* 31:795–825.

Sousounis, P.J., and J.M. Bisanz. 2000. *Preparing for a Changing Climate: Great Lakes.* A Summary by the Great Lakes Regional Group for the US Global Change Program. Ann Arbor, MI: University of Michigan.

US Environmental Protection Agency. 2000. *Great Lakes Ecosystem Report.* Chicago, IL: Great Lakes National Program Office.

US Environmental Protection Agency and the Government of Canada 1995. *The Great Lakes: An Environmental Atlas and Resource Book.* 3d ed. Chicago, IL: Great Lakes Program Office.

Vallentyne, J.R., and A.M. Beeton. 1988. "The Ecosystem Approach to Managing Human Uses and Abuses of Natural Resources in the Great Lakes Basin." *Environmental Conservation* 15(1): 57–62.

Weller, P. 1990. *Fresh Water Seas: Saving the Great Lakes.* Toronto, ON: Between the Lines Press.

4

Toronto: Putting the Georegion in Perspective

LUCY M. SPORTZA

SUMMARY

York, Toronto, Toronto Region, Metropolitan Toronto, Greater Toronto Area, Golden Horseshoe, Greater Golden Horseshoe ... The evolving nomenclature of the Toronto-based segment of southern Ontario calls to mind an ever-growing creature, slowly eating up the surrounding landscape and trapping countless people within the congested concrete jungle. Certainly this is the view of many from outside, and even inside, the region. Often denounced as having too much influence on provincial and national politics, Toronto is seemingly easy to revile. The Toronto Georegion is, on the one hand, one of many important, unique, and special places within Ontario. On the other hand, the Toronto Georegion is in many ways dominant among Ontario's georegions. The reasons for this dominance – in terms of population and economy – are largely explicable when one understands the historical development of the region. One result has been decades of urban sprawl, an issue that we continue to struggle with and that has led to Toronto extending its reach into surrounding communities and georegions, altering them in innumerable ways (Kalbach 1999). However, Toronto has also emerged as one of the most diverse, safe, and respected cities in the world, wonderfully positive characteristics that deserve appreciation and celebration.

This chapter aims to explain, in part, how and why Toronto evolved to its present size and dominance, focusing on the period from roughly 1900 to the present (for earlier information see chapters 1 and 2). One theme will be the role of parks and protected areas in extending the reach of Toronto into surrounding georegions, and the present and future challenges brought forth by globalization will be introduced, as will Toronto's continuing evolution as a "World City." Fostering a positive image has become increasingly important as the region becomes integrated with other places through the processes of globalization, many of which are centred on the City of Toronto but have implications for the Greater Golden Horseshoe and beyond as the region continues to evolve as a functionally interdependent conglomerate.

A BRIEF NOTE ON DEFINING THE REGION

"Toronto" can mean many things, and the administrative boundaries of the Toronto urban system have changed markedly over time as the population and spatial extent grew and surrounding communities were subject to amalgamation (box 4.1).

BOX 4.1   URBAN SYSTEMS AND CENTRAL PLACE THEORY

Urban systems are interdependent sets of urban settlements within specified regions. Within an urban system, an interdependent hierarchy of cities and towns exists.

Central place theory attempts to explain the relative size and spacing of towns and cities. Larger urban areas tend to be rarer and relatively far apart and offer a greater variety of services and amenities. They cater to both local residents and those of more distant cities, towns, and rural areas.

The implication is that in any given region, there is a need for only a limited number of large central places offering a wide spectrum of services, goods, and amenities. Toronto provided the central place functions for the rapidly growing rural and remote areas of Ontario, offering specialized services and amenities. Over time, this role continued to solidify.

By the author, with permission

The core of the region is the City of Toronto, formerly the Regional Municipality of Metropolitan Toronto (1954 to 1999). The Greater Toronto Area (GTA), the next outward expansion, comprises the City

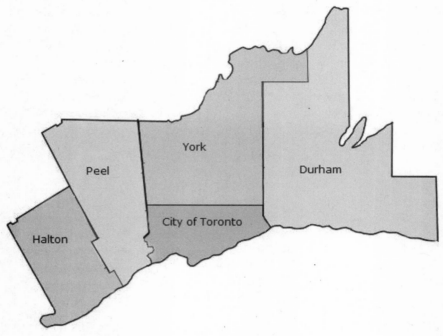

Figure 4.1 Location map. By the author, with permission

of Toronto and the surrounding east to west regions of Durham, York, Peel, and Halton. The Golden Horseshoe includes the GTA and the cities of Hamilton, Niagara Falls, and St Catherine's, while the Greater Golden Horseshoe extends even further to Waterloo Region in the west, Barrie in the north and Peterborough in the northeast.[1] All these names will be used throughout this chapter, depending on the initiative and time period under consideration. As a general rule, the Toronto Georegion/Region is seen as roughly synonymous with the GTA (figure 4.1).

THE LAND USE COMPLEX

The concept of a "land use complex" will be used throughout this chapter to discuss the outward spread of Toronto. A land use complex can be seen to comprise the densely built, urbanized areas fading out into less intensely built suburban areas and eventually the surrounding countryside and farmland. This pattern has existed almost since the inception of York in 1793, expanding rapidly as population grew and technology and particularly transportation allowed. Over time, the extent of this land use complex has increased significantly. As this chapter will show, there have been a number of attempts to modify or control its growth but the continued expansion of the land use complex has had tremendous impacts on the natural and social environments of the peoples and places in and around the region and continues to present a significant challenge for surrounding georegions.

DEMOGRAPHICS

The Toronto Georegion is characterized by high population levels, high rates of growth, and significant cultural diversity (Statistics Canada 2007). The population of the City of Toronto is roughly 2.5 million, while the GTA houses roughly double that number and the Greater Golden Horseshoe about 8.1 million. This latter figure is expected to exceed 11 million by 2031. Currently, the GTA houses two-thirds of the provincial population and one-quarter of the national population. Trends suggest population and spatial growth will continue to be robust.

Forty-three percent of the City's population consists of visible minorities. Roughly one-half of residents are not native to Canada, and 20 percent arrived in Canada during the 1990s. International immigrants continue to be attracted to the Toronto Georegion in large numbers. In addition, the economic and educational opportunities in the Toronto Georegion attract many young adults from other places in Canada.

THE ABIOTIC AND BIOTIC SETTING AND EARLY SETTLEMENT

The abiotic and biotic setting of Toronto, particularly the City of Toronto and the Greater Toronto Area, are well covered in Roots et al. (1999), as well as in chapters 1 and 2 of this volume, and will not be discussed in further detail here. As with any highly settled region, there has been significant land use change and habitat loss in the georegion, although some significant ecological features remain. The Golden Horseshoe Greenbelt area, which is discussed in more detail later, is home to roughly 70 officially recognized species at risk out of a provincial total of 180. Planning to protect this diversity and the associated ecological functions continues to be a challenge and has important implications for the georegion.

The abiotic and biotic setting of Toronto was critical to its early use and occupation both by Native groups and by Europeans (Careless 1984; Heindenreich and Burgar 1999; West 1967; Middleton 1923). Chapters 1 and 2 discuss the early settlement of the Toronto and other georegions and the impacts of evolving transportation technology. In brief, at the time of European settlement, Toronto became a government centre. This, along with a rich terrestrial and marine hinterland, fuelled rapid population and economic growth (Careless 1984; Lemon 1985, 1989; West 1967). As the region's economic power grew, so did its social and cultural relevance as it came to house important institutions and organizations, solidifying its role as a central place (box 4.1). Until the advent of the railroad, however, spatial expansion was limited by the relatively simple transportation systems available (Kalbach 1999). However, early transportation developments such as horse-drawn and, later, electric streetcars did allow for some outward expansion and land use differentiation. Before the turn of the twentieth century, early suburbs such as Yorkville began to emerge, while business and commercial activities were increasingly focused in the city core.

TORONTO IN THE EARLY 1900S

Events leading up to the turn of the twentieth century, including railroad development, population growth, and industrialization, combined with ineffective governance, resulted in a city under significant stress. Living conditions were crowded and deteriorating, in part because of the lack of adequate infrastructure. Similar situations in Great Britain and the United States resulted in a new comprehensive-planning movement (box 4.2) that in subsequent decades would have a profound influence on the growth and development of the Toronto Georegion.

BOX 4.2   THE EMERGENCE OF URBAN PLANNING

The turn of the nineteenth and twentieth century was a significant one for the development of professional planning in Canada and elsewhere. At the end of the nineteenth century, the Canadian population was becoming increasingly urban, and most people settled into already organized municipalities. Industrialization was having a noticeable effect on urban areas, as well as on rural and hinterland areas. In Ontario's case, advocacy on the part of the city of Toronto, which was struggling with uncontrolled suburban development, was an important contribution. The City and Suburbs Plans Act of 1912 and the much-expanded Planning and Development Act of 1917 both aimed at providing municipal authorities with some control over land use decisions. Several iterations were released in the ensuing century.

As urban areas were coping with rapid population growth and uncontrolled development, there was a growing realization that urban planning was a complex task, involving more than a simple design of a town's form. New and more orderly planning was called for, resulting in a municipal reform movement and the development of the planning profession. Two views on the nature of city problems, along with proposed solutions, emerged:

1 Urban problems were the result of deteriorating *living conditions*, leading to the *garden city* movement.
2 Urban problems were the result of the deteriorating *appearance* of cities, leading to the *city beautiful* movement.

The garden city concept was originated by British reformer Ebenezer Howard in 1898. The utopian urban form included considerable green space in the form of both internal greenbelts and a surrounding agricultural greenbelt.

The city beautiful concept involved the redesign of major streets and public areas in existing cities. City beautiful-inspired plans included monument plazas and squares at the heart of cities to foster civic pride and the use of open space as an architectural extension of monuments and important public buildings.

By the author, with permission. *Sources*: Richardson (1989); Hodge (2001).

POPULATION GROWTH AND URBAN STRUCTURE

Population growth continued to be robust in the new century (figure 4.2). An increasing number of immigrants were arriving from

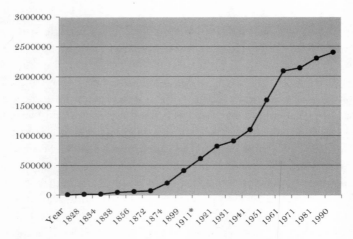

Figure 4.2   Population trend in Toronto. Note that the figures for 1911 and after are for the area of what would become Metropolitan Toronto. The jump in population noted in 1899 is due in part to annexation of neighbouring villages.

continental Europe rather than from Great Britain, changing the face of Toronto.[2] Some of these new arrivals spoke different languages and followed unusual customs, and they were poor – all characteristics that were not common in earlier immigrants. Social unrest was the result, and ethnic enclaves began to appear as part of the urban structure.

By the end of the First World War, Toronto's population was almost 500,000. With the original city already densely populated, growth in the region increasingly occurred in the suburban areas, extending the land use complex outward (figure 4.3). In 1911, about 33,500 people lived beyond Toronto's borders, within what would become Metropolitan Toronto in the 1950s. By 1931 this number had ballooned to 190,000.

The spatial expansion of Toronto occurred mainly through boundary extensions as adjacent communities were annexed. Significant expansions took place in the 1880s and between 1904 and 1912. After this time, the city was no longer so interested in annexing new territories, since the economic costs of extending public services to newly annexed areas were too high. The desire of adjacent, poorer, regions to be sheltered under the Toronto umbrella resulted in various proposals for a metropolitan county governance structure in the 1910s and 1920s, but little was done to implement these proposals because the city and province were not interested.[3]

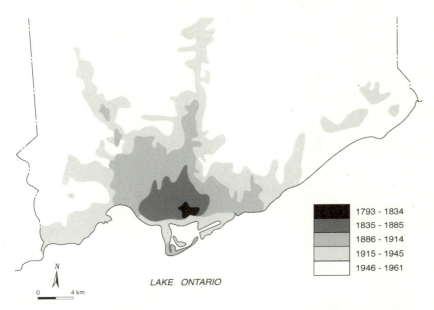

Figure 4.3 Expansion of Toronto Metro Area Population, 1793–1961. From Roots et al. (1999), with the permission of UBC Press

TRANSPORTATION TECHNOLOGY

Changes in transportation technology were at the heart of the city's ability to expand spatially. The streetcar system continued to grow, allowing residential development to expand further from the city core. By the end of the First World War, bus and automobile use was growing rapidly, making longer-distance travel and thus the separation between workplaces and living places increasingly feasible. This resulted in the growth of suburban areas, places that today are part of the City of Toronto but that were then separate communities.

The adoption of this evolving transportation technology not only permitted rapid expansion of the urban complex but also required the reworking of previously developed areas. Street widenings, new street connections, and the installation of traffic lights all became important endeavours, and planning documents in the 1920s and beyond focused heavily on developing the roadway system to accommodate private auto use. In fact, many early organized-planning efforts were centred on meeting the needs of emerging technology and moving an increasing number of people to and from suburban areas. Another significant element of early planning interest was the development of park systems. This dual focus can be seen in such early plans as the Civic Guild of Art's (1909) *Comprehensive Plan for Systematic Civic Improvements in Toronto*, among others (Lemon 1989).

An interesting facet of the Civic Guild's plan was that it not only dealt with lands inside the city's boundaries, but extended its interest into adjacent regions. It reflected a need or perceived need to reach beyond the legal extent of the city in planning for the growth of Toronto, implying that Toronto's interests and residents were more important than those of its neighbours.

While work on developing a comprehensive park system was slow, work on improving the transportation system continued apace, allowing for the increased spread of the land use complex, encroaching on and enveloping neighbouring communities and countryside areas. This trend became especially prominent in the post-Second World War period.

TORONTO IN THE MODERN ERA

Population and Spatial Growth

Toward the end of the Second World War the need for planning became an increasing concern. The region's population continued to grow rapidly, and urban development, particularly in the suburbs, remained largely unplanned. The suburbs added nearly two hundred thousand new residents between 1940 and 1953. Population growth within the city was lower, since it was already heavily urbanized and left little room for new residential development, especially in the form of single-family homes, which gained prominence in this era. The developing sprawl, which enveloped other georegions, was made possible by an increased emphasis on private-automobile use.

In addition to housing an increased proportion of residents, suburban areas increasingly became the location of economic activity, particularly manufacturing development. Changes in transportation technology (e.g., trucking) and the availability of abundant, less expensive land pushed manufacturing activities to the outer edges of the city and into adjacent regions. The result was rapid spatial expansion of the Toronto Region in the 1940s and 1950s. The need to coordinate planning continued to arise, particularly as stark differences between places in the region – for example, in terms of economic development and the provision of social services – became evident.

Planning to the 1990s

The 1943 *Master Plan for the City of Toronto and Environs* (TCPB 1943) represents an early attempt at comprehensive planning. It focuses on transportation, greenbelt development, and the creation of nodes of economic activity throughout the region. The greenbelt was to shape urban form, buffer present and future development, and provide recreational opportunities, but implementation of the greenbelt was impeded by regional differences.

To accommodate the growing prominence of private automobiles, an extensive network of expressways was proposed (box 4.3). Some would be implemented (e.g., the Gardiner Expressway), while others were vehemently opposed by local residents (the Spadina Expressway). The plan clearly demonstrated the prominent role of transportation technology in shaping the urban structure. Expressways were to create a quick and easy way to shuffle people from the suburbs (living space) to the city core (work space). With the growing popularity of private automobiles and the freedom of movement they afforded, the city could and did expand ever further outwards, encroaching on surrounding georegions and transforming countryside lands into part of the metropolis.

With little legislative authority, only bits and pieces of the 1943 plan were implemented. After a major revision to the provincial *Planning Act* and the creation of the Regional Municipality of Metropolitan Toronto (Metro Toronto) in 1953, however, official planning became a legislated requirement.

BOX 4.3   THE SPADINA EXPRESSWAY

The Spadina Expressway was proposed in the 1960s as part of a series of expressways cutting across Toronto. It was intended to run from Highway 401 to the downtown core, following the path of the Cedervale Ravine and Spadina Road and cutting through the well-established communities of Forest Hill and The Annex. Hundreds of homes, including some of historic significance, were slated to be demolished to accommodate the expressway. Local activists, supported by Jane Jacobs, were ultimately successful in preventing construction of the Spadina and other cross-city expressways.

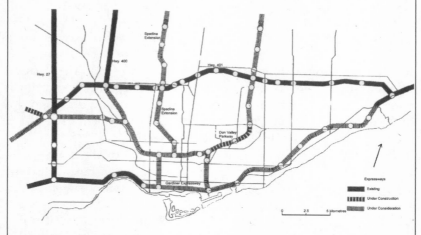

Map showing the proposed expressway development in Toronto, 1959. By the author, with permission. *Sources*: Reeves 1993; Lemon (1985); Sewell (1993)

Planning the New Region

By 1956 the regional population was 1.6 million and expected to rise steadily. The first official plans for Metro Toronto focused on planning for growth (MTPB 1959), including moving forward with expressway development in the expectation that the spatial spread would continue unabated. The creation of a regional parks system was also prominent on the planning agenda, particularly after Hurricane Hazel and its floods struck Toronto in 1954 and turned attention to the control of development and floodplains protection (Fulford 1995). Since providing for the recreational needs of the growing urban population was a key consideration, numerous land areas were excluded from development for this reason, as well as to protect from possible flood hazards (MTPD 1966). The effect was to exclude a considerable land base from urban development. With the ever present thrust for development, however, lands on the fringe rapidly urbanized.

Provincial Interest

The Ontario government did attempt to address Toronto's spatial spread and the challenges posed by the rapid rise in automobile use. The provincial Metropolitan Transportation and Region Transportation Study ultimately led to the Toronto-centred region concept of 1970. Alternative land use scenarios were developed for the Greater Golden Horseshoe area. In effect, the concept allowed Toronto to extend into its hinterland, continuing to absorb surrounding regions, as it had been doing for decades. Subsequent proposals recommended a linear development pattern to facilitate efficient, high-capacity transportation facilities along the lakeshore (Reeves 1993). The "parkway belts" included a greenbelt to control urban sprawl and provide for recreation. The complexity inherent in the proposal, combined with developers and land use pressures and a lack of development by local municipal governments, contributed to the demise of the concept. Urban sprawl in the Toronto Region continued unabated, and little provincial action was taken to control this trend for some time.

The 1980s generally saw a growing concern with the human impact on the environment, including the consequences of urban sprawl (MTPC 1980). One emerging approach to lessen the impact of sprawl was to develop new land use categories such as Environmentally Significant Areas (ESAs) (box 4.4). Most development was to be excluded from these areas. A side effect, however, was that environmentally significant lands excluded from development within the city core meant greater development pressure on lands located on the urban fringe, contributing to further sprawl.

BOX 4.4   URBAN PROTECTED AREA CLASSIFICATIONS

In the 1970s, ecological science was increasingly used in planning urban protected area systems, evolving over time to include increasingly complex greenland and natural heritage systems.

Environmentally Significant Areas
Areas possessing unique or unusual natural features and/or functions essential for ecosystem health and sensitive to human activity.

> *Provincially Significant Wetlands*
> Categorizes wetlands based on biological, social, hydrological and "special" features such as rarity, ecological age, or habitat. Wetlands designated class 1, 2, or 3 are considered "provincially significant wetlands."
>
> *Areas of Natural and Scientific Interest (ANSIs)*
> Areas of land and water that contain natural landscapes or features that are significant on natural, scientific, and educational grounds. ANSIs may focus on physical features (Earth Science ANSIs) or outstanding landscapes and ecological communities (Life Science ANSIs).
>
> *Natural Heritage Systems*
> Systems of core natural habitat areas with existing or potential connectivity. Often seen as an ecological "blueprint" for a region's natural ecosystems, other "green land" areas include greater consideration of recreation and aesthetics.
> Environmentally Sensitive Landscapes (SSLs)
> Geographically and ecologically definable landscapes distinguishable from surrounding areas by the concentration, proximity, and/or overlap of significant ecological features and functions. Stewardship and restoration of regional biodiversity and ecosystem health are central objectives
>
> By the author, with permission.

TORONTO IN THE 1990S AND BEYOND

As noted at the beginning of this chapter, population growth has continued to be robust in the Toronto Georegion, a trend that is expected to continue. At the same time, environmental awareness continued to grow in the 1990s and in the new century. Our understanding continues to evolve and is reflected in a number of initiatives in the 1990s, initiatives focusing frequently on new goals such as biodiversity and sustainability and more recently on combating global climate change.

Planning the Region in the New Environmental Age

A significant revision to how the region could or should be conceived was put forth by the 1992 Royal Commission on the Future of the Toronto Waterfront (RCFTN 1992). The task force introduced the concept of the Greater Toronto Bioregion (GTB), a redefinition of the Greater Toronto Area based on ecological boundaries. Thinking at this scale, it was suggested, was necessary to truly address environmental concerns (e.g., the waterfront). Drawing heavily on the concept of sustainability, it noted that the health of the natural environment was closely linked to that of its economy and society. This scale of attention recognized that the region was now so interconnected and interdependent, reaching beyond Metro Toronto into surrounding georegions, that joint planning was essential.

A key feature of the GTB is the Oak Ridges Moraine (ORM), toward the fringe of north Toronto (box 4.5). Provincial planning efforts for the ORM have focused on protecting significant natural features and functions from development (ORM-TWC 1994). As in the past, the reach of Toronto into surrounding areas and the primacy of Toronto-based interests were notable. Because planning was couched in terms of Toronto's needs, conflict with other municipalities that did not want to be defined in terms of their value to the Toronto Georegion or see their communities annexed (formally or informally) naturally arose. There has been some effort to modify thinking about countryside areas, from considering it as land in reserve for future expansion to considering it as consisting of places with their own uniqueness and importance. Recognizing this would mean realizing that highly urbanized and rural parts of the georegion must work cooperatively to create a more livable and sustainable future for all inhabitants.

BOX 4.5   THE OAK RIDGES MORAINE

One of Ontario's most significant landforms, the Oak Ridges Moraine performs several essential functions, including providing significant natural habitat, surface and ground water resources, and scenes and recreational resources. Population growth has had a significant impact on the Moraine, and development pressures are likely to continue in the coming decades.

> The Moraine provides an example of a "peri-urban" area, still largely rural in that it is at risk of being subsumed by sprawl. Urban residents rely on water resources from the Moraine as well as aggregates extracted for construction materials, while also enjoying the recreational opportunities that the Moraine affords. Planning for the Moraine has often resulted in conflict, since local residents feel their needs and desires are dismissed in favour of those of urban residents.
>
> Provincial plans for the Oak Ridges Moraine were released in the early 1990s and early 2000s. These plans divided the Moraine into different land use categories. The 2001 *Conservation Plan*, for example, divided the Moraine into four areas as follows:
>
> - Natural core areas (38 percent) to maintain and, where possible, improve or restore ecological integrity.
> - Natural linkage areas (24 percent) to maintain, improve, and restore ecological integrity and to maintain or restore regional-scale open-space linkages between core areas and along river valleys and stream corridors.
> - Countryside areas (30 percent) to provide an agricultural and rural transition and buffers between core and linkage areas and the urbanized settlement areas.
> - Settlement areas (8 percent): urbanized areas that reflected the range of existing communities and their needs and values.
>
> By the author, with permission. Information from OMMAH (2002)

At the root of the continuing spread of the urban complex and the conflicts that ultimately arise, however, is our continued overreliance on private automobiles. We as a community have continued to support this addiction, often to the direct detriment of public mass transit systems. The GTA is in many ways a fractured georegion with multiple jurisdictions that are often in conflict over the costs and benefits of action, including curbing sprawl and protecting the natural environment. Jurisdictional arguments have included concerns in the suburban area about being saddled with Toronto's financial issues, many of which benefit residents of surrounding areas or the most vulnerable, but they have dampened any attempts to deal with the GTA in a comprehensive manner. At a provincial scale there has often, apparently, been a reluctance to enforce any new governance structure – or to provide any significant powers to bodies such as the now-defunct Greater Toronto Transit Authority – in order to maintain the political support of those in the areas lying outside the city itself in the "905" region. Interestingly, the October 2007 provincial election suggests that political attitudes in the 905 region and in the 416 (city) area may be merging. However, it is much too soon to reach any firm conclusions.

TORONTO TODAY

Today, the City of Toronto and the GTA house 21 and 42 percent of the provincial population, respectively. The cultural, economic, and political power structures laid down in Toronto since the 1700s have continued to evolve and harden. With current rapid rates of globalization, Toronto's position as the central city in the provincial and national urban systems has continued to strengthen, and it has become increasingly linked to the international economic system.

Urbanization now extends almost to Lake Simcoe, the Niagara Escarpment, and Pickering. With large portions of the GTA and the Golden Horseshoe still rural in nature, there is significant opportunity for further sprawl. Automobile use is increasing steadily, which will only prompt expansion to and beyond the current limits of urbanization, resulting in the spread of the land use complex. The increasing interconnectedness of places on the "fringe," for example Guelph, Kitchener-Waterloo, Barrie, and Peterborough will be aided by transportation changes beyond the use of private automobiles. Recent and planned future expansions in the provincially supported GO Transit service, for example, will make commuter travel within the greater region ever more viable.[4] Furthermore, although lands continue to be excluded from development to protect special places for recreational and environmental purposes, this will increase pressures on other areas (box 4.4).

Recently, the provincial government has developed two plans to try to control growth in the Toronto Region (OMMAH 2005, OMPIR 2005). *Places to Grow* and the *Golden Horseshoe Greenbelt* aim to direct future growth in a way that will protect significant environmental features and functions, agricultural resources, and rural areas. If fully implemented, they would place a presumably permanent limit on the spatial extent of the Toronto Georegion. Together, the two programs are intended to create a more livable and economically prosperous community. Many of the challenges facing the Toronto Georegion are expected to be addressed, including traffic congestion, environmental degradation, infrastructure planning, and sprawl.

It is far too early to judge the effectiveness of these recent efforts. Both initiatives have been criticized as being too timid in their targets to increase density and for excluding too many important areas from protection. "Leapfrog development," that is, increased urbanization pressures to the north and west of designated lands, is a real concern, as is interjurisdictional conflict. However, the value of this planning has to be acknowledged. History has shown that efforts to control sprawl on a regional municipality-by- municipality basis are not sufficient. Until recently – and this is still largely true – there has been no effort to coordinate public transit systems to allow users to travel around the region in an efficient manner, enhancing reliance on private automobiles and feeding the sprawl cycle.

A possible problem is that there is no governance regime for the GTA or the Golden Horseshoe outlined in either plan, meaning that the existing fragmented government structures will be retained, thereby placing emphasis on provincial control with, in turn, a likely continued emphasis on planning for Toronto much more than for other georegions of Ontario. Although this does bypass significant barriers to implementing a regional governance structure, care will have to be taken to ensure that important issues such as uneven economic performance and social equity within the region are also addressed, since the central city, the suburbs, and rural areas continue to struggle over economic and political rights and responsibilities. Conflict continues to exist. Region-wide issues such as waste management and uneven rates of taxation and service provision, as well as providing for the needs of newcomers, all require ongoing attention and seem best dealt with on a GTA or wider basis.

Of particular relevance is developing an efficient, effective public transportation system for the georegion. Recent action in this regard – a new regional transit authority (Metrolinx) – has potential. However, this project is seen as having a significant impact on the environment and farmlands, for example, in the case of the proposed Toronto–Niagara link atop and west of the Niagara Escarpment. The Metrolinx proposal will require sustained political support locally and provincially, as well as a firm, enforceable mandate. It is not inconceivable that unpopular measures, such as congestion charges and tolls on major roadways will be necessary to truly curb dependence on automobiles. The October 2007 re-election of the McGuinty Liberal government means support for Metrolinx, the growth plan, and the greenbelt will continue into 2011, but beyond that, political change could have uncertain outcomes.

TORONTO IN THE POST-INDUSTRIAL GLOBAL ERA

With current globalization, major urban areas in affluent countries face a variety of new and emerging challenges, and central cities such as Toronto tend to be the focus of stress (Hodge 2001). Significantly, globalization has resulted in the loss of many second-order economic activities, such as manufacturing in the region and the province, as the economy moves toward tertiary activities (the exchange of goods and services) and quaternary activities (knowledge and information-based activities). Toronto, which is focused on the city but with implications for other places within the region, has become increasingly integrated into the global economy and represents Canada's best claim to housing a global city, or "World City."[5]

Although it is not a new trend, the Toronto Region continues to be the chosen home of a significant proportion of newcomers to Canada. Over the last several years, the Toronto Region became home to about 107,000 international immigrations each year (City of Toronto 2006), and more than one hundred languages and dialects are spoken in the region. All of this points to a wonderful diversity that is having an impact on the look of the region. Many newcomers are settling into "ethnic burbs" or "ethnoburbs," concentrations of ethnic groups in more central and suburban areas. With uncertain overall implications, related services and amenities follow, changing and shaping the urban structure.

Economic change is also altering the urban structure. Old industrial lands are increasingly being redeveloped for tourism, recreation, housing, and other uses. New housing, notably in the form of condominium high-rises, is rapidly being built within the city core. New communities are planned or are in the process of being built on former "wastelands," such as the West Donlands area of the waterfront. In addition, amenities are being redeveloped to enhance the "livability" of the city and its attractiveness to international investors and workers. Key examples include the recently completed Opera House, redevelopment of the Royal Ontario Museum and the Art Gallery of

Ontario, and the growing array of restaurants and lifestyle-oriented services. Mayor David Miller was active in trying to raise Toronto's profile, attempting to position Toronto as a world leader in dealing with climate change, for example. These moves are intended to increase the attractiveness of Toronto in terms of economic, social, and environmental sustainability but also in the emerging and evolving era of globalization. While this draws new people and businesses to Toronto, it also puts much pressure on the georegion to continue to grow, to continue to exert its influence into surrounding georegions in order to "make the city work." This is often couched in terms of the idea that "what is good for Toronto is good for Ontario/Canada," since Toronto is the "economic engine" of the province. Concern over Toronto's prominence all too frequently follows.

CONCLUSION

Where does all of this leave Toronto as one region among many in Ontario? Certainly some features of Toronto place it in a unique position. The Toronto Region continues to grow relatively quickly in terms of population, employment, and spatial extent, the latter driven in large part by the current transportation system. The interconnectedness and interdependencies between the Toronto Georegion and the Greater Golden Horseshoe are becoming increasingly complex, blurring the definition of the Toronto Georegion.

There are important equity challenges that are difficult to reconcile. The Toronto Region is often accused of demanding and receiving too much attention, often at the perceived or real expense of other places. On the one hand Toronto does house a disproportionate amount of the provincial and national population. From a political standpoint, no politician can afford to ignore the Toronto Region and its needs. The region – especially when considering the Greater Golden Horseshoe – is also very complex, comprising multiple jurisdictions ranging from small rural areas struggling to succeed, to the largest city in Canada. Creating some fairness within this disparate area requires support from an upper level of government, one that can work through local jurisdictional conflicts.

On the other hand, many regions within Ontario have their own unique histories, characteristics, and challenges, as the chapters in this book demonstrate. These georegions, which are home to about 50 percent of the province's population, have value in and of themselves. The portions of the province closest to the Toronto

Georegion understandably are concerned about being melded into it and losing their identity as unique places. The critical challenge is to balance the needs of the Toronto Region and other places, addressing them in a fair and equitable manner (whatever that may be), while continuing to celebrate the diversity and opportunities – social, economic and environmental – of all of Ontario's georegions. At the core of this will be finally halting the excessive spatial spread of the megalopolis, thereby allowing smaller communities and currently distinct places to thrive without threat of engulfment by the big city. Perhaps then Toronto can be appreciated for the wonderful place it is, rather than often being reviled as a monstrous, uncontrollable entity.

POSTSCRIPT

Since completion, elections have been held in Toronto and the province. A new mayor and council were elected for the city whose policies, among other things, could affect transit, the role of the automobile, and sprawl. The Liberals remain in power provincially, but as a minority government whose effects are uncertain.

NOTES

1 The Greater Golden Horseshoe includes the cities of Toronto, Hamilton, Kawartha Lakes, Guelph, Peterborough, Barrie, Orillia, and Brantford, as well as the regional municipalities of Halton, Peel, York, Durham, Waterloo, and Niagara and the counties of Haldimand, Brant, Wellington, Dufferin, Simcoe, Northumberland, and Peterborough.
2 This trend would continue throughout the century, accelerating after 1967, when Canada adopted an official, non-racist immigrant-selection policy, ultimately creating the great diversity that is seen in the Toronto Georegion today.
3 In contrast, by later decades the city would favour annexation, while surrounding communities resisted and pursued a new tier of regional governance as an alternative. By the late 1990s, when amalgamation was imposed by the provincial government, it was opposed by many both in the city and in surrounding communities.
4 Increases in the price of gasoline, as happened in the summer of 2008, may be pushing some people away from the private automobile, which could in time change this trend and result in increased support for mass

transit and denser city living. Certainly, there has been a significant influx of population into the city core in recent years. However, no decrease in the rate of sprawl is yet apparent.

5   A world city is one that exerts disproportionate control over other areas through information, finance, and cultural products. It acts to sustain and spread globalization.

REFERENCES

Careless, J.M.S. 1984. *Toronto to 1918: An Illustrated History.* Toronto: James Lorimer & Co.

City of Toronto. 2006. *Toronto's Racial Diversity.* Accessed online. www.toronto.ca

Frisken, F. 1983. *Conflict or Cooperation? The Toronto-Centred Region in the 1980s: Symposium Proceedings.* Urban Studies Programme. Toronto: Faculty of Arts, York University.

Fulford, R. 1995. *Accidental City: The Transformation of Toronto.* Toronto: Macfarlane Walter & Ross.

Heidenreich, C.E., and R.W.C. Burgar. 1999. "Native Settlement to 1847." In B.I. Roots, D.A. Chant, and C.E. Heidenreich, eds, *Special Places: The Changing Ecosystems of the Toronto Region,* 63-75. UBC Press: Vancouver and Toronto.

Hodge, G. 2001. *Planning Canadian Communities: An Introduction to the Principles, Practice, and Participants.* 4th ed. Toronto: Thomas Nelson.

Kalbach, W.E. 1999. "Spatial Growth." In B.I. Roots, D.A. Chant, and C.E. Heidenreich, eds., *Special Places: The Changing Ecosystems of the Toronto Region,* 77-89. UBC Press: Vancouver and Toronto.

Lemon, J. 1989. "Plans for Early Twentieth-Century Toronto: Lost in Management." *Urban History Review* 18(1): 11-31.

Lemon, J.T. 1985. *Toronto since 1918: An Illustrated History.* Toronto: James Lorimer.

Metropolitan Toronto Parks Department (MTPD). 1966. *A Twenty-Five Year Development Concept for Regional Parks.* Toronto: Metropolitan Toronto Parks Department.

Metropolitan Toronto Planning Board (MTPB). 1959. *The Official Plan of the Metropolitan Toronto Planning Area.* Toronto: Metropolitan Toronto Planning Board.

Metropolitan Toronto Planning Committee (MTPC). 1980. *Official Plan for the Urban Structure: The Metropolitan Toronto Planning Area.* Toronto: Municipality of Metropolitan Toronto.

Middleton, J.E. 1923. *The Municipality of Toronto: A History*. Toronto: Dominion Publishing Co.
Oak Ridges Moraine Technical Working Committee (ORM-TWC). 1994. *The Oak Ridges Moraine Area Strategy for the Greater Toronto Area: An Ecological Approach to the Protection and Management of the Oak Ridges Moraine*. Maple, ON: Ontario Ministry of Natural Resources.
Ontario Ministry of Municipal Affairs and Housing (OMMAH). 2002. *Oak Ridges Moraine Conservation Plan*. Toronto: Ministry of Municipal Affairs and Housing.
— 2005. *Greenbelt Plan*. Toronto: Ministry of Municipal Affairs and Housing. www.mah.gov.on.ca
Ontario Ministry of Public Infrastructure Renewal (OMPIR). 2005. *Draft Growth Plan for the Greater Golden Horseshoe.* www.pir.gov.on.ca
Reeves, W.C. 1993. *Vision for the Metropolitan Toronto Waterfront, II: Forging a Regional Identity, 1913–1968*. Major Report 28. Toronto: Centre for Urban and Community Studies, University of Toronto.
Richardson, N. 1989. *Land Use Planning and Sustainable Development in Canada*. Ottawa: Canadian Ecological Advisory Council.
Roots, B.I., D.A. Chant, and C.E. Heidenreich, eds. *Special Places: The Changing Ecosystems of the Toronto Region*. Vancouver and Toronto: UBC Press.
Royal Commission on the Future of the Toronto Waterfront (RCFTW). 1992. *Regeneration: Toronto's Waterfront and the Sustainable City: Final Report*. Toronto: Queen's Printer of Ontario.
Sewell, J. 1993. *The Shape of the City: Toronto Struggles with Modern Planning*. Toronto: University of Toronto Press.
Statistics Canada. 2007. *Portrait of the Canadian Population in 2000: Population and Dwelling Counts, 2006 Census*. Statistics Canada Catalogue no. 97-550-XIE. www.statscan.ca
Toronto City Planning Board (TCPB). 1943. *Master Plan for the City of Toronto and Environs*. Toronto: Toronto City Planning Board.
Toronto Guild of Civic Art. 1909. *Report on a Comprehensive Plan for Systematic Civic Improvements in Toronto*. Toronto: Toronto Guild of Civic Art.
West, B. 1967. *Toronto*. Toronto: Doubleday Canada.

5

The Carolinian Canada Georegion: Farmland, Forests, and Freeways in Conflict

STEWART HILTS

SUMMARY

The Carolinian Georegion in southwestern Ontario is the heartland of Canada in more ways than one. It is the urban-industrial heartland, with dynamic cities, manufacturing, and important transportation facilities; it is the agricultural heartland, with a greater diversity of crops than any other region in Canada owing to its relatively warm climate; and it is the ecological heartland, with more rare and endangered species than any other region in the country. The long settlement history of this region has led steadily to the growth of an urban-industrial economy centred in Toronto but extending increasingly southwest across the Carolinian region; urban-industrial "progress" and economic growth are still dominant values in an "open society."

But the Carolinian region is also a climatic region, the warmest region in Canada, supporting both unique natural habitats and the most diverse agricultural production in the country. The unique ecological values of the region have spawned a dedicated long-term conservation effort, spearheaded by the non-government organization Carolinian Canada Coalition. Among other things, this effort has pioneered stewardship programs to recognize and reinforce the conservation commitments of private landowners. The warm temperatures and adequate rainfall have fostered intense and diverse agricultural production.

The policy challenge in this region is to balance the economic values of "progress" with support for conservation and agriculture, to design a balance of policies that will support farmland, forests, and freeways together. We must move beyond the metropolitan dominance of urban-industrial growth, driven by the economic forces centred in Toronto, to a more balanced future in the Carolinian region, as in other rural regions of Ontario, including a flourishing future for agriculture and conservation.

THE REGION

In Ontario, the Carolinian Region is too often seen merely as the empty space between Toronto and the key American border crossings at Buffalo and Detroit. It has a vague agricultural character, but it is mainly a "blank slate" available for highways, rail lines, and sometimes auto plants, all essential to keep our economy growing. No other georegion of Ontario epitomizes this blank slate model of landscape planning as does the Carolinian region.

In fact, the Carolinian region is anything but an empty space and has unique and important values of its own (table 5.1). Agriculture here is the most diverse of any region, not just in Ontario, but in Canada. Natural habitats and species populations are equally significant in their own right, with a greater concentration of species diversity, including rare and endangered species than any other georegion, again not just in Ontario, but in Canada (Johnson 2007).

While the City of Toronto sees itself as the economic heartland of Canada, whose continued "growth and progress" is essential for the health of the country, the Carolinian region in southwestern Ontario is in fact the agricultural and natural heartland of Canada. Here we grow more grapes, peaches, tomatoes, cucumbers, and numerous other vegetables, as well as flowers, than any other place in the country. Here we find tulip and Kentucky coffee trees, rare warblers and turtles, the blue racer snake, and many other unusual species in the remaining natural habitats. Some of Canada's rarest prairie habitats are here, as are unique shoreline habitats such as sand dunes, protected in provincial and national parks like Point Pelee and Long Point.

Both the productive agriculture and the unique ecology of the Carolinian region result from its relatively warm climate (it is the

Table 5.1
Characteristics of the Carolinian Georegion

Location	The Carolinian Georegion stretches from Toronto southwest to include all of the area north of Lake Erie, approximately along Highway 401 to London and then along Highway 402 to Sarnia.
Urban dominance	Defined by natural characteristics of soil and climate, the Carolinian region includes Toronto and areas west along the north shore of Lake Ontario, but most of the natural characteristics of this part of the region have been destroyed by urban growth. Toronto is considered a separate region in this volume.
Size	24,300 square kilometres
Population	Approximately 6,163,000 including the Greater Toronto Area, but only 3,682,000 not including Toronto. The overwhelming majority of this population is urban.
Major urban centres	Toronto, Oakville, Burlington, Hamilton, Niagara Falls, St Catharines, Brantford, London, Sarnia, Windsor.
Economic activities	Manufacturing, especially auto parts and autos, although this sector is currently in decline; agricultural production including a wide variety of crops and livestock.
Landscape fabric	With the exception of the Niagara Escarpment and major river valleys, a relatively flat landscape with high-quality soils and the warmest climate region in Canada.
Environmental quality	The best farmland with the greatest diversity of crop production in Canada, and the greatest concentration of biodiversity and rare species in Canada.

Source: By the author, with permission.

warmest region in the country) and its high-quality soils (it has the most class 1 soil for farming of any part of Canada). Unfortunately, these natural characteristics are in direct conflict with the highways, industry, and urban growth that Toronto demands and fosters in the region.

As discussed in chapter 1, the Carolinian region is one of the oldest settled regions in Ontario, with numerous small towns and villages, most first established during European settlement before 1850. The "age of progress" from 1850 to 1900, discussed in chapter 2, brought the spread of railroads and then industry, reinforcing the growth of major urban centres, above all Toronto. Today this has created the Golden Horseshoe around the west end of Lake Ontario, a circle of light easily visible from space. Now highways 401, 402, 403, 407, and the Queen Elizabeth Way (QEW), all linking the Toronto region to the United States at Buffalo and Detroit, are a magnet for major industries such as auto plants and suppliers. Urban growth continues apace regardless of the agricultural and ecological uniqueness of the landscape that is being devoured.

The physical landscape, the history of settlement on that landscape, and today's pattern of land use are a complex story. The combination of rapidly growing urban centres and industrial development, excellent farmland with intensive agriculture, and of forests and wetlands with rare species presents a huge set of policy challenges. Conflicts among farming, forests, and freeways underlie all the land use choices we face in this region. Today government policies struggle to recognize a balance between protecting the environment and farmland and also allowing urban and economic development. In the long-run future, how we as a society handle these choices will determine the ecological, agricultural, and economic health of the province and the nation.

In this chapter we look first at what defines Carolinian Canada as a georegion and then examine the physical and ecological landscape. Following this we consider its settlement history and patterns of land use today. The story of conservation efforts is of special interest, since private stewardship programs now widely used across Canada were pioneered here. Finally we examine a group of key policies influencing the future of the Carolinian region.

THE CAROLINIAN GEOREGION

What makes the Carolinian Georegion a georegion? Location, location, location! From an economic point of view, location in the urban-industrial heartland of North America defines this georegion. The highways and rail lines that tie Toronto and the GTA to the northeastern United States support economic growth at all costs. From an agricultural and ecological viewpoint, the high-quality soils and warm climate define the Carolinian Georegion, supporting diverse agricultural production and unique ecological habitats. These values are in direct conflict.

Economic Growth and Industry

The first element of location that defines the Carolinian Georegion is that it sits in the urban-industrial heartland of eastern North America, between Toronto, Buffalo, and Detroit. The region exhibits major centres of urban and industrial growth, including Toronto, Hamilton, and London. It is the core of Canada's auto industry, served by both railways and highways, especially the 400 series, highways 401, 402, and 403 and others that are planned. In terms of this dominant economic development, the unique natural heritage and the prime agricultural land are both almost irrelevant to this burgeoning region.

The history of this settlement growth is well told in chapter 2, and won't be repeated here. Even in the 1850s southern Ontario's economy depended on exports, in that era primarily of wheat, unlike the exports of lumber of the regions further north and east. Early railways were built to facilitate the movement of these exports, and ports were important to carry them to Europe by ship. Although the focus has changed from exporting wheat to Britain to exporting cars, auto parts, and agricultural products to the United States, transportation facilities and urban growth have been continuously reinforced over the past two hundred years.

Because of the close economic linkages between the United States and Canada, the Carolinian Region is once again often seen simply as an undeveloped space between Toronto, Detroit, and Buffalo, a space merely awaiting future urban and industrial growth. Because both auto production and agriculture, our number one and number two industries, depend heavily on exports, transportation to the US markets is critical, and highways and rail lines, always under pressure for expansion, provide the routes. In the past twenty years we have come to depend heavily on the trucking industry particularly, for "just-in-time" delivery of virtually all consumer goods. Anyone who regularly drives our highways knows the dominance of transport trucks in our manufacturing system.

Urban growth is directly linked to this industrial economy. The Greater Toronto Area (GTA) dominates, but Hamilton, London, Windsor, St Catharine's, and other urban centres provide nodes of urbanization throughout the region. Auto plants have spread to smaller cities like Woodstock and Ingersoll. Servicing these urban communities, especially the many suburbs and commuters in the Golden Horseshoe of urban areas that circles the west end of Lake Ontario, puts an enormous strain on public financial resources. The provision of the drinking water supply and of sewage treatment without degrading surface and groundwater resources is as costly as the constant improvements to transportation facilities.

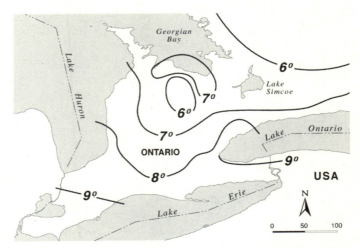

Figure 5.1 Mean daily temperatures. From Hilts and Mitchell (1998), with permission

Natural Habitats and Ecosystems

Given that the Carolinian Region is the urban-industrial heartland of Canada, we often forget that it is also essentially an ecological region defined by climate. As mentioned in the summary, it is the warmest climate region in all of Canada and has some of the best soils in the country. As a result, it has both the best farmland and growing conditions for crops and the greatest range of biodiversity in terms of trees, plants, birds, insects, and other species. It is an ecologically defined region, the northernmost (Canadian) part of the great deciduous forest region of eastern North America.

Comparing a map of the Carolinian region to a map of mean daily temperatures in southern Ontario reveals that the boundary of this area corresponds closely to the eight-degree (Celsius) isotherm (figure 5.1). South of this line the climate is on average significantly warmer than to the north.

It so happens that this eight-degree line corresponds to the northern limit of many tree and shrub species, so that ecologically, this region differs significantly from areas further north. Dr James Soper and Dr Sherwood Fox, botanists from the University of Toronto and the University of Western Ontario, did the pioneering work on mapping these species in the 1940s and 1950s and developing

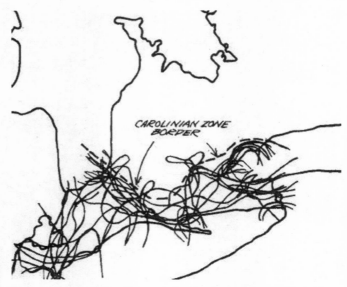

Figure 5.2 Sample Carolinian species distributions. From Hilts and Mitchell (1998), with permission

maps showing the northern limits of many tree and shrub species. A sampling of these limits is shown on figure 5.2, illustrating how clearly many species are restricted to the Carolinian zone of southern Ontario.

This Carolinian forest region is commonly known as the southern deciduous forest region. In southern Ontario sugar maple, white ash, American beech, black cherry, walnut, and other trees collectively known as hardwoods are dominant. Species restricted to the Carolinian Region and much less common or rare include tulip, sassafras, Kentucky coffee tree, blue ash, flowering dogwood, and several oak and hickory species.

Trees are not the only rare Carolinian species. Birds such as the Acadian flycatcher and herptiles such as Fowler's toad, the blue racer, the Lake Erie water snake, and the eastern spiny softshell turtle are all restricted to this ecological region of Canada. Nationally, because these species are found nowhere else in Canada, a large number of them are designated as rare or endangered species under species at risk legislation.

Small patches of rare prairie and oak savanna habitats remain. These associations often arose from aboriginal village and cultivation patterns, since early native communities found farming on lighter (sandy) soils easier. The Carolinian zone was not a continuous forest when white settlers arrived on the scene; it was a mixture of forests, clearings, wetlands, and prairie meadows.

Beyond the unique mixture of species characteristic of the Carolinian are the unique physical features, especially hundreds of miles of Great Lakes shoreline with important wetlands and three major sand spits that extend into Lake Erie as well as the Niagara Escarpment. The Escarpment curves through the Niagara region, around Hamilton, and northwards. It is governed by the Niagara Escarpment Plan, a conservation-oriented program that has now been in place for many years, as discussed in chapter 13 of this volume. Long Point, Rondeau, and Point Pelee, which are major sand spits along the north shore of Lake Erie, all bring a unique assemblage of plants and habitats along the shoreline. As well, they are subject to intense recreational pressure, since they have beautiful sand beaches. All three are largely conserved as parks, though management challenges related to surrounding land use and recreational pressure continue.

Since together these species and habitats rank the Carolinian Georegion of southern Ontario as having the greatest biodiversity of any region of similar size in Canada, it has been the focus of unique and innovative conservation programs, in particular the Carolinian Canada program described in greater detail below. Obviously the potential conflict between this rich natural heritage and dominating urban-industrial growth provides one important dimension of land use policy.

Agriculture

But to the conflict just described we must add a third layer that defines this georegion, the dominant agricultural land use of the rural parts of this area. The same climate that leads to high natural biodiversity enables the growth of a wide variety of crops that will not survive elsewhere in Canada, not only because temperatures are warm but also because rainfall is adequate and water is available for the irrigation of specialized crops.

Agricultural meteorologists have developed an index called the Agro-Climate Resource Index, or ACRI, to measure the adequacy of warm temperatures and rainfall for farming. ACRI values across Canada clearly show how unique the Carolinian region of southern Ontario is. The ACRI 2.5 isobar includes all of the Carolinian zone and extends around the Great Lakes throughout southern Ontario; all of the area with an index greater than ACRI 3 is in the Carolinian Georegion. No place else in Canada provides such good conditions for farming.

The outstanding climate is matched by excellent soils. Over half the class 1 soil in Canada, soil with no limitations for agriculture, is located in southern Ontario. Together, this soil and the excellent climate provide for a diversity of crops unheard of elsewhere in the country. Within the Carolinian region there are specific local soil and climate differences that are also significant. The climate provides a particularly long frost-free season, which has attracted greenhouse agriculture to the Niagara Fruit Belt. The combination of the soil and the local micro-climate caused by the proximity of the Niagara Escarpment to Lake Ontario allows for the growth of tender fruits like peaches, as well as grapes for the wine industry.

Farmland, Forests, and Freeways

The total acreage of the land that has been cleared for agriculture represents the largest portion of land in the Carolinian Georegion. Forests and other remaining natural areas occupy much less acreage, perhaps 20 percent of the region overall. The warm climate and the excellent soil that, as we have seen, underlie the natural biodiversity also underlie the agricultural diversity, but while these two land uses compete for the same resources, both also compete with urbanization. Our devotion to materialistic values and economic growth ensures that urbanization will always win. Land use policies to support the continued existence of farmland and forests will always be the major challenge in this region.

Many specific examples of land use conflict illustrate this three-way land use dilemma. A highway up the Red Hill Creek Valley along the Niagara Escarpment in Hamilton, the loss of Warbler Woods in London to a subdivision, the new Toyota auto plant on excellent farmland in Woodstock, a proposed new highway from Niagara Falls to Hamilton, these and many other examples pose difficult land use choices for local politicians. Often it is these specific local conflicts that draw citizens into awareness of the land use pressures, the policies we use to balance them, and the tough choices we must make between them. Behind them all is the dominating influence of the Greater Toronto Area and its demands for urban growth. The province's growth-management strategy, calling for urban growth centres, simply reinforces this pattern.

Before considering some recent innovations in land use policies, we turn to a review of the settlement that has occurred in this region, in order to make the origins of the current situation clear.

SETTLEMENT HISTORY

As discussed in the first two chapters of this volume, the human history of Ontario is a fascinating story. Beginning with the retreat of the glaciers approximately ten thousand years ago, the landscape of southern Ontario moved through stages such as arctic tundra and northern forest to eventually become the mixed and deciduous forest that characterized the region in about 1800. Native peoples changed over these millennia along with the environment, from hunters of caribou and gatherers of indigenous plants to settled agriculturalists, and the Carolinian region was critical for this transformation. Native farming activities began as early as 500 BC and were concentrated on areas of lighter sandy soil, which was more readily worked by simple wooden tools. The cultivation of corn, beans, and squash in a semi-cleared landscape around villages was typical. The best story of this culture is told and demonstrated at Crawford Lake, west of Toronto. Here, just south of Highway 401 on the edge of the Niagara Escarpment, a reconstructed Indian longhouse village tells the story of native agriculture ending about four hundred years ago when this particular village was abandoned.

The first contact between European society and native peoples occurred with the fur trade. While the later fur trade era spread throughout the north and west of Canada, very early trade routes did include the Carolinian Region, and both Niagara and Detroit were important French outposts in the 1700s (see chapter 1). The valleys of both the Grand River and the Thames were important travel routes. By the time settlement by Europeans began in earnest, as the nineteenth century arrived, native people of Iroquoian heritage lived in several locations in the Carolinian Region. Unfortunately white settlers brought diseases such as small-pox that wiped out many people in the native communities.

Throughout the eighteenth century, until the War of 1812, native peoples had been drawn into the larger conflict between the French and the English in eastern North America, much of which took place in the Carolinian Region. There had been some earlier French settlement in association with the fur trade, especially along the east shore of the Detroit River, around Windsor and Lake St Clair, but with the establishment of the United States of America in 1783, following the American Revolution, the lines were drawn, and many Loyalists (loyal to the British Crown) migrated to southern Ontario, including the Carolinian Region, crossing the border at Niagara and finding land in Upper Canada, often along the Lake Erie shore.

For the most part, native peoples sided with the British during the War of 1812, and a number of famous battles were fought against American invasions in the Carolinian Region. Many know the story of Laura Secord walking through the night to warn the British troops before the Battle of Beaver Dams, near present-day Thorold, in 1813. The war ended in stalemate between Britain and the United States, peace was restored, and settlement continued as the flow of Loyalists increased, augmented by Europeans driven off their land in "the Highland Clearances" (box 5.1). This white settlement cleared the land for homesteading and the beginning of Euro-American agriculture starting in the early 1800s; the Carolinian forest has been declining ever since. Native communities were gradually relegated to reservations, where they still live today.

> BOX 5.1   COMMENT ON CLEARANCES AND SETTLEMENT
>
> The contrast between the clearances in Scotland and pioneer settlement in southern Ontario, and elsewhere that Scots settled, is one of the great ones in the modern interpretation of historic events. The clearances, from a Scottish point of view, were an entirely negative, almost evil process; pioneer settlement once the same people got off the boat in North America, has been seen entirely positively.
>
> By the author, with permission.

Transportation in this era was most commonly by boat, not by road, so ports on Lake Ontario and Lake Erie were early nodes of settlement. As settlement expanded, the land in crops increased, and wheat was exported through these ports to England. The early building of the Welland Canal across the Niagara Peninsula to bypass Niagara Falls reflected this shipping emphasis of the times.

This was very much the Short Distance Society postulated by Fuller and discussed in chapter 2. Settlers lived in relatively small areas, and each village was a service and industrial centre for local supplies. Although the arrival of the railways started the process of change, this pattern persisted culturally well into the twentieth century for many people who rarely travelled far from home. It was well developed in the Carolinian Georegion, but it began to end when the railway suddenly arrived in southern Ontario, heralding the beginning of the "Age of Progress," starting in 1850. By this time much of the present agricultural landscape of the Carolinian Georegion had been claimed by early settlers and cleared for crops; the frontier of white pioneer settlement had moved further north into the Huron Tract and then into Grey and Bruce Counties. The small village of York, which had grown into the large town of Toronto, dominated the province economically. Both of these regions are discussed elsewhere in this volume.

THE INDUSTRIAL ERA

It is impossible to capture in words the revolution brought by the railway. Municipalities fell over each other offering bonuses to ensure that railway lines were routed through them. The mighty locomotive was virtually worshipped. The railway was going to bring "progress" such as civilization had never known before (Hilts 1981). Construction on the Great Western Railway, the first in Ontario, was completed in 1853. The line ran west from Toronto through Hamilton to London and Windsor, cutting east to west across the entire Carolinian region. It changed the orientation of the transportation system fundamentally, cutting off all the north-south routes to Lake Erie ports and drawing export traffic inevitably east to Toronto.

This was the first step toward industrial dominance by the Toronto Georegion, which drew off and controlled trade throughout southern Ontario for its own benefit, a recurring pattern among the georegions that continues to dominate our thinking and planning today, a pattern that the great historian Maurice Careless labelled "metropolitanism." The railway has today been replaced by Highway 401.

The Great Western was merely the first. By the 1880s a network of railways criss-crossed the region, and the export of wheat to Britain boomed, along with towns and villages throughout the region. It takes only a brief look at the old main streets and downtown residential areas of these communities to see the optimistic urban growth that resulted during the 1850s, 1860s, and 1870s. Two- and even three-story buildings of Victorian architecture in the downtown and matching large red-brick houses nearby were all paid for by the booming local economy based on wheat exports carried by the new railways. On farms log cabins were replaced by brick houses, and the large barns that dot the rural landscape were built. All this epitomized the "Age of Progress." Nationally it was highlighted by Confederation and the building of the Canadian Pacific Railway from coast to coast.

Locally much of our built heritage in rural and urban Ontario dates from these decades. Along with the railway came factory manufacturing. In town after town textile mills, lumber mills, and other factories sprang up in the second half of the nineteenth century. Too often torn down and redeveloped today, these sites form the built heritage of the industrial economy. Typically in many communities, this industrial development was dominated by Toronto, just as the railways gradually converged on Toronto. Over time it came to be the central metropolitan influence over much of southern and northern Ontario, as discussed elsewhere in this volume – a pattern that continues today in highways, industry, and government policy.

By the 1870s the clearance of the landscape was at a maximum. In places where settlers had cleared their entire one hundred acres, there was actually a shortage of firewood, in a time when most homes were heated by wood. Trees had not yet grown up along fence rows, so the farming landscape had a bare, open appearance that would surprise us today. The earliest conservation efforts, directed at tree planting, arose during this time, and Arbor Day, now usually forgotten, was established before the end of the century to encourage tree planting. Beautiful lines of sugar maples lining rural roads, farm lanes, and city streets date from this era.

The twentieth century brought a long, gradual shift from railways to automobiles and trucks for transportation. By the 1930s the first four-lane expressway was under construction, the Queen Elizabeth Way, built around the developing Golden Horseshoe of urban centres surrounding the west end of Lake Ontario. Unfortunately the highway was built straight through the Niagara Fruit Belt, arguably the best region of farmland in all of Canada. The urbanization that followed the highway has destroyed large parts of this unique resource. The second expressway was Highway 401, built in the 1950s from Toronto to Windsor (and east to Montreal). This highway, followed later by Highways 402 and 403, has captured most of the traffic that formerly moved by rail. Highways now carry many of Canada's exports to US markets and are attracting industry such as auto plants, in turn reinforcing urbanization through the Carolinian Region and often putting rural land values beyond the reach of agriculture.

While the growth of many smaller towns and villages has faltered, major urban centres are growing rapidly. Urban and industrial development has shifted away from sites with natural advantages, such as mill sites along rivers and the Lake Erie ports, to be based largely on the transportation advantages provided by today's highways. Situated between Toronto, Buffalo, and Detroit, the region is increasingly a magnet for large industrial development such as auto plants and associated support industries. Concentration of urban and industrial development in the Greater Toronto Area continues as the GTA urban conurbation eats up more and more of the eastern end of the Carolinian Georegion, continuing the pattern of Toronto domination.

On the better soils, agricultural production today is intensive and diverse, with the greatest range of crops produced of any region in Canada. The farm economy is actually comprised of several interrelated economies of livestock production; international commodity production, especially of grains, specialty crops including the greenhouse and wine industries; and food grown for local markets. The historic built heritage of nineteenth-century barns is steadily disappearing as farms modernize.

While farming has taken over most of the rural landscape, forests, wetlands, and other natural habitats have suffered and are now reduced to a very fragmented patchwork of remnant natural vegetation. The conservation of natural habitats and the associated rare and endangered species of flora and fauna in "Carolinian Canada" has become an important issue over the past twenty years, and the conservation of water resources will be an increasingly important issue in the future. The sudden growth of conservation efforts in the past thirty years reflects the growth of broader environmental concerns in society.

CONSERVATION IN CAROLINIAN CANADA

The first of many conservation efforts developed in response to an International Joint Commission survey in the late 1960s that declared that "Lake Erie is dead." With no sewage treatment in major US cities like Chicago, Detroit, and Buffalo, this was not surprising; the lake was suffering from severe eutrophication. When the Cuyahoga River in Cleveland caught fire and burned, something had to be done. The eventual outcome was the joint US-Canada Great Lakes Water Quality Agreement. Research and agricultural stewardship programs flowed out of this, with an important influence on the landscape in the Carolinian Georegion.

A critical discovery in the research that was done to determine the causes of declining water quality was the recognition of pollutants, especially phosphorous, carried in the soil sediments that had eroded from farm fields. This "non-point source pollution" was much more challenging to treat than pollution from individual sewage pipes ("point sources"), and it led to the widespread testing of various forms of conservation tillage on farms, including the now widely adopted "zero-till" approach, minimizing the exposure of soil to erosion. Zero-till has been a successful twenty-year revolution in agriculture, fostered by strong and continuing agricultural stewardship programs and grants to farmers. This particular form of pollution has been largely reduced, though the water quality in Lake Erie is still declining from invasive species and toxic chemicals.

Parallel to the recognition of the need for improved land and water stewardship on farms was the recognition of the need to protect the remaining unique ecological features of the Carolinian Georegion, which has led to an intensive and widespread conservation effort over the past twenty years known as the Carolinian Canada program. Begun in 1984, this effort identified thirty-eight unprotected but significant natural areas throughout the region and developed strategies to encourage their conservation. From the beginning the program has been based on an active partnership between an extensive list of government agencies and non-government organizations.

Conservation authorities, the provincial Ministry of Natural Resources, and NGOs such as the Nature Conservancy of Canada have all played important roles. While some of the significant natural areas have been purchased as parks or conservation areas, an important innovative aspect of the program was the Natural Heritage Stewardship Program.

Carolinian Canada is, and always will be, a largely privately owned landscape, with both farmers and non-farmers who live in rural areas owning most of the remaining natural wetlands and woodlands. Many of these natural areas are now fragmented into small remnants of original habitats. Beginning in 1985 a landowner contact approach was developed to visit every private landowner of the identified significant sites. Over seven hundred owners were visited, informed about the ecological significance of natural habitats on their land, and asked if they would voluntarily agree to protect this habitat.

The representatives who conducted these visits, over the kitchen table, so to speak, were given focused training in advance, and explicitly tried to negotiate a verbal "handshake" agreement for conservation during the visit. Landowners who did agree received the Natural Heritage Stewardship Award, in the form of a plaque signed by the premier of Ontario, to recognize their agreement. Surprisingly, in particular to many of the government agencies involved, 467 agreements covering over fifteen thousand acres of natural habitats were negotiated among 716 landowners who were successfully visited.

Once they had learned about the unique features of their land, both farm and non-farm rural landowners proved to be very interested in conservation. Although the agencies sometimes viewed private landowners as antagonistic before this program began, they came to see them as important partners in conservation by the end of the contact process. A large number of positive, friendly, and supportive landowners had emerged.

The Carolinian Canada program was the first in Canada to develop and use such a landowner contact process to encourage private stewardship of natural habitats on a widespread basis. This stewardship approach was later adopted by many groups across Canada: more than twenty similar landowner contact programs were modelled after the program in the Carolinian Georegion. By the year 2000, there had been more than ten thousand landowner visits in the many programs across the country, leading to over five thousand stewardship agreements covering three hundred thousand acres of natural habitats. This result represents on average a 59 percent agreement rate among private landowners owning these sites.

The significant stewardship effort in Carolinian Canada has underscored the fact that rural landowners are generally agreeable people who have chosen to live in rural areas and usually love their land – whether they are farmers or not. Their support is essential for conservation, whether through stewardship or regulation. Since the stewardship program was initiated, the farm community has developed and extensively participated in the Environmental Farm Planning program. In 1998 I produced a stewardship handbook oriented to rural non-farm landowners entitled *Caring for Your Land* (Hilts and Mitchell 1998), and today a coalition of agencies and organizations is developing a separate stewardship program similar to the Environmental Farm Planning program for such non-farm owners.

There have been many spin-offs of the intensive Carolinian Canada conservation effort over the past twenty years. One has been the development of advanced conservation tools such as conservation easements and an increased prevalence of land donations by interested landowners. Another has been the improvement of conservation policies in local and regional official municipal plans, supported by the increased public awareness of the unique ecological features of the region.

However, a remaining question mark is the governance of stewardship programs and the future of non-government conservation organizations. The conservation effort has always been fostered by a partnership of government and non-government agencies and has relied largely on voluntary conservation. Carolinian Canada is now an incorporated non-profit organization devoted to continuing this conservation effort. But there are only scattered ongoing programs to support a voluntary stewardship effort by private landowners and non-government groups, including land trusts (box 5.2) and all are in a constant search for funding to enable their own work to continue. In competition with both urban-industrial development and agriculture, natural habitats and rare species in the Carolinian Georegion remain at risk.

> BOX 5.2 LAND TRUSTS:
>
> Land trusts are community-based, non-government conservation organizations that have emerged in the past twenty years to promote local environmental stewardship. They often focus on the preservation of specific significant natural habitats, seeking donations or conservation easements or sometimes purchasing land. Several important land trusts now exist in the Carolinian region, growing out of the public interest generated through the Carolinian Canada programs. These include the Thames-Talbot Land Trust, the Lower Grand River Land Trust, the Long Point Basin Land Trust, Lambton Wildlife Inc, and the Canada South Land Trust.
>
> The emergence of land trusts as another focus of citizen effort in the conservation sector reflects the broader environmental concerns in society and the growth of stewardship programs, which are seen as positive, constructive things that landowners and citizens can contribute to, in contrast to protesting environmental policies. Initial voluntary, "handshake" stewardship agreements as pioneered in the Carolinian region have led to a wider variety of approaches to stewardship, including the use of conservation easements. The non-government land trust sector has provided much of the leadership in this direction.
>
> By the author, with permission.

AGRICULTURAL STEWARDSHIP IN THE CAROLINIAN GEOREGION

As described above, the same soil and climate that make for the Carolinian zone's unique natural habitats make it Canada's best farmland. Over 52 percent of Canada's class 1 soils for farming are in Ontario, and much of this is within the Carolinian region. All of Canada's top Agro-Climatic Resource Index ratings are in the Carolinian zone. The combination of our best soils in our best climate zones supports the unique diversity of agricultural production in this region.

At the same time we have been losing farmland in Ontario steadily for seventy years. In 1951 10.4 million hectares were being actively used for agriculture, but by 2001 we were down to 7.5, an annual loss of 60,000 hectares of our farmland as defined by the census. Most of this loss has been outside the Carolinian Georegion, in northern and eastern Ontario, where poor land on the Canadian Shield should never have been cleared for farming in the first place. However, there has also been continuing loss of class 1 soils in the eastern portion of the Carolinian region, in the GTA and the Niagara Peninsula, and there has been a small but steady loss to urban and industrial land uses in the rest of the Carolinian region. At the county or regional level, attempts have been made to create strong land use planning policies to protect high-quality farmland from urbanization. Within the Carolinian region, Oxford County in particular has always emphasized the preservation of a flourishing agriculture. Unfortunately, its location has also brought two major auto plants, in turn pushing land values to such a level that farmers are finding it very hard to expand their operations at a reasonable price.

The challenge is that transportation facilities, especially the 400-series highways, reinforce urban growth in a self-perpetuating pattern. Urban sprawl has extended nearly continuously for ninety kilometres west from Toronto in the past forty years. With the location of major auto plants such as the new Toyota factory further southwest in Woodstock along this transportation corridor this pattern of urbanization is reinforced and extended. This continuing reinforcement of urban and industrial growth to the southwest from Toronto through the heart of Carolinian Canada is our biggest land use policy challenge. The land values involved in industrial development simply make it impossible for agriculture to compete.

The establishment of the Greenbelt, which includes parts of the eastern end of the Carolinian Georegion, appears to have exacerbated this pattern, since development pressure has leapfrogged over the protected Greenbelt onto good farmland further west, for example around Brantford. Government policies such as the Greenbelt Act and the growth management strategy both reinforce this trend. Although the Planning Act in theory protects prime agricultural land, urban pressures inevitably extend onto land zoned agricultural, and prices skyrocket. Toronto-centred urban growth pressure wins again! This pattern, wherein the economics of the marketplace determine land values, rather than its inherent agricultural or environmental value, reflects our commitment to globalization and a materialistic living standard as the underlying ethic of our times.

The prospect of a major new highway through the eastern end of Carolinian Canada, sweeping from the Niagara region past Hamilton and around to the GTA via highway 401, 403, or 407 raises further serious conflict with agriculture and conservation. Such a highway would not only take many hectares of land directly but it would also raise land values and attract industrial development, causing further loss of high-quality farmland in our best climate region, as well as of significant natural habitats.

A land use problem specific to agriculture is the disruption caused by rural severances – individual residential lots in the middle of land otherwise devoted to farming. Often such residences place limits on farming through generating extra traffic and complaints about farming operations such as complaints about smells associated with livestock. In significant parts of the Carolinian zone, especially Niagara and Essex, rural residential severances have been spreading rapidly, fragmenting not only natural areas but farmland itself, although recent revisions to the Planning Act have now limited this land use.

Farmers themselves face enormous pressure through unfair international subsidies by the United States and Europe, subsidies that leave many Canadian farmers in a losing position economically before they start. In turn, their economic position makes it more likely that farmers will sell out to urban land uses if they get the chance, since they can't make a reasonable income from farming. However, today interest is spreading in growing fruits and vegetables for local food markets as a means of adding value to production. The climate of Carolinian Canada is ideal for farmers who wish to pursue this option.

In the face of continuing large subsidies for agriculture in Europe and the United States, farmers continue to argue for similar subsidies here – but now with an environmental twist. The European and American subsidies are increasingly shifting to supporting good environmental practices on farms, and Canadian farmers are promoting "payments for ecological goods and services" as a similar mechanism. Should such a shift in Canada's agricultural policy occur, there will be a significant impact throughout this region, one supporting the conservation of natural habitats as well as the farming industry (box 5.3).

BOX 5.3   WETLANDS, SHORELINES, AND OTHER
RESOURCES

In this chapter we have focussed on urban/industrial growth, the conservation of the unique natural features of Carolinian Canada, and agriculture. But a closer look at the region would reveal many other resource issues than these.

Wetlands have been steadily lost to farming over the past two centuries, especially in the southwest part of the region. Only the larger or deeper wetlands remain today, too difficult to drain for agriculture. However, there has been a strong reversal of attitudes in the past thirty years, and policies to support wetland conservation are now in place. An environmental inventory system enables the ranking of wetlands in terms of their significance, and land use planning policies are doing a much-improved job of protecting them.

Water quality along Great Lakes shores and in many streams and rivers is another key issue. The increase in livestock pressure and the spreading of manure has too often led to contamination with bacteria or nutrients; swimming opportunities are lost; and fish populations are degraded. Today the Environmental Farm Plan provides a framework through which farmers can develop improved manure handling systems, in turn protecting water quality.

Shoreline development is a third major issue around the shores of Lake Erie and elsewhere. In some cases, cottages built on hazardous shorelines have been damaged or even washed away during severe storms. In spite of this, the demand for second and retirement homes is today putting extreme pressure on shoreline properties.

By the author, with permission.

THE POLICY CHALLENGE

Within the Carolinian Georegion, farmland, forests, and freeways will continue to be in conflict in the future. The jobs and land values associated with urban-industrial growth are so overwhelming in comparison to the natural heritage and agriculture that urbanization is widely seen as inevitable. It is easy to conclude that dominant patterns of urban growth; the demand for economic growth, employment, and transportation routes to the United States; and the government subsidies and policies that encourage these will define the future of Carolinian Canada at the expense of agriculture and natural habitats.

In fact, the Carolinian Georegion is a distinctive region, with its own very significant values for agriculture and the natural heritage. It is well worth planning for its future in its own right, not merely as an "empty development field for the Toronto growth engine." There are some encouraging signs of a more balanced future, both in terms of public attitudes and government policy, though it remains to be seen how effective these will be in the future.

Currently, land use is governed by the Planning Act and its associated Provincial Policy Statement. Along with public opinion, these have been strengthened considerably over the past twenty years to provide greater long-term protection for farmland and the environmental, especially in rural areas. Nevertheless, urban development continues steadily at the urban edge, along with ripple effects throughout the rural areas. At the regional level municipalities vary considerably in their commitment to protecting agricultural and natural resources.

THE GREENBELT AND GROWTH MANAGEMENT

Two significant land use policy innovations that influence the eastern portion of the Carolinian region are the Greenbelt and the province's growth management plan for the Golden Horseshoe. The Greenbelt is a wide belt of land including the Niagara Escarpment and the Oak Ridges Moraine that circles the west end of Lake Ontario and extends along the northern edge of the GTA. Much of the land is either natural forest cover or farmland. It is essentially to be preserved as a rural area, with no urban development. Officially included in the Greenbelt is the Niagara Escarpment region, extending all the way to Tobermory. One of the most innovative land use planning initiatives in Ontario's history, it predates the Greenbelt by twenty years and is discussed in detail in chapter 13 of this volume.

The passing of the Greenbelt legislation in 2005 was opposed by many farmers, who felt that their land value had been removed unfairly. Now that the legislation is in place in spite of this opposition, there remains a serious question of whether farming can be supported to flourish in this near-urban landscape. Local food advocates say it can; farmer's markets are expanding rapidly, and at least some urbanites appear to be reconnecting with agriculture, hoping to buy at least some of their food from a farmer they know. Only the future will tell whether the Greenbelt succeeds in protecting this area.

A number of farmers and groups such as the Ontario Farmland Trust have argued that the Greenbelt should have extended across all of southwestern Ontario in order to restrict urban growth to designated areas and protect the entire rural landscape for farming – some areas should be for "farming first." As noted earlier, development is leapfrogging over the Greenbelt to create new pressures for urban growth in such centres as Brantford and Woodstock, further reinforcing urbanization southwest of the GTA into the Carolinian Region. The dominance of the Toronto-centred economic core continues to expand.

When the Greenbelt was established, the province introduced a growth management plan for the central part of southern Ontario. Behind it lies the fact that the provision of highways and new water supply and sewage treatment to service miles and miles of urban sprawl is simply too expensive for the provincial treasury. Part of the answer is to increase the density of development, to support transit, and to restrict other services to more compact areas, thus reducing cost. The growth plan designates downtown areas both in the GTA and in larger urban centres such as Hamilton as areas where the density of urban development must be increased. It is hoped that in the long run this will reduce the extent of urban sprawl. While developers are arguing that people do not want to live in high-density neighbourhoods, this plan can only be a good thing if it protects more of the rural landscape.

A key element of the plan will be the choice between transit and highways. In theory, growth management will foster transit instead of highways, but in practice, new 400-series highways will likely still be approved because of the continuing growth in private automobiles and trucking. The proposed mid-peninsula highway from Niagara Falls through the Niagara Peninsula, around Hamilton to join Highway 401 (euphemistically labeled an "economic corridor") will have the largest impact on Carolinian Canada.

However, if planners declare that they need to continue building suburbs on prime farmland in order to meet the province's growth targets, the entire purpose of the growth management strategy will be undercut, Toronto's need for more sprawl will simply spread further, and the unique agricultural and environmental features of the Carolinian Georegion will be further degraded.

All these policy initiatives, along with the history of settlement, conservation, and agriculture described above, reflect the long-term trend toward an "open society" and the continuing dominance of the idea of "progress." To a great extent they also reflect the steadily growing metropolitan dominance of an urban-industrial future, driven by the economic forces centred in Toronto. A balanced future in Carolinian Canada, as in other relatively rural regions of Ontario, will require policies that protect agriculture and conservation, as well as smaller rural communities, and help them to flourish. Such balanced policies will be critical if the uniqueness of the Carolinian Canada landscape is to still be with us in the future.

CONCLUSION

The Carolinian Canada region *is* a georegion. It is defined by its climate, which is relatively warmer than the climate anywhere else in Canada, and the resulting diversity of both natural habitats and native species and of agricultural production. But it is also defined by its critical location in eastern North America, between Toronto, Buffalo, and Detroit. This makes it the Canadian focus of urban-industrial growth and key transportation routes.

The future of agriculture in this region will depend heavily on changes in international subsidies, diversification, and environmental services, as well as land use policies that protect farmland and keep land prices at an agricultural level. The future of conservation will depend heavily on continued environmental protection through land use planning and the commitment of conservation organizations and private landowners to good land and water stewardship. Non-government organizations and individual active citizens will both play key roles in fostering such policies in the future.

Looking back, it is easy to conclude that the building of the Great Western Railway in the early 1850s was the first step toward Toronto's economic domination over this entire region, a pattern that has never stopped. Only if we agree that issues such as food security and environmental conservation are also essential services will the future of this region change. But is also easy to conclude that government policies will support urbanization, transportation, and employment at the expense of agriculture, conservation, and private landowners. The challenge will be to design a balance of policies that will support farmland, forests, and freeways together, to promote a new, more balanced idea of progress that includes a new rich new localism to balance our open society.

REFERENCES

Allen, G., P. Eagles, and S. Price. 1990. *Conserving Carolinian Canada: Conservation Biology in the Deciduous Forest Region*. Waterloo: Waterloo University Press.

Eagles, P., and T. Beechey. 1985. *Critical Unprotected Natural Areas in the Carolinian Life Zone of Canada*. Toronto: Nature Conservancy of Canada, Ontario Heritage Foundation and World Wildlife Fund Canada.

Hilts, S. 1981. "The Idea of Progress: Attitudes to Urbanization in Southwestern Ontario, 1850–1900." PHD diss., Department of Geography, University of Toronto.

Hilts, S., and P. Mitchell. 1998. *Caring for Your Land: A Stewardship Handbook for Carolinian Canada Landowners*. Guelph, ON: Centre for Land and Water Stewardship, University of Guelph.

Johnson, L., ed. 2007. *The Natural Treasures of Carolinian Canada*. Toronto: James Lorimer.

Simpson-Lewis, W., et al. 1979. *Canada's Special Resource Lands*. Ottawa: Environment Canada.

Troughton, M., and W. De Young. 2007. "Natural-Human Interactions." In *The Natural Treasures of Carolinian Canada*, ed. L. Johnson. London: Carolinian Canada.

6

The Huron Georegion: Rurality in an Urbanizing Province

WAYNE CALDWELL

SUMMARY

While Ontario is expected to grow by nearly four million people over the next thirty years, the Huron Georegion, which includes most or all of Bruce, Huron, and Perth Counties and parts of neighbouring counties, will struggle to maintain its existing population. This georegion is located to the immediate east of Lake Huron and comprises one of the most important agricultural regions in Canada. By virtue of its rurality, it will require policies (federal, provincial, and local) that respect the different needs and aspirations of rural Ontario relative to the rapidly urbanizing regions of the province. Special issues, such as a probable shortage of skilled labour and the special needs of agriculture, present challenges that will need to be addressed to ensure that this region remains a vibrant and vital rural community.

IDENTIFYING THE GEOREGION

The Huron Georegion could be identified using a number of criteria. The Merriam-Webster dictionary defines a region as "a large tract of land" and "an indefinite area." Individual areas, from a geographic context, tend to share criteria that link places together into something approaching a single or common entity. From this perspective a region could be identified using characteristics of the physical or human landscape, which may include individual traits ranging from soils and climate to history, culture, and economic activity. Ontario, of course, consists of a nesting of regions, the boundaries of which may vary depending on the criteria or scale used to delineate the region. In this context the delineation of the Huron Georegion will vary accordingly, but from the perspective of this chapter it shares the following characteristics. It

- consists of much of the western part of the province in proximity to Lake Huron,
- shares climate and soils supportive of an intensive agricultural industry,
- has an interconnected history related to settlement patterns, culture, and ethnicity,
- shares common and overlapping institutions related to education and health care, among others,
- includes a network of towns, villages, hamlets, and farmscapes, but excludes large urban centres such as London and Kitchener-Waterloo,
- comprises a rural economy dominated by agriculture, manufacturing, and related service activities,
- includes the political entities of Huron, Perth, and Bruce Counties (excluding the Peninsula), as well as the western parts of Grey and Wellington Counties and the northern parts of Middlesex and Lambton counties (figure 6.1), and
- shares many overlapping institutions including school boards, church conferences, political ridings, and training boards, which plan and promote local labour market strategies.

This chapter reviews the defining characteristics of this region, identifies key planning concerns, and concludes by offering policy initiatives to help address ongoing issues. It asserts that this area will play a critical and ongoing role in the economy, social fabric, and natural amenities of Ontario. It is not the intent of this chapter, however, to suggest a single vision of this area nor to relegate it to the status of a "rural museum." As a working landscape this area will evolve and change as the needs of the individuals who comprise its numerous communities also evolve.

THE HURON GEOREGION: ESSENTIAL BACKGROUND

Although the vast majority of Ontario's landscape is rural, it is not a single, homogenous entity. The basic human and physical geography

The Huron Georegion

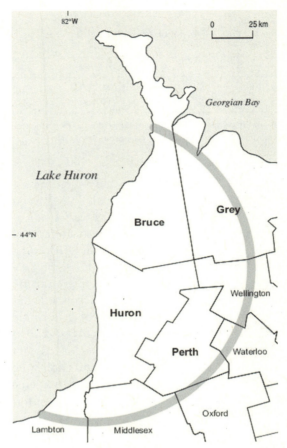

Figure 6.1 The Huron Georegion. By the author, with permission

of rural areas varies across the province. More fundamentally, residents of a given rural area tend to develop an affinity for the community of communities that comprises "their" region. In this context a single definition of the Huron Georegion is illusive. Within the Huron Georegion some people will affiliate with the boundaries of the Huron Bruce provincial and federal riding and others with the Huron Perth School Board. In addition, there will be residents who identify the Maitland Valley watershed, which spans large portions of Huron and Perth Counties, as part of the georegion, and still other residents will look to the local network of sports teams that

Table 6.1
Current and Projected Population: Southern Ontario versus the Huron Georegion, 2006 and 2031

	Southern Ontario	Huron Georegion
Total Population, 2006	11,881,200	205,300
Total Projected Population, 2031	15,719,400	234,600

Source: By the author, with permission. Data from Ontario Ministry of Finance http://www.fin.gov.on.ca/english/economy/demographics/projections/2007/demog07+6.html.

will be shared across the area. Others may use economic criteria to delineate their region – areas from which they buy and sell commodities. Furthermore, as discussed in chapters 1 and 2, residents of the Huron Georegion may also identify with shared historical roots, since the area was settled in the early 1800s predominately by individuals who had emigrated from the British Isles.

Whatever the basis for an individual's regional affiliation, this constant state of interactions with others across a broad geographic area helps to define what can be thought of as a single entity – the Huron Georegion. The boundaries of this area are also fixed in the sense that Lake Huron to the west, the Niagara Escarpment to the north and the cities of London, Waterloo, and Kitchener to the south and east, respectively, exert their own influence, which constrains or compresses the Huron Georegion as is illustrated in figure 6.1.

THE HURON GEOREGION: POPULATION AND POPULATION FORECASTS

In 2006 the population of the Huron Georegion was 205,300, compared to 204,800 in 2001. The population is distributed across a network of towns, villages, and hamlets, and it includes a dispersed farm and rural non-farm component. Stratford, with a population of approximately thirty thousand, is the largest single urban centre. The Lake Huron shoreline is a destination for cottagers and retirees. As shown in table 6.1 the population of the Huron Georegion is small compared to the population of Southern Ontario.

Table 1 compares population forecasts for the Huron Georegion with those for Southern Ontario to the year 2031. The significant projected population growth in Ontario relative to the Huron Georegion is even more evident when existing and projected population densities are compared (figure 6.2). In 2001, the population density of Southern Ontario as a whole was 112 people per square kilometre, but the comparable number for the Huron Georegion was 21 people per square kilometre. In 2031 it is anticipated that the population density for Southern Ontario will increase to 148 people per square kilometre in contrast to 24 people per square kilometre within the Huron Georegion. This increased density of 36 people per square kilometre in Southern Ontario is driven by anticipated growth in the Greater Toronto Area (GTA), but for many rural areas, including Huron, there will be only minimal growth if present trends continue.

Obviously the population of the Huron Georegion is much more rural than that of the province as a whole (figure 6.3). While 85 percent of the province's population lives in urban communities, there is nearly an even split between the rural and urban population in the Huron Georegion; 60 percent of the 2001 Huron County population was rural, in contrast to 53 percent and 35 percent in Bruce and Perth counties respectively. As mentioned, Stratford has 30,000 residents, while the next largest settlement is Goderich, with 7,500 residents. The remaining urban population is scattered across the region in a number of small towns and villages. With its low population density and high proportion of residents living in non-urban areas, the population of the Huron Georegion is one of the most rural in Ontario.

The population of the Huron Georegion is also older than the Ontario average. Huron County, for example, has the third oldest population in Canada (Huron Community Development Corporation 2005). In 2001, less than 40 percent of the population was of prime working age (25–54), and almost 30 percent was 55 years of age or older. There is also a larger proportion of youth in this region than the provincial average.

THE HURON GEOREGION: THE ECONOMY

At first glance, because of the dominance of agriculture on the landscape, there is a tendency to equate rural with agricultural. Making that assumption, however, would fail to recognize other key aspects of the local economy. In Huron and Perth counties, for example, the three major categories of agriculture, manufacturing, and wholesale and retail trade support approximately equal proportions of the workforce. The major employment sector, however, consists of

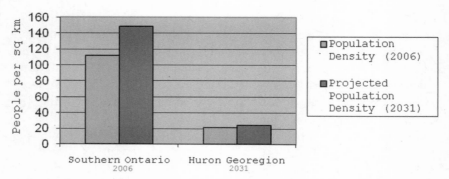

Figure 6.2 Current and projected population densities from 2006 to 2031. By the author, with permission

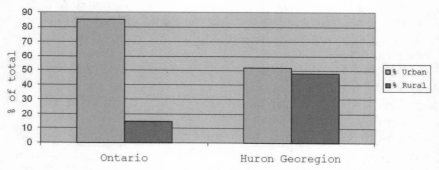

Figure 6.3 Rural versus urban population distributions for the Ontario and Huron Georegions. By the author, with permission

services (personal, business, health education, and government services) (figure 6.4).

Within the Huron Georegion manufacturing has grown dramatically in recent years. In fact, according to the Huron Business Centre, manufacturing shipments increased by 172 percent between 1992 and 2002. The value of manufacturing shipments in Huron County, for example, now exceeds the value of gross farm receipts, placing manufacturing in the lead role as a revenue generator for this county's economy.

The importance of agriculture to the local economy, however, should not be underestimated (figures 6.5 and 6.6). In 2006, gross farm receipts in the Huron Georegion totalled nearly $1.8 billion, a 16 percent increase over 2001 levels (figure 6.6). The area's continued role as a national leader in agricultural production is obvious.

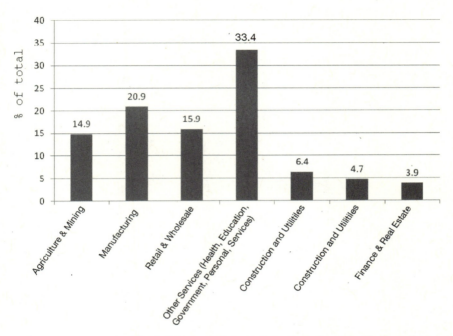

Figure 6.4 Huron and Perth County, employment percentages by sector. By the author, with permission

Whereas the Huron Georegion has 13 percent of the province's farms it tends to account for a higher proportion of provincial production. In livestock, for example, the region produced 38 percent of the province's pigs and 21 percent of all cattle, and it had 20 percent of all poultry and egg farms. In crops, the area produced 21 percent of all the province's corn, 18 percent of all the wheat, and 59 percent of all the white beans. Over the same period, however, the number of census farms in the region dropped from 7,795 to 7,435, a 4.6 percent decline caused by farm consolidations (figure 6.5).

The impact of agriculture on the local economy goes beyond farms. The analysis of the economic impact of agriculture completed by Cummings (2005), for example, demonstrates that in Huron and Perth counties there are nearly 10,000 jobs directly connected to agriculture and more than 17,000 indirect jobs, as well as 6,500 induced jobs, for a total of nearly 34,000 agriculturally related jobs. Incredibly, at a county level this represents nearly 80 percent of Huron County's total employment and 29 percent of Perth's.

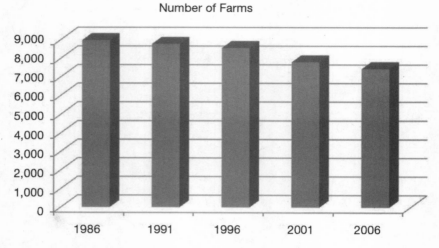

Figure 6.5 Total number of farms for the years 1986 to 2006. By the author, with permission

Clearly many of the manufacturing- and service-related jobs within the region have a strong connection to agriculture.

The pervasive impact of agriculture is evident in more than local employment. It dominates the landscape and defines a way of life. In total, 77 percent of the entire area is active farmland (61 percent in Bruce, 86 percent in Huron, and 93 percent in Perth). Support for agriculture and what it represents is also evident in local planning policies. The following quotation is from the Huron County Official Plan:

> Agriculture in Huron is of national significance. Huron leads all counties and regions in Ontario in total value of production; and it also exceeds the production totals of several provinces. Huron has the advantage of an informed and progressive farm community, a supportive service sector, high capability soils, a diversified agricultural industry, a favourable climate, and limited non-farm intrusion. Agriculture has an economic impact in the County that goes beyond the farm gate and rural areas into Huron's towns and villages. The continued health of agriculture is important not only from an agricultural perspective, but also from a broader community and economic perspective. (Huron County Department of Planning and Development, Official Plan 1999)

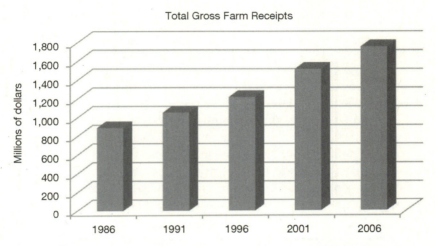

Figure 6.6 Total gross farm receipts for the years 1986 to 2006. By the author, with permission

ISSUES AFFECTING THE HURON GEOREGION

The identification of issues affecting the Huron Georegion is important from several perspectives. First, the issues and their implications vary significantly from those of much of the rest of the province. Recognition of these issues is fundamental for policy development. Sometimes they will be recognized at the provincial or the federal level, but other times these locally important issues will fail to garner the attention of either level of government. Consequently, a thoughtful local response by the municipality or a non-governmental organization (NGO) is required. Moreover, an intentional "scan" of issues could identify key issues at an early stage, thereby contributing to the potential for enhanced effectiveness of policy. Gunton (1985), for example, identified the lag that often exists between the development of issues and the concomitant development of policy. Finally, and perhaps most importantly, these issues matter. They matter to individuals and their livelihoods, as well as to the social, economic, and environmental well-being of communities. From a planning perspective, addressing these issues is fundamental to the practice of rural and community planning.

Across rural Ontario local issues and their relative significance vary by region. As outlined earlier, agriculture is a dominant activity in the Huron Georegion, and the nature of agricultural production is largely a reflection of climate, soils, markets, and the overriding historical, economic, and political structure. Combined, these elements produce a varied agricultural industry with respect to such details as the distribution of crops and livestock and farm productivity and finances. For example, the sale of agricultural products in the Huron Georegion in 2006 amounted to $2,290 per hectare of farmland in contrast to $459 in Renfrew County, in Eastern Ontario (Statistics Canada 2001, 2006). The result of these differences is the generation of regionally specific issues. In Huron and Perth County, for example, the significant concentration of the hog industry has led to various concerns related to air and water quality that have led to concerted municipal action, particularly within Huron County. This response will be discussed in more detail later.

In order to deal with the wide range of issues that affect the Huron Georegion, broad types of issues have been identified under six general headings. They include economic (employment, income, investment), social (education, welfare, health, housing), environmental (ecology, conservation), land use (goods and services, population distribution), political (government, administration, decision making), and quality-of-life (variety, richness) issues. Each of these categories includes numerous subcategories and individual issues.

Economic Issues

The Canadian agricultural industry has been shedding farmers for decades, and the Huron Georegion is no exception. Over the twenty-five-year period from 1981 to 2006 the number of farms in Bruce, Huron, and Perth counties dropped from 9,860 to 7,435. Many communities now receive fewer and fewer benefits from agriculture, since the traditional role of many small towns and villages as service centres for agriculture has changed. Input expenditures (for fertilizer, equipment, and so forth) are often sent directly out of the community. A continuing cost-price squeeze has made earning a livelihood from agriculture very difficult. In the attempt to maintain profitability, there is a constant pressure to increase efficiency in order to keep the cost of production below the value of commodity prices. This drives the ongoing pressure for increased size in order to benefit from economies of scale. Farmers who cannot compete are either forced out of production or they subsidize the farm with off-farm employment. Despite these issues, however, the business of agriculture in this area, has in some ways thrived. As of 2009, farm productivity and gross farm sales were at record highs, and investment in the agricultural sector continues unabated: the number of census farms with farm capital value exceeding $1 million increased from 2,229 in 2001 to 2,946 in 2006, despite a reduction in farm numbers. There does, however, appear to be an interesting parallel trend. While the numbers barely create a blip in census data, local food production and related consumption has an increasing profile. In Huron and Perth counties, for example, there is an active Buy Local, Buy Fresh campaign.

As in the rest of the province, the manufacturing and tourism sectors are closely tied to broad economic factors. The high value of the Canadian dollar, access to American markets (border crossings), and competition with emerging economies (particularly in China and India) create issues that are not easily addressed. The 2008 closure of the road-grader-producing Volvo plant in Goderich and the related loss of five hundred jobs, for example, represented the end of an industry that had existed for generations. Business revenues have, however, increased significantly. In Huron County manufacturing shipments have increased at an annual rate of 17 percent since the mid-1990s, to $933 million in 2002. Tourism and retail activities are also an important part of the region's economy. Entrepreneurial activity is prevalent: nearly 20 percent of the workforce is self-employed, compared to 11 percent provincially. In the future, however, the issue for the Huron Georegion is not likely to be a shortage of jobs but, rather, a shortage of people to occupy key positions. With impending retirements, continued youth out-migration, and the attractiveness of large urban centres to immigrants, it will be difficult to find skilled labour for many professions. This problem is likely to continue to be particularly challenging for the health care field, where an aging population, combined with incoming retirees destined for lakeshore areas, will in time place enhanced demands on the health care system. An opportunity, albeit a challenging one, is for rural communities to target immigrants as a means to maintain essential services.

Environmental Issues

The Huron Georegion is subject to various local, regional, and continental environmental pressures, and the implications of broad issues such as climate change are as yet unknown. It can be expected, however, that agriculture will witness greater weather extremes and fluctuating patterns. Peak oil, while equally an economic issue, speaks to the long-term implications of increasingly expensive and scarce conventional energy supplies. The impacts on a dispersed rural population and an energy-dependent agricultural sector are equally uncertain. A related trend, with corresponding economic and land use implications, is the increasing search for alternative energy supplies. The trend towards large-scale solar and wind energy projects is often the source of local debate and conflict. Large-scale wind turbines, in particular, have proliferated in proximity to Lake Huron, where many towers have recently been constructed. Moreover, the move towards large-scale ethanol production (based largely on corn) is changing cropping patterns and leading to a corresponding increase in corn acreage, along with the potential for higher levels of erosion and fertilizer use.

Concerns over water quality have been the source of ongoing tensions, particularly between the farm and the lakeshore communities. The intensification of the livestock sector and, in particular, the construction of new hog barns has contributed to near-shore water quality concerns along Lake Huron. Faulty septic systems, discharges from municipal sewage treatment plants, and birds (seagulls and geese) have also contributed to this problem, which has led to municipal action. The approach of the Huron Water Protection Steering Committee will be profiled in further detail later in this chapter.

Increased public awareness is creating new demands and pressures on farms and businesses to be environmentally accountable. There are local concerns over water quality, environmental protection, and habitat preservation, and government is being pressured to regulate in these areas, while it is increasingly recognized that stewardship can also play an important role. Agriculture needs to adopt best management practices, and nutrient management planning (provincial and local) provides an example of government attempts to force environmental responsibility.

Political Issues

Politically, the Huron Region faces a number of issues. First, at the national and the provincial levels, the overall reduction of the rural population as a proportion of the total population has resulted in a loss of political influence. As the country has shifted from a nation of farmers to a nation of urban dwellers, the programs and the attention of governments have shifted in response. Second, at the opposite end of the spectrum the composition of many rural communities is changing and, as a result, the direction of rural municipal councils is also shifting because the reduction in the farm population relative to the non-farm population has changed the composition of municipal councils and introduced a series of competing interests. Addressing these interests requires significant attention on the part of municipal councils.

The Quality of Life

"Quality of Life" is an encompassing yet ambiguous term. The concept, however, is of long-standing importance to residents of many rural communities. It speaks to an adequate standard of living, access to appropriate services, and the availability of cultural and social services that contribute to a meaningful, rewarding life. For many it implies close ties within the community, such as regular contact with family and friends. Although rural communities offer much that enhances the quality of life, there are challenges. Traditional institutions ranging from the post office to the local church to schools have been threatened in various ways. Sometimes they are driven by changing priorities and in other instances by the search for elusive economies of scale (such as the application of provincial funding formulas that value efficiency and cost savings).

Much has been written in the sociological literature concerning the increasing loss of the rural way of life. Rural communities are inherently dependent on larger centres, but their distance limits access to certain services that contribute to the quality of life. Areas with a small population base will not be able to support, for example, certain medical, cultural, and commercial services. As a result, rural people will sometimes need to look elsewhere for many necessary services and amenities. In the Huron Georegion the continued provision of quality services will be an ongoing issue. In addition to health care, attracting and retaining members of other professions will create new challenges. Furthermore, the region's relatively low population density will put pressure on the ability of local governments to maintain acceptable public services related to roads and infrastructure. Likewise, it may be difficult for the private sector to provide certain services, such as high-speed internet, to an area with a relatively low population density.

Social Issues

Related to the aforementioned concerns, two overriding social issues affect the Huron Georegion: the availability of institutional services and the presence of stress in the rural community. The change in the nature of the rural community combined with variable growth rates across the region, including depopulation in some areas, has negatively affected the viability of rural institutions such as schools, churches, and other community organizations. Furthermore, the overall state of many rural economies and the combined effects of distance and dependency outlined above have created considerable stress at the community and family level. There is evidence in the numbers of farmers that have left the profession over the last twenty-five years and the ongoing transition within the manufacturing sector of the challenges many people face in securing an adequate livelihood. This is also reflected in significantly lower average family income and education levels relative to the province as a whole.

It is noteworthy that while the total population of the Huron Georegion has changed very little over the past thirty years, the composition of the population has changed. There is a much lower percentage of the population in the 25-to-54 age cohort than for the province as a whole and a higher percentage in the 10-to-19 age cohort. As well, there is a larger percentage of population aged 55 and over. While education levels are improving, relative to the province as a whole a disproportionate percentage of the working-age population lacks a high school certificate. In Perth County for example 41.1 percent of residents do not have a high school degree, in contrast to 33 percent for the province as a whole. This has implications for employment and the need for labour training. A number of these trends contribute to poverty and the loss of youth through out-migration.

Land Use Issues

Land use and related issues are of particular importance in the Huron Georegion, since there is a strong connection between land use, the environment, and the economy. Since agriculture and tourism, for example, rely heavily on the natural resources and amenities of the area, appropriate land use and environmental policies that protect these resources for future generations are important. Significant land use issues within the area may be considered under four separate but related headings:

- The changing function of rural communities. The changing nature of agriculture (larger farms, a reduced farm population, and a reduced reliance on local services) contribute to the declining importance of and changes in many traditional service centres. While some towns and villages within the region are thriving, others have an overabundance of commercial space evident in abandoned or underutilized buildings.
- Agricultural land and rural non-farm development. The growth of a non-farm population in rural areas has created its own set of issues. While people may argue the merits of this trend, it has affected rural communities by increasing land costs and creating issues of land use compatibility. Although the counties of Perth, Huron, and Bruce have some of Ontario's most supportive agricultural policies limiting rural non-farm development (Caldwell and Dodds-Weir 2007), there continue to be ongoing issues and conflict between industrial agricultural practices and other uses located within the countryside, particularly along the Lake Huron shoreline.
- The role of natural areas. The role of natural areas within the rural community has been an evolving land use issue. Ongoing concerns over water quality and habitat preservation (wetlands and upland forests), for example, point to the need for appropriate policies, including the protection of natural areas and improved farm practices. Moreover, Lake Huron as a resource deserves more attention from both the provincial and the federal government. Locally, for example, in reference to the lack of attention received from upper levels of government Lake Huron is sometimes referred to as the "forgotten lake."
- New land uses. Wind turbines and the potential for solar farms introduce a new use to the Huron landscape that has potentially significant implications. Wind towers, which have an imposing presence on the rural landscape, have generated much local debate. At a minimum they speak to the need for policies that are

sensitive to the protection of "heritage landscapes" and ensure that in the future those landscapes, which are highly valued by residents and tourists alike, are retained. Moreover, renewed investment in Bruce Power (the Bruce Nuclear Plant), north of the town of Kincardine, contributes to evolving local land use patterns.

This outline of the major issues affecting the Huron Georegion, does not try to identify their relative importance. Depending upon individual perspectives, any one of the issues could be considered the most important. As mentioned at the outset of this chapter, it is important that the development of policy be sensitive to the local community's needs. In this context two separate approaches will now be presented, largely promulgated by local authorities as a response to both environmental and economic issues.

KEY ISSUES AND A LOCAL POLICY RESPONSE

What does it mean for the Huron Georegion to be rural and small-town at a time when the population of the province of Ontario is forecast to grow by more than four million people over the next thirty years? Clearly, the provincial government needs to devote resources to planning for this anticipated growth, and yet there is a risk that in this process the needs of rural Ontario will go unrecognized. Some observers suggest that the urban agenda of the province is leading to the creation of inappropriate policy for rural Ontario (see, for example, the website for the Rural Revolution, a protest group established in Eastern Ontario, at www.ruralrevolution.com/website/). Some residents will point to real or perceived reductions in the budget of the Ministry of Agriculture, Food, and Rural Affairs or to certain articles in the urban media as an indicator of a prevalent urban agenda. While some of these concerns are not surprising, given the province's overwhelming urban population, the role of rural issues remain critical for the province's future.

While those who live within rural communities and advocate for them may not embrace the demographic shift from rural to urban, the challenge is for rural communities to acknowledge that the political landscape of the province has changed permanently and that a focus on rural and agricultural needs is a relatively small part of the provincial government's agenda. Recognizing this change will provide a renewed opportunity to respond and reposition the rural agenda. While a thoughtful rural policy is required at both the provincial and the federal levels, there is an opportunity for local communities to partner with upper levels of government and to show leadership in developing an appropriate response. The following two sections focus on approaches to two diverse issues: the work of the Huron Water Protection Steering Committee in responding to water quality issues along the Lake Huron shoreline, and the Huron Business Development Corporation's response to economic issues.

An Approach to Environmental Issues: The Huron Water Protection Steering Committee

The Huron Water Protection Steering Committee was established in the spring of 2004 by Huron County Council as a municipal and community-based response to the ongoing concerns over water quality in Lake Huron. The county had also completed various ground-water studies that pointed to the need to carefully manage and protect ground and surface water quality across the region. Leadership and direction has been provided by the Huron County Planning Department, but the composition of the myriad of partnerships testifies to the community-based approach that has been utilized. This example is profiled because it demonstrates effective local action in response to a regional issue.

The Water Protection Steering Committee (WPSC) was established with three key goals:

- To bring together representatives of agencies, groups, and municipalities, including planning agencies, health units, municipalities, conservation authorities, the Ministry of the Environment (MOE), the Ontario Ministry of Agriculture, Food and Rural Affairs (OMAFRA), and representatives of agriculture, manufacturing, tourism, cottage associations, and watershed groups.
- To prioritize and recommend implementation measures to participating agencies.
- To coordinate activities at a broad level, subject to the resources of the participating agencies.

The committee has approximately twenty-five members who meet four times a year. They represent a diversity of interests ranging from cottage associations, to agriculture, to tourism. There are also representatives from key government ministries and agencies: the Ministries of Agriculture, Food and Rural Affairs, Natural Resources, and Environment, as well as Environment Canada and Conservation Authorities. Several water protection projects have emerged since the WPSC was launched. As projects are identified and discussed, resource needs are identified, and other organizations and agencies step forward to collaborate and share staff, volunteers, contacts, funding, meeting rooms, and so forth. Under the auspices of the committee nearly $2 million has been leveraged to pursue water-quality work. Numerous projects have been pursued, including the following:

1 Huron Clean Water Project. This project is funded by the County of Huron and is delivered by area conservation authorities. It provides financial assistance to landowners to adopt improved management practices ranging from fencing cattle out of watercourses to wellhead improvements. There is also a category for community mobilization.
2 Septic System Re-inspection Program. This project involves working with volunteer landowners to raise awareness of proper septic system operation and maintenance and to identify faulty systems. It is funded by the County of Huron and the Ministry of Environment and is delivered by the Huron County Health Unit.
3 Stewardship Manual for Cottagers and Non-farm Rural Residents. Modelled on the Environmental Farm Plan, this manual educates residents about how they can reduce their environmental impact. Workshops accompany the manual. This is a joint project with the University of Guelph, Friends of the Bayfield River, Ausable Bayfield Conservation Authority, the Planning Department, and the Lake Huron Centre for Coastal Conservation. Funding for the project is from the OMAFRA, the Grand Bend Community Foundation, the Municipality of Central Huron, TD Canada Trust, the Ministry of the Environment, and a number of additional community groups.
4 Anaerobic Digester Pilot Project. Anaerobic digesters are built to use manure to create fuel. In doing so, they reduce the pollution created by manure. A subcommittee was established to investigate the possibility of a farm-based anaerobic digester as a pilot project. Representation on the subcommittee is drawn from a number of sectors.

5 Payment for Ecological Goods and Services (PEGS). PEGS is a stewardship model that pays farmers lost opportunity costs for retiring farm land into stewardship projects. A sub-committee including representatives from the Conservation Authorities, Stewardship Council and the Planning Department has developed a $50,000 pilot project.

While this list provides just a sample of the types of projects that have been initiated by the Water Protection Steering Committee it does point to the impact that this group has had. Several factors have contributed to success:

- A diverse membership within the group brings an equal diversity of issues and approaches to discussions. The views range from those of members with a strong environmental ethic to those of members who are immersed in manufacturing and agriculture.
- The approach builds on multiple partnerships between various organizations. Many groups work on environmental issues, but there is a risk that actions may proceed with little co-operation or communication. The steering committee creates a platform for dialogue and coordinated activities.
- The group is engaged in action. While strategic planning is visited on a regular basis, group members can look to ongoing tangible action, and they recognize their contributions in making this happen.
- There is a broad definition of "planning." The group has supported the notion that while the regulatory side of planning has a role to play, there is much to be gained by focusing on diverse non-regulatory approaches that support voluntary action, education, and community development.

The Huron Approach to Economic Issues

Just as partnerships have been key in the response to environmental issues within Huron, they play an equally important role in the response to economic issues. Beginning in the mid–1980s the County of Huron began to implement a community development approach to economic development that received a significant boost in 1992 when the federal government gave preliminary approval to the establishment of a Community Futures Development Corporation and a corresponding Business Development Corporation. This eventually culminated in the creation of the Huron Business Development Corporation, which has for more than ten years provided a diversity of programs that strive to tackle the key underlying issues that hamper the Huron economy.[1]

While the federal government's involvement has been critical, it is the local community-based approach to development that is noteworthy. Most recently this has led to renewed and enhanced partnerships between the county (the Warden's Task Force), The Huron Business Centre, and federal and provincial partners. Under the rubric Community Matters, there are the Economic Development Matters, Youth Matters, and Employment Matters initiatives.

Community Futures remains key to the Huron Business Development Corporation. The Community Futures program is administered locally by a board of directors that represents the various communities and economic sectors in the area and provides a range of options to communities designated under Community Futures. These options include loans and management consulting to assist businesses during start-up and expansion, training programs to develop new skills and maintain the competitiveness of local businesses, and financial support for community development projects that strengthen, diversify, or expand the local economy. Examples include entrepreneurial development seminars and programs, business training and counselling initiatives, market research on new business opportunities, and information for partnerships for new business development, as well as joint marketing and promotional activities. Recent (and ongoing) successes include workshops, youth engagement, specialized training programs, loans amounting to nearly $1.5 million dollars, and work with multiple community partners.

There are many motives for coordinating community economic development activities. Efficiency, the delivery of a quality product, public support and confidence, and the goal of providing "one-stop access" represent some of the more important reasons. Because of jurisdictional and political differences between levels of government, however, there are few Canadian examples of concerted efforts to achieve coordination and integration that have been attempted or have succeeded. It is this combined effort at the coordination and integration of federal, provincial, and municipal community development initiatives that distinguishes the Huron Business Development Corporation (HBDC).

Several reasons help to explain the organizational and administrative accomplishments of the HBDC. They can be summarized as follows:

- Local leadership and support. The HBDC benefits from the leadership provided by a local board of directors and community support. The board is drawn from a diversity of interests from across the county. Local government participation and linkages are fundamental to its success.
- County-wide focus. The staff of the HBDC has attempted to maintain a single county-wide focus in their programs. From the perspective of the public, the HBDC, even though it represents an integration of certain federal, provincial, and municipal programs, represents a single service.
- Community development model. The adoption of a "community development model" has been a significant factor in the operation of the HBDC. To this end, there has been an attempt to focus on direct public initiative, participation, education, action, and a process based on local community leadership. This has the benefit of frequently leveraging access to other resources, while at the same time helping to enhance the community's capacity for self-help.
- Support for diversity and small business. The activities of the HBDC have focused on development and support for small business, as opposed to a focus on "mega-projects." The HBDC has also recognized the extent and diversity of the rural labour market across Huron County. Programs have been designed and delivered to all sectors of the local economy, thereby avoiding the potential tendency to treat the Huron economy as a "single industry."
- Building on provincial and federal initiatives. The HBDC has brought the resources of the provincial and federal government to Huron County, but local officials have been and are seen to be in the driver's seat. This is not a program developed in Ottawa or Toronto, and, furthermore, it is seen and understood to be a locally directed initiative.
- Strategic planning. Strategic planning has been an integral tool in helping to establish the priorities of the constituent parts of the HBDC. Priorities are identified, worked with, and amended on a regular, ongoing basis. As clear statements of community aspirations, these strategies have been accepted and widely endorsed as an appropriate mandate.

- The concept. The appropriateness of the organizational and administrative concept of the HBDC has contributed to the acceptance politically and within the community of the notion of a one-stop, integrated, and efficient community-based service.

This discussion of a local approach to environmental and economic issues is intended to highlight the importance of understanding the need for rural communities to identify with, understand, and respond to challenging local issues. While enhanced provincial and federal engagement is ideal, the political reality reflects shifting demographics and priorities within the country. The Huron Georegion, for example, remains one of the most rural areas of the province. Moreover, it will likely stay this way. As we have seen, while provincial population projections forecast some growth, it will be minimal relative to the rest of the province.

ISSUES IN THE HURON GEOREGION: THE BASIS FOR A RESPONSE

There is no single solution for the myriad of issues that face the Huron Georegion. Responses are required from upper levels of government, but they are also required from municipalities, community groups, and individuals. There is a role for regulation, but there is also a role for stewardship, education, and community development.

In some respects governments at all levels need to show a sensitivity to rural needs relative to the burgeoning urban mandate at the provincial level. While it is difficult to identify what this may mean in practice, at a certain level it simply means that governments need to be open to the stories and concerns of rural regions and open, as well, to a philosophy that holds that one solution does not fit all. The federal Community Futures program, for example, integrates local leadership and governance with federal program criteria and federal dollars (for local business start-ups and community economic development) in a way that has been supportive of rural communities across the country. Likewise, Ontario Stewardship provides resources and a structure that builds local capacity to engage in environmental issues.

This chapter does not identify specific actions for the Huron Georegion but, rather, concludes with a series of overriding strategies or principles that can help to direct the region's future:

- Sensitivity to ruralism. Rural communities have their own set of challenges and issues. Governments need to listen, be sensitive to rural issues, and avoid the temptation to impose a single solution on many rural communities that differ among themselves just as they differ from urban Ontario.
- Vision. A vision developed by and shared among members of a community can lead to concerted action. Perth and Huron Counties, for example, are leaders in the province in planning for agriculture. For more than thirty years Huron and Perth have resisted non-farm development in agricultural areas. This vision continues to establish a land use pattern that supports the agricultural industry. There are, however, conflicting visions concerning the future role of agriculture in the region.
- Political commitment (cumulative impact). Political commitment to the implementation of a community vision is an absolute necessity. While decisions in the interest of an identified public interest may be difficult, especially when it affects a private interest, the impact of individual incremental decisions can lead to an overall negative cumulative impact.
- Equitable access. Adequate health care and education, broadband access, and safe roads are examples of services that should be available to the residents of the Huron Georegion just as they should be available to residents across the province. A relatively low population density may compromise some of these essential services.
- Community/human resource capacity. This is a broad topic area, but it speaks to the need to ensure that there are appropriate programs to assist with the development of employment skills, leadership, and community support.
- Environmental responsibility. From many perspectives – water quality, agriculture and its connection to the environment, and the actions of individual residents – there is a need to take responsibility and action to address environmental issues.
- Community-based planning. Community-based planning implies a willingness to engage the local community and to listen and act on the direction that the community helps to identify. It presupposes that actions are most effective when they are endorsed, supported, and implemented by individuals in their day-to-day activities.

In summary, the differences between the Huron Georegion and the rest of the province, especially those areas experiencing significant growth, will increase. The challenge provincially and locally will be to acknowledge this change, to assess its impacts realistically and to engage the community in long-term planning.

NOTE

1 Community Futures Development Corporations also exist in Bruce and Perth Counties. See, for example, http://www.bruce.on.ca/ and http://www.perthcfdc.ca/Home.aspx

REFERENCES

Caldwell, Wayne, and Claire Dodds-Weir. 2007. "Rural Non-Farm Development and Ontario's Agricultural Industry." In *Farmland Preservation: Land For Future Generations*, ed. Wayne Caldwell, Stew Hilts, and Bronwynne Wilton, 229–52. Kitchener, ON: Volumes Publishing.

Cummings, Harry. 2005. *Socio-Economic Impact Study of Agriculture in Ontario*. Guelph, ON: Sustainable Rural Communities Program, University of Guelph.

Gunton, T. 1985. A Theory of the Planning Cycle. *Plan Canada* 25(2): 40–4.

Huron Community Development Corporation. 2005. *The State of the Huron County Economy*. Seaforth, ON: Huron Business Development Corporation.

Huron County Department of Planning and Development. 1999. *Huron County Official Plan*. Goderich, ON: County of Huron.

Statistics Canada. 2001. *Census of Agriculture*. Ottawa: Statistics Canada.

– 2006. *Census of Agriculture*. Ottawa: Statistics Canada.

7

Peterborough: A Georegion in Transition?

ALISON BAIN AND JOHN MARSH

SUMMARY

The idea of the region is foundational to the practice of geography. The division of space into regions has a long history within the discipline of geography, and the properties of regions, their core areas, and their boundaries have long been the subject of intense debate (Gregory 1978; Massey 1984; Gilbert 1988; Pudup 1988; Thrift 1990a,b; MacLeod and Jones 2001; Paasi 2002). Regions in their traditional interpretation by geographers could be defined as parts of the earth's surface that have discernibly distinctive climatic, landform, vegetation, land use, demographic, and other characteristics that differentiate them from other regions (Bone 2000). Yet geographers have also wanted regions to be studied in less descriptive and more theoretically informed ways. Nearly three decades ago Derek Gregory (1978, 172) argued that "we need to know about the constitution of *regional* social formations, of *regional* articulations and *regional* structures." Geographers such as Nigel Thrift (1990a,b; 1991) and Ron Johnston (1983, 1991) followed Derek Gregory's lead and called for a "new" reconstructed regional geography that recognized that regions are a product of social change and are not fixed divisions of territory. That regions are territorial entities that are produced, reproduced, and transformed through human agency will be explored in this chapter through an examination of the Peterborough area in central Ontario.

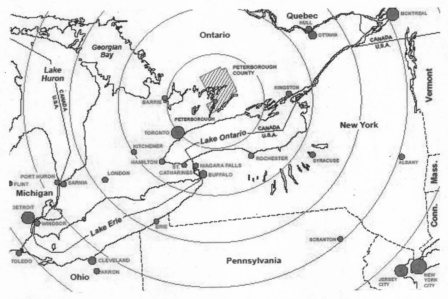

Figure 7.1 Location map. From Adams and Taylor (1992), with permission

BIOPHYSICAL CHARACTERISTICS: PETERBOROUGH AS A TRANSITION ZONE

Peterborough is located at the northern edge of the St Lawrence Lowlands, adjacent to the southern edge of the Canadian Shield (figure 7.1). The identification of a St Lawrence Lowlands region hints at the potential for an alternative physical regionalization based on watersheds. The upper Trent River watershed incorporates a very small southerly portion of the Canadian Shield, while the lower Trent River watershed comprises part of the St Lawrence Lowlands watershed. However, whereas the main river in the watershed south of Rice Lake is called the Trent, the main river in the watershed above Rice Lake is called the Otonabee. Why one name was not used for the full length of the river remains to be determined. Peterborough lies on the edge of the Otonabee approximately halfway down the Trent watershed. It is important to note that there are many lakes in the upper watershed and a few, notably Rice Lake, in the southern watershed (figure 7.2).

It is also possible to identify a hierarchy of regions based on geology and geomorphology. Thus, major regions such as the St Lawrence

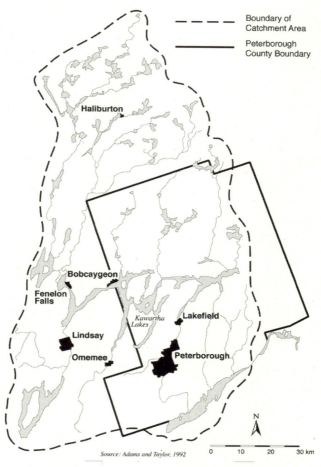

Figure 7.2 The Ottonabee Watershed. From Adams and Taylor (1992), with permission

Lowlands can be further subdivided based on the geological and geomorphological variations within them, as demonstrated by Chapman and Putnam. In their classic work, *The Physiography of Southern Ontario* (1966, viii), they state that "Southern Ontario was regarded as being composed of three physiographic divisions, eastern, south-central, and south-western Ontario." Peterborough is located in the south-central division. The three major divisions are then divided into fifty-two "physiographic regions." One of these is the Peterborough Drumlin Field, which is described as "a rolling till plain with an area of about 1,750 square miles." Extending from Hastings County in the east to Simcoe County in the west and including the drumlins south of the moraine in Northumberland County, this belt contains approximately three thousand good drumlins, in addition to many other drumlinoid hills and surface flutings on the drift cover. The name of Peterborough has been used to designate the whole group because the city occupies its geographical centre and because the drumlins are most typical in form and most densely distributed in Peterborough County.

Regions may be characterized by their physical and vegetational characteristics or even primarily by their distinctive vegetation. The vegetation around Peterborough is a mixed coniferous-hardwood forest. But since it is part of the Great Lakes-St Lawrence Forest region, one of the great forest regions of Canada, the vegetation does not provide a local regional identity.

Recently, the Peterborough area has been considered part of "The Land Between," a corridor stretching west-east from Georgian Bay through to the Frontenac Arch near Kingston, with rather indeterminate northern and southern boundaries (www.thelandbetween.ca). This corridor has been described as an ecological transition zone between Southern Ontario and the Canadian Shield characterized by changes in elevation and climate. Targeted for greater population settlement and resource extraction with recent habitat conservation and provincial planning initiatives, there is ongoing concern that land management and planning practices could threaten the future of this area.

In sum, no obvious biophysical characteristics serve to bound a region around Peterborough. One might ask, then, how the area around Peterborough has been bounded as a result of its settlement and development. In what follows, we discuss how boundaries have been drawn around Peterborough both for administrative purposes and in the public imagination.

SETTLEMENT AND EARLY DEVELOPMENT

While various native peoples moved through and lived in the Peterborough area for thousands of years, it was not until the 1800s and the arrival of European settlers that large-scale landscape modification, farming, industry, and urban development occurred.

As Nelson and Troughton discuss in chapter 1, the first settlers were mainly from England, Ireland, and Scotland, arriving individually or in groups. According to Coleman (n.d., 27), "the sudden influx of nearly 1900 settlers into an area with a scattered population of only 500 was a dramatic event which was the turning point of the development of the Otonabee." The attraction of this frontier area at that time was the availability of land with potential for logging, farming, and water power, as well as trade with developing cities around Lake Ontario, the United States, and Europe.

In chapter 2, Nelson and Troughton outline how by the late nineteenth century Peterborough had become an industrial, service, and administrative centre serving a considerable hinterland. Railway linkages were being developed in all directions, especially with cities on the shore of Lake Ontario, such as Port Hope, Belleville, and Toronto, thus facilitating trade.

In the early twentieth century the population of the Peterborough area increased. Industrial concerns such as Quaker Oats, Canadian General Electric, the Peterborough Canoe Company, and Outboard Marine Corporation flourished, and tourism based on the Kawartha Lakes resorts and cottages developed (Jones and Dyer 1987). The road system was expanded and improved, while the just-completed Trent-Severn Canal became obsolete, and the railway system of less importance. The City of Peterborough had established itself perceptually and functionally as a significant centre drawing on and servicing an extensive region, albeit with ill-defined and evolving boundaries.

Over the last two hundred years various Peterborough-focused regions have been "constructed." They have been more or less persistent, evolving, and overlapping in the public mind. They include regions focused on the river and the Trent Canal, the Kawartha Lakes, the Otonabee Watershed, administrative regions such as Peterborough County, and promotional regions such as the Peterborough-Kawarthas tourist region and the Greater Peterborough Area Economic Development Corporation region.

A RIVER AND A CANAL

The Trent and Otonabee Rivers have been a focus and linking feature of the area for thousands of years. Various First Nations peoples travelled along and lived beside the rivers and lakes, and some still do. Early European explorers, such as Champlain, penetrated the area north of Lake Ontario along these waterways. In the nineteenth century, they were used by European settlers, such as the Peter Robinson group, to gain access to their land around Peterborough. Catherine Parr Traill and Susannah Moodie also used the Otonabee River to reach their land near Lakefield (Traill 1971). However, the small size of the Trent and Otonabee Rivers, their numerous rapids, circuitous route, winter freezing, and variable flows impeded their use as a major transportation corridor and prevented the Trent Valley as a whole from emerging as a region. To improve navigation along this route, proposals were made in the early 1800s for a series of canals. Surveys were undertaken in 1833, and the first of a series of recommended locks was built in 1835 in Bobcaygeon. Progress was slow. The Peterborough Liftlock was completed in 1904, but it took until 1930 to finish the whole canal. After nearly a century of construction, the canal was no longer needed for commercial purposes; instead it became popular with recreational boaters (Tatley 1978). The canal route is now known as The Trent Severn Waterway. Designated as a National Historic Site, it is managed by Parks Canada. The City of Peterborough, which lies near the midpoint, has the most significant cultural feature, the liftlock, the head office for managing the waterway, and the most substantial visitor centre for interpretation purposes (Brunger 1987). The waterway has gradually been recognized as a corridor of recreational, tourism, and heritage significance, rather than as a region, with Peterborough being but one centre along its route (Ontario Ministry of Culture and Recreation 1981).

CONSTRUCTING THE KAWARTHAS

Today Peterborough is frequently associated with the Kawartha Lakes. Weather forecasts for the regions of Ontario refer to "Peterborough and the Kawarthas." The association of Peterborough with surrounding rivers and lakes is also made in the title of the book, *Peterborough: Land of Shining Waters* (Borg 1967), published to celebrate Canada's centennial, yet its introductory map shows Peterborough County as the area being considered. In another book on the Peterborough area, published by the Geography Department at Trent University and entitled *Peterborough and the Kawarthas*, the editors state: "Even in the tourist industry, where the concept of the 'Kawarthas' as a lake district has been promoted for a long time, the Kawartha Lakes are not yet as clearly defined in the public consciousness as other regions" (Adams and Taylor 1985, 1). The City of Peterborough is variously described as being "the Gateway to the Kawarthas" or the "Heart of the Kawarthas." For a hundred years, the Kawartha area has been variously defined as a few lakes, many lakes and Peterborough, and an even larger swath of territory.

In order to determine the boundaries of the region named Kawartha, or the Kawarthas, a cartographic and brochure analysis was conducted (Marsh 1981, 2004). Any descriptions, for example in tourism brochures, of the spatial extent of the Kawarthas were used to produce maps delimiting the region. They were combined with existing maps that had "Kawartha" in the title to give a total of twenty-two maps indicating varying spatial extents of the region labelled Kawartha from 1901 to 2002.

What is probably the first brochure delimiting the Kawartha Lakes was published in 1901 by the Grand Trunk Railway. Here the Kawartha Lakes District referred to a chain of ten lakes located north of Peterborough and Lindsay. In 1909 a brochure entitled "Fenelon Falls: The Prettiest Summer Resort on Kawartha Lakes" stated that there were eleven Kawartha lakes. *The Reference and Guide Book to the Trent Canal* (Department of Railways and Canals of Canada 1911, 27) stated that the Kawartha Lakes comprised fourteen lakes: Scugog, Sturgeon, Cameron, Balsam, Pigeon, Bald, Sandy, Buckhorn, Chemong, Deer, Lovesick, Stony, Clear, and Katchewanooka, "to which might be added Rice Lake, twenty-one miles (thirty-two kilometres) down the Otonabee River below Peterborough." Subsequent descriptions and maps have included a larger region. For example, Mallory (1992, 34–5; see also 1991) states that "other lakes, like Sandy, Big Cedar, Catchacoma, Mississauga, Anstruther and Four Mile Lake are considered part of the Kawartha region." In recent years, all the territory from Lake Ontario north to the northern boundary of Peterborough County has been labelled the Kawarthas. However, an examination of these maps and descriptions reveals a core area identified in all of them that includes Cameron Lake, Sturgeon Lake, Buckhorn Lake, part of Pigeon Lake, part of Chemong Lake, and Clear Lake.

Various relatively new administrative units have further complicated the identification of a boundary for the region. For example,

North Kawartha Township extends up to thirty-five kilometres north of Stony Lake and includes lakes such as Anstruther and Chandos. Most recently, the naming of an area north of Peterborough as the Kawartha Highlands Signature Site has not only confirmed that this area is part of the Kawarthas but suggested that the region has "highlands" (Ontario Ministry of Natural Resources 2003). The controversial amalgamation of local governments in Victoria County as the City of Kawartha Lakes has also extended the region to the east and implied a more urban landscape.

THE OTONABEE REGION

The Otonabee Region Conservation Authority created another water-based approach to defining a region focused on Peterborough. It was established in 1959 as the Otonabee Valley Conservation Authority. Then, in 1960 at the request of the authority and certain municipalities adjacent to its boundaries, the authority was enlarged to include the watershed of the Indian River, and its name was changed to the Otonabee Region Conservation Authority. A further enlargement took place in 1961, when the watersheds of the Ouse River and a number of smaller streams were also placed under the jurisdiction of the authority. More recent concerns about source-water protection have fostered some new collaborative initiatives by conservation authorities in the region; for example the Trent Conservation Coalition for source-water protection includes the following authorities: Otonabee, Crowe Valley, Ganaraska, Kawartha, and Lower Trent.

PETERBOROUGH COUNTY

In the lobby of City Hall in Peterborough is a mosaic map showing Peterborough, the Kawartha Lakes, and an area extending to Haliburton, Bancroft, Trenton, and Oshawa. Before 1841 the Peterborough area was part of the old Newcastle District of Upper Canada. In October of that year, the area north of Rice Lake was separated from the Lake Ontario region known as "The Front," and the Colborne District was proclaimed, its border containing all of the later Peterborough and Victoria Counties and part of Haliburton. In 1850, the Colborne District was divided in two, the northern part being called Peterborough County, and then in 1851, the United

Counties of Peterborough and Victoria. "For a further ten years this political union continued with most of its power centred in Peterborough. But in 1861 Victoria was granted independent status. Haliburton County separated in 1874, drawing its twenty-three townships from both of the formerly United Counties, leaving the composition of townships in Peterborough County as it is today" (Cole 1988, 1). Since then, various townships and villages have been amalgamated, often contentiously. Peterborough County, then, is not synonymous with a geographically defined region or with all the Kawartha Lakes.

A TOURISM REGION

For the purpose of tourism promotion, the Ontario government over the last decade has subdivided the province into varying regions. In 2000 it identified six major tourism regions. Peterborough was located in the "Eastern Ontario" region, which extended from Toronto east to the Quebec border and from Lake Ontario north to include Algonquin Provincial Park. This region was further subdivided, one part including Peterborough, Victoria, and Haliburton Counties, and other areas to the southwest, southeast, and northwest. It was labelled Getaway Country – Central Ontario and featured on the website www.getawaycountry.com. The name, of course, begged the question, from where and what tourists were being encouraged to get away?

In 2004, the government divided the province into seven tourism regions. Peterborough is now located in the Central Ontario region, which extends from Georgian Bay east to the eastern boundary of Hastings County and from Lake Ontario north to Algonquin Provincial Park. This "region" is further subdivided, one subdivision being the Kawarthas and Haliburton. This so-called region exhibits a wide range of topography, vegetation, cultural landscapes, settlements, recreation opportunities, and economic activities. Although the government maintains that this is a "region renowned for its recreational opportunities," it is doubtful that it constitutes a region in the public mind, and it has not been conceived as a region for any other purpose than tourism promotion (Peterborough and the Kawarthas 2004, 62).

The Kawartha Lakes and Peterborough, though variously bounded, have been promoted as a tourist region for over a century. Today, the Peterborough and the Kawarthas Tourism and Convention Bureau promotes part of this area as "Peterborough and the Kawarthas." Its travel guidebook is entitled *Peterborough and the Kawarthas*, and its website is www.thekawarthas.net. The bureau urges visitors to "Let our region embrace you with its own brilliant personality" (2007, 2). It is thus implied that all of the Kawarthas and Peterborough are in a region. Yet the maps of the region in the guide show only Peterborough County and the City of Peterborough. Furthermore, the western part of the Kawarthas, west of Omemee and Bobcaygeon, is promoted by a separate organization, the City of Kawartha Lakes, based in Lindsay. There is thus a competitive, contested, and confusing approach to defining and promoting the Peterborough area as a tourist region.

A SERVICE REGION: THE GREATER PETERBOROUGH AREA

The Greater Peterborough Area as defined by the Greater Peterborough Area Economic Development Corporation (GPAEDC) is comprised of the City of Peterborough and eight municipalities within the County of Peterborough (the Township of Asphodel-Norwood, the Municipality of Cavan-Millbrook-North Monaghan, the Township of Douro-Dummer, Smith-Ennismore Lakefield, Galway-Cavendish-Harvey, Havelock-Belmont-Methuen, the Township of North Kawartha, and the Township of Otonabee-South Monaghan). In economic development terms, the Peterborough region has been defined by the political boundaries established for the County of Peterborough. The County of Peterborough is located between Ottawa and Toronto and has an estimated population of 133,000 that can increase by 30,000 with summer cottagers.

At the heart of the county is the mid-sized City of Peterborough with a population in 2006 of 74,898, an increase of 4.8 percent from 2001. Accessed from the southwest by a four-lane divided highway, the city is positioned fifty kilometres north of a central focus of population and economic activity in Canada – the Windsor–Quebec City corridor embracing Highway 401 (figure 7.1). This location sets the City of Peterborough somewhat apart from the main flow of people and commerce and has allowed it to become a regional service centre for East Central Ontario. It is the largest urban centre in the Kawartha Lakes Region. To use the language of American geographer Richard Hartshorne, who laid out the philosophical basis of regional geography, Peterborough could be described as a "functional region" in the sense that the City of Peterborough is the common node around which the county is unified and organized.

As a former industrial town, Peterborough capitalized on a cheap supply of hydroelectricity and accessible railway networks. Today, metal fabrication, automobile parts, and food processing remain key industries. However, the City of Peterborough's economy has gradually become much more diversified, with a shift from manufacturing to service industries. The city now specializes in regional education, health, and government-related functions; Trent University, Sir Sandford Fleming College, the Peterborough Regional Health Centre, and the Ontario Ministry of Natural Resources serve as major employers. The city has a joint service agreement with the County of Peterborough covering social services, land ambulances, and provincial legal offenses. The city and county also share the Greater Peterborough Area Economic Development Council, an arm's-length economic development and tourism agency formed in 1999.

The GPAEDC has played a central role in raising the profile of the Peterborough area through strategic marketing initiatives. With a mission to create and to sustain wealth within the Greater Peterborough Area, it employs a private/public, non-profit governance model of regional economic development that has the support of the Peterborough City and County Councils. Such a partnership between the city, the county, and the private sector is said to allow for greater marketing and administrative efficiency that reduces the duplication of services and permits the promotion of the area on a regional scale with a singular political voice. In the late 1990s, the GPAEDC ran community consultation sessions with the intent of "identifying regional strengths for future economic progress and expansion. The goal was to define a 'personality' for the region that would move the area towards an identity of excellence" (www.gpaedc.on.ca). In its 2000 strategic plan, the GPAEDC prioritized "the formation of a unique Peterborough-based regional cluster with the goal to harness local innovation and develop a stronger regional economy" (www.gpaedc.on.ca). DNA research was identified as the new strategic area of economic growth and development, and a plan for the Peterborough DNA Cluster was developed in 2002 (www.innovationcluster.ca).

Despite being based on the Trent University campus within the City of Peterborough, the DNA project has been re-framed in regional terms as the Greater Peterborough Region DNA Cluster (GPRDC). This initiative to advance biotechnology research in DNA and forensic science is a private-public educational partnership between Trent University, Sir Sandford Fleming College, the Ministry of Natural Resources, and Maxxam Analytics. The Ontario provincial government has invested $1.3 million in the project through the Rural Economic Development program in order to develop the infrastructure needed to advance the commercialization of technologies related to DNA profiling, forensics, robotics, geomatics, and bioinformatics. The project is expected to provide 1,000 to 2,500 direct jobs and 3,000 indirect jobs over the next ten to twenty-five years and has been strategically positioned as a key component in a developing "Ontario biotechnology corridor," with other nodes in Guelph, Toronto, Kingston, and Ottawa. With continued investment, the DNA Cluster is expected to function as an "innovation resource" at the regional, provincial, and national scales, while the research process and the application of findings are expected to provide a boost to what is interpreted by economic development officers and politicians as the "regional economy." It remains to be seen whether, as the promotional material touts, "the DNA cluster [will] become the focal point, the catalyst for a bright new future in our region" (www.dnapeterborough.ca).

Concomitantly with the identification of the DNA Cluster as a regional economic focus for Peterborough, the cultural sector has received increasing attention from politicians as a significant source of economic development and quality of life. Cultural workers represented 2.2 percent of the local labour force in the City of Peterborough in 2006. On 22 April 2005 Peterborough hosted one of five provincially funded regional forums in Ontario on Municipal Cultural Planning (Driscoll 2005). The forums were targeted at municipal staff, elected officials, and local cultural, business, and community leaders in an effort to encourage a broader vision of culture and cultural resources as pillars in new economic development and community-building strategies. At the forums, municipal cultural planning was presented through case studies of good planning practice. The stress was on cultural planning as inclusive, integrated, and ongoing. The objectives of the forum were to present cultural planning as an activity helping to break down silos separating different forms of local cultural activity, to build and strengthen local and regional networks, to share experiences and good practices, and to identify needs and opportunities to advance cultural planning in Ontario.

The Peterborough forum drew participants from a wide catchment area including Bancroft, Cobourg, Haliburton, Kingston, Lakefield, Lindsay, Oshawa, Picton, Port Perry, Tweed, and Whitby – demonstrating that in practice the region can have quite loosely defined boundaries. The councillors, mayors, municipal staff from planning, tourism, and recreation departments, representatives from cultural organizations, and local community leaders who attended the event had never before been in the same room together discussing the importance of culture. Yet at the forum the city of Peterborough was presented as a leader in municipal cultural planning. The city was one of the first Ontario municipalities to conduct a cultural-mapping exercise, in 1995, that identified 134 arts, culture, and heritage organizations, 185 cultural businesses, and hundreds of cultural workers all within a fifteen-kilometre radius of the city. This mapping exercise helped to shift attitudes to culture within the city by documenting the breadth and diversity of cultural activity in the area.

Culture is now accepted as a core municipal function and responsibility. This political acceptance is reflected in the 60 percent increase in the municipal cultural budget between 1995 and 2001. Per capita investment grew from $27.45 to $42.56. While the focus of the good-practices case studies was on the city of Peterborough, one of the key ideas to emerge from this forum, as highlighted at meetings of the Local Advisory Committee, was the notion that cultural planning cannot be undertaken just at the municipal level. There is increasing recognition among local politicians and community leaders that in the Peterborough area cultural planning needs to be undertaken on a county-wide basis with a focus on joint service provision and partnerships between townships and municipalities. The urban-rural and city-county divisions are seen by many as too divisive, exclusive, and neglectful, and the concept of the region enters the political discourse as the appropriate scale for undertaking cultural planning.

THE RELATIONSHIP WITH TORONTO

Since road and railway connections were established in the mid-nineteenth century with Toronto, Peterborough, however its region is defined, has been in a dynamic relationship with that growing city and smaller cities east of it, from Scarborough to Oshawa. This relationship has proved both beneficial and problematic for the Peterborough area. Certainly, a significant challenge to maintaining connectivity was the gradual abandonment of freight and passenger rail service to Port Hope, Hastings-Belleville, Ottawa, Lakefield, and Lindsay. Of all past rail services, only a freight service between Havelock, Peterborough, and Toronto remains. Occasionally, negotiations have been initiated to restore passenger-train service between Peterborough and Toronto, either as a GO train or VIA Rail service. In the 2008 federal budget, $500 million was set aside for transit projects nationwide. While VIA Rail has no immediate plans to reinstate the service between Toronto and Peterborough, politicians maintain that this rail link is well positioned to have its service restored at a cost of $150 million. In the meantime, people wishing to travel between Peterborough and Toronto will continue to have to use a slow and infrequent bus service or to drive to Oshawa and take the GO train from there.

The deterioration of rail links from Peterborough to Toronto has to some extent been offset by improvements over the last twenty years in the highway linkage. These include the twinning of Highway 115 from Highway 401 to Peterborough, the widening of Highway 401 between Oshawa and Toronto, and the opening of the Highway 407 toll road. The extension of Highway 407 to Highway 115 is expected and welcomed by commuters and trucking-based industries. Highways to the east, north, and west of Peterborough have also been improved in the last decade, facilitating economic development through reinforced transportation and communication linkages.

Good highway connections between Peterborough and Toronto strengthen the linkage between the two regions. Many residents of the Peterborough area work in the car factories in Oshawa, and some in Toronto. The shopping facilities in Oshawa, Scarborough, and Toronto have long been deemed attractive and better than those in Peterborough. Many residents of the Peterborough region go to Toronto for entertainment, especially that provided by theatres, concert halls, art galleries, and sports venues. Travelling in the opposite direction are people from the Toronto region seeking recreational, educational, and retirement opportunities.

Peterborough's proximity to Toronto is a double-edged sword. While Toronto provides significant retail, cultural, and recreational opportunities, it also draws away customers, audience members, and audience dollars from local retailers and events, fostering a relationship of consumption dependence. This relationship reinforces the mistaken assumption that local activities are not as significant or not of as high quality as those in Toronto. Although the Peterborough region is linked with the Toronto region through networks of retail, cultural, and recreational consumption, understanding the complexity of these linkages and planning for them, above and beyond a concern for expanding highway capacity, has been weak. It remains unclear whether the Peterborough region or the Toronto region will benefit more from improved transport links between them.

In summary, the Peterborough region wants better access to the Toronto region for employment, business, shopping, and entertainment. It also wants more people from the Toronto region to access Peterborough, especially for recreation and its economic benefits. However, the Peterborough region also wants to maintain its independence, viability, and growth. A healthy relationship between the two regions requires linking the planning for the Peterborough region with that for the Toronto region, and for the province as a whole.

CONCLUSIONS

In this chapter we have demonstrated how the concept of the region in the Peterborough area has been variously defined, delimited, and deployed at different periods in history and in response to different political-economic circumstances. The boundaries of the Peterborough region are fluid, contracting and expanding in the public imagination, symbolizing an effort to construct a collective representation of a place. The concept of the region, then, can be interpreted as a social construct – a product of socially contingent practices and discourses (Paasi 2002). It is a concept created by people in response to global economic forces that demand that a place must be distinctive in order to survive and compete with other places; created to foster attachment to place, to encourage people to learn a culture, and to contribute to its continuation.

Our preliminary exploration of the application of the concept of the region to the Peterborough area has raised a number of potential research questions. What are the implications of the evolving confusion surrounding the definition of a Peterborough region? Should administrative and promotional regions be more related to biophysical regions? How do various people define the regions of Ontario, and the Peterborough "region" in particular? Are the concepts of corridors and transition zones more useful than regional concepts? How does the perception of the Peterborough area influence, for better or worse, provincial land use and economic planning and its relationship with Toronto? Questions such as these deserve ongoing critical and sustained attention by scholars and policy-makers.

REFERENCES

Adams, Peter, and Colin Taylor, eds. 1992. *Peterborough and the Kawarthas*. Peterborough: Heritage Publications.

Bone, Robert M. 2000. *The Regional Geography of Canada*. 3d ed. Don Mills, ON: Oxford University Press.

Borg, Ronald. 1967. *Peterborough: Land of Shining Waters*. Peterborough: City and County of Peterborough.

Brunger, Alan. 1987. *By Lake and Lock: A Guide to Historical Sites and Tours of the City and County of Peterborough*. Peterborough: Heritage Publications.

Chapman, L.J., and D.F. Putnam. 1966. *The Physiography of Southern Ontario*. 2d ed. Toronto: University of Toronto Press.

Cole, A.O.C., ed. 1988. *Peterborough Historical Atlas, Abridged Edition*. Peterborough: Corporation of the County of Peterborough.

Coleman, J. n.d. *The Settlement and Growth of the Otonabee Sector*. Peterborough: Trent-Severn Waterway, Interpretive Program Publication.

Department of Railways and Canals of Canada, The. 1911. *Reference and Guide Book: The Trent Canal*. Ottawa: The Department of Railways and Canals of Canada.

Driscoll, John. 2005. "Raising the Profile of Municipal Cultural Planning." *The Grassroots Review*. http://www.grassrootsreviewpeterborough. ca/2005/April/April25.php (accessed 5 May 2005)

Fenlon Falls. 1909. *Fenlon Falls: The Prettiest Summer Resort on Kawartha Lakes*.

Gilbert, Anne. 1988. "The New Regional Geography in English- and French-speaking Countries." *Progress in Human Geography* 12: 208–28.

Grand Trunk Railway. 1901. *Kawartha Lakes–Grand Trunk Railway System.*

Gregory, Derek. 1978. *Ideology, Science and Human Geography.* London: Hutchinson.

Johnston, Ronald J. 1983. *Philosophy and Human Geography: An Introduction to Contemporary Approaches.* London: Edward Arnold.

– 1991. *Geography and Geographers: Anglo-American Human Geography since 1945.* 4th ed. London: Edward Arnold.

Jones, Elwood, and Bruce Dyer. 1987. *Peterborough: The Electric City.* Peterborough: Windsor Publications.

MacLeod, Gordon, and Martin Jones. 2001. "Renewing the Geography of Regions." *Environment and Planning D: Society and Space* 19: 669–95.

Mallory, Enid. 1991. *Kawartha: Living on these Lakes.* Peterborough: Peterborough Publishing.

– 1992. "The Kawartha Lakes." In *Kawarthas Nature*, ed. Peterborough Field Naturalists, 33–5. Erin: Boston Mills Press.

– 1994. *Countryside Kawartha.* Peterborough: Peterborough Publishing.

Marsh, John. 1981. "Early Tourism in the Kawarthas." In *Kawartha Heritage: Proceedings of the Kawartha Conference*, ed. A.O.C. Cole and J.M. Cole, 43–52. Peterborough: Peterborough Historic Atlas Association.

– 2004. "Kawartha: The Origin and Application of a Name." *The Heritage Gazette of the Trent Valley* 9:23–4.

Massey, Doreen. 1984. *Spatial Division of Labour: Social Structures and the Geography of Production.* London: Macmillan.

Ontario Ministry of Culture and Recreation. 1981. *Heritage Studies on the Rideau-Quinte-Trent-Severn Waterway.* Toronto: Historical Planning and Research Branch.

Ontario Ministry of Natural Resources. 2003. *Kawartha Highlands Signature Site, Charter.* Peterborough: Ministry of Natural Resources.

Paasi, Anssi. 2002. "Place and Region: Regional Worlds and Words." *Progress in Human Geography* 26:802–11.

Peterborough and the Kawarthas. 2004. *Peterborough & the Kawarthas, 2004 Travel Guide.* Peterborough: Peterborough Kawartha Tourism and Convention Bureau.

Peterborough and the Kawarthas Tourism. 2007. *Peterborough & the Kawarthas, 2007 Travel Guide.* Peterborough: Peterborough & the Kawarthas Tourism.

Pudup, Mary B. 1988. "Arguments within Regional Geography." *Progress in Human Geography* 12:369–90.
Tatley, Richard. 1978. *Steamboating on the Trent-Severn*. Belleville: Mika Publishing Company.
Thrift, Nigel. 1990a. "For a New Regional Geography, 1." *Progress in Human Geography* 14:272–9.
– 1990b. "Doing Global Regional Geography: The City of London and the South-east of England." In *Regional Geography: Current Developments and Future Prospects*, ed. Ronald Johnston et al., 180–207. London: Routledge.
– 1991. "For a New Regional Geography, 2." *Progress in Human Geography* 15:456–65.
Traill, Catherine Parr. 1971. *The Backwoods of Canada*. Toronto: McClelland and Stewart.

8

Georgian Bay, Muskoka, and Haliburton: More Than Cottage Country?

NIK LUKA AND NINA-MARIE LISTER

SUMMARY

The mere mention of "cottage country" anywhere in Canada almost inevitably conjures up images of Ontario's forested lakelands. This chapter explores the georegion stretching from Georgian Bay eastward to Haliburton and the Algonquin Highlands and from the French River southward to the Trent-Severn waterway and the Oak Ridges Moraine. Although permanently if sparsely settled since the early nineteenth century, it is widely considered a summer leisure destination, and indeed it is now dominated by second homes numbering in the tens of thousands. Residents of major cities in the lower Great Lakes basin have extended their everyday life-spaces to include its waterfront settings for over a century. What once were modest summer cottages are increasingly elaborate permanent dwellings occupied part-time by Greater Toronto Area residents, entrenching the quasi-suburban role of this georegion. This chapter highlights the confluence of abiotic, biotic, and cultural components in the Georgian Bay, Muskoka, and Haliburton georegion, suggesting how its hybrid cultural-natural ecologies stem from its transitional qualities – both symbolically, as an exurban cultural "edge" to the Toronto-centred metropolis, and functionally, in geophysical terms.

Contemporary Ontario folklore tells us that the province contains a space known as "cottage country" – a swath stretching from Georgian Bay through Muskoka into Haliburton and the Kawartha Lakes. History shows that this space is more than a hinterland of second homes for an increasingly (sub)urban society. However, at various junctures in human history it has been the home territory and hunting ground for various aboriginal groups, a place of ruthless resource extraction, and (with varying degrees of success) an agricultural area. As a georegion, it is a compelling case study in the confluence of abiotic, biotic, and cultural forces – the ABC paradigm that underlies the case studies in this volume. This chapter traces how these forces have generated a hybrid of cultural-natural ecologies in this georegion. We pay close attention to the georegion's transitional quality. This is true both symbolically – for this area has become an exurban cultural edge to the Toronto-centred metropolitan region – and functionally, in terms of its geology and ecology, since it straddles the transition from the arable St Lawrence Lowlands to the rocky Canadian Shield, with its proliferation of freshwater lakes. By markedly shaping settlement possibilities and patterns, this "edge" has also tended to be self-reinforcing. It has embedded itself in the materiality of its own cultural landscapes. Its haunting beauty and ruggedness, in concert with the Anglo-American compulsion to "escape" the city, have, moreover, elevated parts of central Ontario to near-mythic status as the very archetype of a summer playground.

In this chapter, we present the Georgian Bay/Muskoka/Haliburton georegion through a summary of its physiographic and ecosystem characteristics, along with an account of its human settlement and activity patterns over the past several thousand years. We also introduce a set of critical arguments concerning what this "cottage country" georegion now is and what it represents in the Anglo-American world. Key questions are presented as we trace social history from the earliest human settlement through European colonization and the late-nineteenth-century rise of large-scale tourism. Of particular interest is how this georegion has come to be uniquely replete with saleable landscape commodities in the form of waterfront or water-oriented second homes, most of which have been built, owned, and used by city dwellers. Of course, other regions in Ontario, across North America, and around the world are also strongly marked by the presence of second homes. Yet in this area these waterfront settlement patterns represent an especially characteristic intersection of abiotic, biotic, and cultural layers of the landscape of the georegion of which they are part. They are not the only distinguishing features, but they are extremely salient. We suggest ways in which the curious blend of abiotic, biotic, and cultural forces in this georegion can offer hope for the contemporary challenge of getting people to understand the links between "nature" and "culture" in human settlement systems. Our chapter therefore culminates with a set of comments for planning and design practice.

THE GEOREGION

Some clarification is needed to define this georegion, which we call the Central Ontario Lakelands. It is in some ways the centroid of Ontario, and as with all georegions, its boundaries are blurry, overlapping with its neighbours. As shown on figure 8.1, the Central Ontario Lakelands Georegion adjoins the Great Lakes on the west; Manitoulin, the French River, and the Algonquin Highlands on the north; the Peterborough Georegion on the east and the southeast; and finally, the Toronto Georegion and the Niagara Escarpment on the south and southwest.

The Central Ontario Lakelands roughly correspond to the Simcoe and Muskoka/Parry Sound watersheds – a space where different areas of influence overlap. The physiographic heritage of this georegion, combined with its animal and plant geographies as well as human life, make up a landscape mosaic of ecosystems that are nested and intricately interconnected at many scales. Much of the georegion is part of "The Land Between" as described by Alley (2004, 2006): a wide ecotone between the St Lawrence Lowlands and the Canadian Shield stretching from Georgian Bay southeastward to the Frontenac Axis. Jurisdictionally, the Central Ontario Lakelands include the current regional municipalities or counties of Parry Sound, Muskoka, Haliburton, Simcoe, the western parts of Peterborough, and the new "City" of Kawartha Lakes. It is a sparsely settled georegion in terms of its permanent population, with only a few small urban centres. Its most characteristic cultural feature is the dispersed metropolis of waterfront or water-oriented second homes known as "cottages" to their users.[1]

We focus here on the Central Ontario Lakelands using its second-home landscapes as subject while using the ABC approach as object; the two act in concert as a series of illustrative and analytic lenses for making sense of this georegion.

ABIOTIC AND BIOTIC DIMENSIONS

The extremely dense and hard rock of the Canadian Shield dominates the landscape of the Central Ontario Lakelands. This rocky

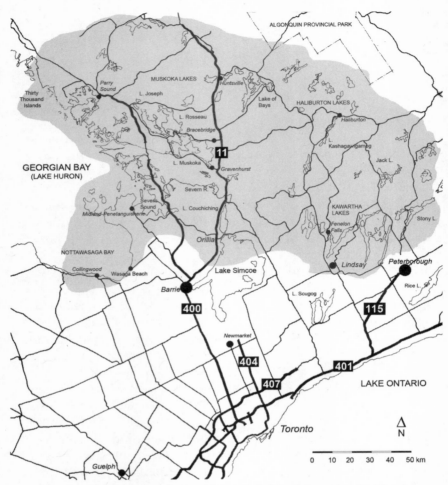

Figure 8.1 General extent of the Central Ontario Lakeland Georegion. By the authors, with permission

foundation is mostly granite (a crystalline formation resulting from the cooling of liquid rock underground) and gneiss, a derivative of granite. Formed 1.0 to 1.6 billion years ago as part of a massive mountainous complex, the Shield has since been transformed by heat, pressure, erosion, and glacial scouring, the latter especially in the past hundred thousand years, when ice sheets covered much of Ontario on at least four separate occasions. Several advances and retreats gouged deep striations, channels, and holes in the hard bedrock. As the glaciers thawed, they became freshwater lakes and rivers, as well as bowl-shaped bogs and wetlands where natural outlets for drainage were lacking. Watercourses now tend to follow faults in the bedrock – paths of least resistance, as it were – creating convoluted and redundant "deranged drainage" patterns (Baldwin et al. 2000; Chapman and Putnam 1984; Marsh 2005).

The final retreat of the great ice sheets, some ten thousand years ago, left two major lake chains in the Central Ontario Lakelands. In the Muskoka-Parry Sound zone to the northwest, the three largest water bodies form a single complex with a surface area of over 250 km^2: Lakes Joseph (57 km^2), Rosseau (67 km^2), and Muskoka (132 km^2). Linked to this complex is the Lake of Bays, with an area of 70 km^2, and a great many smaller lakes. These waterbodies tend to flush rapidly, draining from east to west into Georgian Bay, mainly through the Moon River, a fairly fast-moving channel under 50 km in length. At the southeast end of the Central Ontario Lakelands are the Kawartha Lakes, a chain of narrow, shallow waterbodies found in depressions in front of a large rock escarpment called the Gull River Formation. They drain from west to east by way of the lowest saddles in the bedrock. Following the convoluted courses of the Otonabee and Indian rivers in a southeastwardly direction over approximately 150 km with little change in elevation, they flush very slowly and ultimately reach Lake Ontario at Trenton. Between the two major lake chains is an area called the Haliburton Highlands – the southerly reaches of the topographical dome now forming Algonquin Provincial Park. Here the landscape is marked by deepwater lakes, steep cliffs, and generally rough terrain (Chapman and Putnam 1984). Compared to the St Lawrence Lowlands, much of the Central Ontario Lakelands lies from 100 to 250 m higher (Baldwin et al. 2000; ESWG 1996).

Soils in the Central Ontario Lakelands vary. For the most part, they are acidic and poor in nutrient value, reflective of their glacial parent material and underlying bedrock. From the Kawartha Highlands northwestward through Muskoka to Georgian Bay, the soil cover is very thin with scattered deeper pockets. In the Kawartha Lakes, however, a thicker layer of soil has historically made for viable agriculture. Elsewhere, deposits of peat or rotting vegetation, in layers up to 30 m thick, leach tannins, giving the lakewater a rust-brown hue and high turbidity (Baldwin et al. 2000; Chapman and Putnam 1984; Sanderson 2004).

The climatic and biotic characteristics of the Central Ontario Lakelands differ noticeably from the more densely populated georegions to the south, east, and west. Within the humid continental climate of the Great Lakes region – where summers are hot and humid and winters cold and snowy – the Central Ontario Lakelands tend to be cooler, because of their higher elevations, but also because of the strong westerly winds blowing across central Ontario from the upper Great Lakes. In addition to causing intense thunderstorm activity in June, July, and August, as cold air masses from over Lake Huron meet warmer overland air, this has had a "flushing" effect. Historically, air quality has been superior to that of the urbanized St Lawrence Lowlands, where smog has long been a summertime fact of life, but in recent years, poor air quality has dogged much of the Central Ontario Lakelands, as it does the rest of the Great Lakes basin (McDonald 2003).

The vegetation of the Central Ontario Lakelands clearly reflects the coarseness of this georegion's soils. The mixed wood forest cover is dominated by coniferous species such as hemlock, black and white spruce, balsam fir, Jack pine, and historically the white pine, which often grew in stands reaching 50m or more in height but which was largely logged out (Thompson 2000a). Over the past several thousand years, the mix has diversified with white birch, sugar maple, beech, yellow birch, trembling aspen, and white or red oak. The forest is also rich in animal and bird life, including beaver, muskrat, wolf, black bear, red fox, chipmunk, ruby-throated hummingbird, and the common loon. In all, the Central Ontario Lakelands have, or historically had, about twenty kinds of forest amphibians and reptiles, some fifty species of forest mammals, and over one hundred different species of birds (ESWG 1996; Thompson 2000b; Voigt et al. 2000).

CULTURAL DIMENSIONS

The Central Ontario Lakelands are a space of transition in cultural terms; this has been especially true since the time of European exploration. In effect, this georegion never settled into its roles until sometime in the early twentieth century, as tourism came to dominate its activity patterns.

Aboriginal Life and Settlement Patterns

Aboriginal oral tradition suggests that people moved into areas that had previously been blanketed in ice or under water perhaps thirty-five thousand years before present. While a highly symbiotic relationship was maintained between hunters and animals, a remarkable change took place some fifteen hundred years ago when proto-Iroquoian groups in the lower Great Lakes took up corn cultivation, having learned of it from groups to the south. For at least twelve subsequent centuries, until the arrival of the first Europeans, much of central Ontario cottage country seems to have been a common ground for three groups: the nomadic Algonquin and Ojibway, hunters and trappers coming from the north and west, travelling great distances by canoe; several sedentary Iroquoian groups concentrated around Lakes Erie and Ontario; and the Huron nation, an Iroquoian group that made themselves the hub of an extensive trade network. In their well-established settlements, they grew corn, squash, and beans, moving their villages every ten to fifteen years as the soil nutrients and handy firewood supply were exhausted. What we consequently know as Huronia stretched from the southern part of Georgian Bay around Lakes Couchiching and Simcoe and down to Lake Ontario along the Humber valley. This corridor, known as the Toronto Carrying Place, was an important axis of movement then as now. The Huron were largely on friendly terms with the Algonquin to the north, with whom they traded regularly, but this co-dependency does not seem to have been as true between the Huron and Iroquois Five Nations concentrated to the south of Lake Ontario. Certainly, by the time of first European contact, they were bitter enemies.[2]

European Exploration and Contact with Ontario Aboriginals

The first Europeans known to have visited the Central Ontario Lakelands were the French explorers Étienne Brûlé (ca 1592–1633) and Samuel de Champlain (ca 1570–1635). On the south shore of Georgian Bay they encountered the Huron nation, which then numbered in the tens of thousands and whose well-developed trade network with the Algonquin to the north lent itself extremely well to the fur trade so coveted by the European explorers. Brûlé and Champlain were soon followed by Jesuit missionaries who established a mission at Sainte-Marie-among-the-Hurons, near what is now Midland. Some evidence suggests that they developed a hybrid Christian-Huron faith system and that they were seen as benevolent by the Huron. The arrival of the Europeans brought about the downfall of this aboriginal nation, however, as huge numbers of the Huron succumbed to diseases to which they had no immunity such as smallpox. Within fifty years of their first contact with Europeans, the Huron were forced to abandon and burn many of their villages. Their final migration to Québec, where their descendants still live, was brought on by Iroquois attacks motivated by competition in the fur trade.

Following the almost complete demise of the Huron nation, the Ojibway gradually moved in from the north shore of Lakes Huron and Superior, where they had long taken advantage of the rich Georgian Bay fisheries. With the British conquest of New France, a series of aboriginal land surrenders took place. The once-nomadic Ojibway groups – known collectively as the Chippewa – became increasingly sedentary. By 1850 they had been isolated into some twenty-four designated reserves across the georegion through a comprehensive and no doubt highly coercive set of settlement "agreements."[3] Perhaps ironically, they came to be seen as the "local" aboriginal group (Gray 1994).

Early European Settlement and the Logging Era

Through the early nineteenth century an important axis of European colonization followed the Toronto Carrying Place from Lake Ontario up the west shore of Lake Simcoe to the south end of Georgian Bay. Efforts to settle this area – known as the Ottawa-Huron tract – led the government of Upper Canada to pass a Public Lands Act in 1853 offering one hundred free acres to settlers along designated colonization corridors reaching northward into the Central Ontario Lakelands. Promotional brochures produced in English, French, German, and Norwegian were circulated throughout northwest Europe. Despite conflicting reports about the arability of the land, a good number of eager settlers came, largely from Ireland and England; for instance, Muskoka's population increased from five thousand residents in 1871 to three times that in 1891. Like roads elsewhere in Ontario, the early colonization roads were little more than rough paths, and not much awaited settlers on arrival; some found themselves atop pockets of fairly good soil, but most would-be farmers discovered that their land was hardly arable. It was not long before the Upper Canada government decided to suspend the settlement program in the eastern parts of the Central Ontario Lakelands, where the soil was quite unsuitable for farming and where the number of willing settlers had rapidly decreased (Angus 1988; King 1994; Murray 1963; Øverlund 1989; Parson 1987).

The story of logging and lumber barons looms largest in the nineteenth-century history of the Central Ontario Lakelands. A growing demand for wood products in Britain and the northeastern United States transformed this georegion, since the more accessible old-growth forest areas of the province, such as the Ottawa Valley, had been largely depleted by the 1830s. A US–Upper Canada reciprocity treaty signed in 1854 allowed for the duty-free exchange of sawn lumber, spurring a logging boom until a tariff was introduced in 1866 (Lower 1938). Logging was further facilitated by the speculative building of railways and by policy changes awarding logging rights to Crown land for a fee.

As enormous log booms began to regularly cover the lakes and rivers of the Central Ontario Lakelands, timber barons emerged as the controllers of networks of logging camps upstream from the major water bodies where sawmills could easily be built and maintained. Ruthless logging operations rapidly depleted the old-growth forests of the georegion; production peaked in 1881 and then rapidly declined. The general scarcity of soil and the tendency to remove soil-retaining groundcover vegetation in the logging process meant that many tree species such as red, white, and black spruce, yellow birch, and eastern hemlock were unable to regenerate in many areas. The forest composition was thus significantly transformed by logging. Awareness of how the United States had quickly depleted much of its forest cover and was now hungry for Canadian wood helped bring about a change in attitudes to the forest, which had long been considered merely a barrier to a prosperous agrarian Ontario, if not a latent threat to civilized life writ large. The exportation of sawlogs from Crown land was prohibited in 1898, marking the end of the great logging era in Ontario (Epp 2000; Lower 1938; Nelles 1974; Smith 1990; Thompson 2000a).[4]

Logging was not the only major resource-extraction activity in the Central Ontario Lakelands in the mid-nineteenth century. On the waters of Georgian Bay, a commercial fishery emerged beginning in the 1830s. By the 1870s, railway connections and the development of high-performance fishing craft tailored to the conditions of the Bay, such as the Collingwood skiff and the Georgian Bay mackinaw, meant that whitefish, trout, and pickerel were being shipped through the Great Lakes basin and the northeastern US states. The annual commercial yield peaked sometime in the 1870s at over one million kilograms (Barry 1995). The bountiful waters also attracted sport fishing, which together with hunting soon gave rise to the resort industry and the second-home phenomenon, which was to last much longer than the logging and fishing industries.

The Beginnings of Tourism

The modern history of the Central Ontario Lakelands began in the 1870s with the first deliberate tourists: hunters and sports folk from Toronto and other industrial hubs on the American side of the Great Lakes, particularly Pittsburgh and Cleveland. There are accounts of "shooting boxes" in the southern Kawartha Lakes as early as 1858 (Wolfe 1962). By the 1870s dozens of hunting camps and lodges were established for seasonal use across central Ontario. Lakes Simcoe and Couchiching were colonized by 1863, when a Norwegian tourist recorded this account:

> On the southern shores of Lake Couchiching, a continuation of Lake Simcoe, is the delightful village of Orillia with hardwood forests in the background, beautiful gardens in its midst, and charming islands in the foreground ... No wonder that the merry young people of Toronto liked to spend their summers there. Small rowboats and sailboats had been brought up to Lake Simcoe by rail and every day pleasure boats filled with ladies and gentlemen could be seen passing the village. (Schrøder, in Øverlund 1989, 122).

Before long, the southern reaches of the Canadian Shield had become a choice holiday destination among the well-to-do of the Great Lakes basin, and once again, the liminal qualities of this georegion were key. One late-nineteenth-century advocate of these summer sojourns wrote that "the prudent man flies from all artificial conditions and yields himself to the soothing influences of nature on the shores of the lakes and rivers in the depths of our primeval forests" (Stevenson 1886, 382). There were clear physiological, psychological, and symbolic benefits associated with such an exodus: "by leaving an atmosphere tainted with sewer gas to inhale the tonic perfume of the pine bush," the "weary, over-worked toiler[s] of the city" and their families would find "their cheeks flushing with freshened tints of purified blood" (Hague 1893, 263; Varley 1900, 29). It stands to reason that with the west winds bringing cooler, unpolluted air – a welcome relief before the advent of mechanized air-conditioning – summer sojourns in the Central Ontario Lakelands came to be seen as a necessary coping strategy for regimented, over-structured, and "unnatural" city life.

The "Golden Era," from 1890 to the Great War

Census data reveal that the permanent population of the Central Ontario Lakelands had stabilized by the turn of the century. From that time onward, however, the seasonal population increased dramatically. By the economic boom of the 1890s, the georegion enjoyed a "Golden Era" of trains, steamships, and grand summer resorts. The Kawartha Lakes, at first the destination of choice for well-to-do urbanites, were eclipsed by Muskoka by the turn of the century. Soon an intensely social atmosphere developed in luxurious hotels and rustic lodges or "Houses" that were often owned and operated by land grantees who had all but given up on farming. Guests would spend weeks or even months there, making the round trip only once each season (Wall 1977; Wolfe 1951, 1962).

This Golden Era was made possible only by the further development of passenger railways and small fleets of steamers that met the passenger trains from Toronto at special wharfside stations (figure 8.2). By 1879, there were ten passenger vessels on the Kawartha Lakes. In Muskoka, short canals were built in 1871 to make the three big lakes navigable to large steamers such as the *Muskoka,* the *Segwun,* and later the *Sagamo.* In effect, these canals created a single 250-square-kilometre body of water with countless ruggedly picturesque bays and inlets. Throughout the Central Ontario Lakelands, the increasingly well-developed network of trains and steamers ensured a comfortable and reliable journey for travellers, as well as providing postal service and supplies when roads were often impassable and bad at best (Tatley 1983, 1984; Wall 1977; Wolfe 1951, 1962). On all the major lakes of Muskoka and the Kawartha

Figure 8.2 Late-nineteenth-century postcard showing steamers at one of the many resorts on the Muskoka Lakes. A postcard in the public domain

Figure 8.3 Typical early-twentieth-century waterfront cottage scene in the Central Ontario Lakelands. Archives of Ontario, with permission

districts, these ships were often the social and practical lifelines well into the twentieth century (Angus 2000; Marsh 1983; Marsh and Moffatt, 1979; Wolfe 1962).

If tourism seemed poised to sustain a viable local economy across the Central Ontario Lakelands where both agriculture and logging had failed, it was nevertheless transitional. Even at its zenith, the resort industry helped sow the seeds of its own demise, since hotel patrons sought private properties on which they could build their own retreats. The first waves of private-cottage building were underway by the late 1890s. Sometimes they were built by "locals" and rented out to city dwellers, but land was also being subdivided and sold off, often in large parcels (figure 8.3).

Cultural narratives were being built up by the early twentieth century, not least through the rise of the Group of Seven and Tom Thomson, who painted the rugged Canadian Shield and the dark forests of Central Ontario Lakelands as things of great beauty. Their art strengthened the identity not only of this georegion but of Canada as a whole. Meanwhile, resource extraction continued apace. On Parry Sound on Georgian Bay's north shore, a transportation hub was created in the late 1890s, when Depot Harbour was made the western terminus of the Ottawa, Arnprior, and Parry Sound Railway. It was the shortest route for the shipment of grain, lumber, and other raw materials eastward from the Great Lakes basin to the Atlantic seaboard, with manufactured goods returning westward (Brown 1983). Neither Depot Harbour nor other nascent industrial activities were to last very long, however – just another way in which the Central Ontario Lakelands have historically been a liminal space, always becoming something else.

Early Cottage Life: The Interwar Years 1920s–1930s

The permanent population of the Central Ontario Lakelands remained stable through the 1920s and 1930s, but each summer brought an increasing number of seasonal residents – up to thirty-five thousand by the Second World War.[5] They occupied private summer houses that, while altogether unusable in winter, were often quite large and elaborate (figure 8.3). Shops and services were usually provided on a somewhat ad hoc basis by locals; for instance, until well into the 1920s, provisions were still being delivered by boat on the larger Muskoka lakes (Attfield 2000).

The Great Depression of the 1930s seems to have slowed growth only slightly, although social patterns became more well-defined. Muskoka was increasingly for the well-to-do. In contrast, the Kawartha Lakes and Nottawasaga Bay – notably Wasaga Beach –

were increasingly accessible to the "masses" as private vehicles became more affordable thanks to mass production techniques and as the provincial government had upgraded and paved several main roads into the Central Ontario Lakelands by the mid–1930s. Several provincial parks with designated campsites were also opened (Killan 1993).

If a distinct summer ritual had come to be normalized in the 1940s, it was marked by ethno-religious exclusivity. Protestants were numerous on the major Muskoka Lakes, and evidence suggests that various "others" were excluded. Social histories of anti-Semitism in Toronto reveal that several Muskoka resorts explicitly barred Jewish guests, even boasting through advertisements that they were for "Gentiles only" (Betcherman 1975; Levitt and Shaffir 1987). Lake Simcoe was an exception; Jewish cottagers came to be numerous there by the late 1930s (Wolfe 1951). Anti-Catholic sentiment may have discouraged Catholics from spending time in the major summer holiday destinations of central Ontario, while encouraging the emergence of second-home colonies on the largely bilingual Midland peninsula, where Jesuits had established Sainte-Marie-among-the-Hurons in the seventeenth century.

Tourism, while important, was not the only economic activity and driver of landscape change in the Central Ontario Lakelands. The Georgian Bay fisheries continued to provide whitefish and pickerel to the urban markets of the Great Lakes basin and the northeastern states. Sawmills were still in operation in places, but lumbering was a dead industry on Georgian Bay by 1940 (Barry 1995). Logging did continue on a modest scale through parts of Haliburton, while new industrial activities were established.[6] Something symbolic can be seen in an end-of-war event at the transportation hub of Depot Harbour. Economic activity had slowed considerably in the 1930s and never really picked up again. During the war years, explosive cordite was stored in what were by then disused grain elevators. End-of-war celebratory fireworks inadvertently set the cordite and what remained of the town ablaze in a spectacular inferno. Depot Harbour was never rebuilt, and indeed the scattered resource-extraction and industrial activity of the Central Ontario Lakelands would soon be overwhelmed by a postwar boom in cottage life that was to have the most lasting effects on the landscape of any of its human-induced transformations.

The Postwar Cottage Boom: 1945-72

Second homes increased dramatically in number beginning in the late 1940s, echoing growth and expansion in housing markets across Canada and the United States, following fifteen years of economic hardship and war. Instrumental in the postwar rise of second-home multiple residency across central Ontario was the extensive release of waterfront Crown land – part of aggressive provincial government policies to encourage tourism and recreation. As in parts of the United States, Australia, Norway, Sweden, Denmark, and Finland, cottaging was "democratized" as an affordable and enjoyable way to spend largely new-found leisure time (Halseth and Rosenberg 1995; Wolfe 1965). Rising real incomes, much more widespread automobile ownership, and the extensive construction of highways fuelled growth in second homes. Under Leslie Frost's Conservative government, expressways such as the 400 and Highway 11 provided high-speed, dual-carriageway links from Toronto into Muskoka.

Extensive changes were seen in the waterfront settlement patterns of the Central Ontario Lakelands as cottage lots were surveyed and platted into small standard sizes to be sold off as speculative ventures, sometimes on the very land previously occupied by sprawling resorts. The geographical proximity of this georegion to the growing metropolis of Toronto meant that it bore the brunt of the province-wide increase in the number of second homes. Reciprocally, Toronto dominated the second-home scene; Wolfe (1951) reported that as of 1941, Toronto was home to half the province's cottaging households, in spite of having only one-sixth of its total population. A generation later, Hodge (1970, 1974) estimated that one Toronto-area family in seven owned a cottage, while Census data suggest that there were perhaps as many as a hundred thousand across Ontario by the early 1970s (Halseth 1998).

The folklore of cottage life, fuelled by the mass media, evolved in such a way that summer holidays at a waterfront second home in the Central Ontario Lakelands came to be seen as a favourite summer pastime among residents of the Toronto-centred metropolitan region. Here city folk could sojourn, for the summer cottage was by no means a substitute for the primary urban or suburban dwelling.

A central feature of the widespread cottage phenomenon was what Halseth (1998) calls a "single-tier necklace" along the waterfront,

Figure 8.4 Sketch map showing typical "single-tier necklaces" of cottage dwellings encircling waterbodies with private space (Kennisis Lake). By the authors, with permission

effectively precluding public access to lakes and rivers (figure 8.4). This echoed the broader pattern in which Toronto was separated from its cottage country by a belt of rural land that McIlwraith (1997) has aptly described as the "drive-through countryside." Indeed, where agriculture had failed and logging had been short-lived, tourism had become the mainstay of cottage areas, with the resort industry yielding in time to the provision of goods and services to cottagers as commercial hotels declined in the 1960s.

Contemporary Patterns: A Critical Juncture?

In the early 1970s, there was a slowdown in the processes of transformation across the Georgian Bay, Muskoka, and Haliburton. The postwar burst of commercial activity faded, including the closure of several resorts that had survived the rise of mass cottaging. As for the second-home boom, it waned as vacant waterfront parcels became increasingly scarce (Halseth 1998; Wall 1977), and owing to broader demographic and economic patterns – notably high energy costs and general "stagflation."

The slowdown of the 1970s gave way to a new set of transformations as many owners converted their summer cottages into year-round dwellings. Driven in part by the rising real and effective costs of cottaging, this seems to have intersected with the growing tendency for Baby Boomers to "move to the country" as they approached retirement from the labour force (Bourne et al. 2003). Indeed, there has been a surge in demand for "exurban" living in areas such as Caledon as well as Simcoe, Dufferin, and Grey counties since the 1980s.[7] There is clear evidence of this for the waterfront settings of the Central Ontario Lakelands, notably including the concentration of demand in the small waterfront urban centres of the georegion (Dahms 1996; Dahms and McComb 1999; Halseth 1998; Halseth and Rosenberg 1995). Consequently, there has been a great intensification in the use of waterfront settlements in the Central Ontario Lakelands, as rising numbers of "cottagers" (considered for official purposes to be "seasonal residents") now choose to live by the water year-round (figure 8.5). This has coincided with a more widespread "renaissance" of cottage life in Ontario as revealed by Census data.[8] There has also been a marked non-inflation-related increase in second-home real estate values (Royal-LePage 2001, 2002, 2004), and by 2004, the average cost of a waterfront second-home property in Ontario had exceeded the average cost of a dwelling in Toronto (Wong 2004).

Ever attuned to changing consumer preferences, some Ontario housing developers have picked up on contemporary demand patterns by building other forms of housing on desirable waterfront sites. The archetypal cottage thus now increasingly vies for space with lakeside subdivisions, condominium clusters, and a variant on the timeshare condominium: the fractional-ownership cottage property (figure 8.5). Many of these are built on the sites of now-defunct resorts or summer camps that were simply bought out by land developers. Perhaps paradoxically, several of the remaining long-established luxury resorts have seen a recent renaissance, especially on the sought-after Muskoka Lakes. Older establishments have been expanded and converted into luxurious resorts, such as the Bigwin Inn on Lake of Bays, which was rebuilt after having been closed

Figure 8.5 Contemporary growth patterns: New housing typologies (Peninsula Lake). Photo by the authors, with permission

down and disused for some twenty-five years. It now operates as a luxurious resort with about 150 tourist accommodation units and private dwellings. Often these resorts combine golf courses with spas – the generic sorts of leisure retreats that have become so widespread in amenity areas in recent decades.

Time will tell how readily permanent lakefront housing and emerging genres of the luxury resort will defy the patterns of liminality that have so marked the Central Ontario Lakelands since colonization by Europeans. More urgently, however, some issues arising from these current patterns must be discussed in a critical perspective. We therefore now turn to a synthesis of themes alluded to thus far in our discussion.

SYNTHESIS AND CONCLUSIONS

The Central Ontario Lakelands now face significant challenges. Their powerful attraction as a second-home and recreation area for people from elsewhere now ironically threatens the very qualities that have long made this georegion so alluring. Summer residency and tourism are giving way to permanent waterfront living, and the result – a dispersed metropolis of quasi-exurban settings – feeds its own destruction, as demand for "cottages" pushes real estate values upward and further entrenches a long-established interdependence between the agriculturally productive, densely populated southern part of Ontario and the nearby landscapes of lakes and forests. Compounding the situation is a long-standing social disparity between those who can afford to be on the water and the permanent "local" population, as well as a host of concerns over the ecological impacts of intensified activity patterns across the Central Ontario Lakelands.

The ABC paradigm guiding this chapter is potent, and yet it can be misleading, for it implies that we can parse complex wholes into categorical parts. It highlights differences, disparities, and contradictions – most commonly in terms of how cultural dimensions are at odds with their abiotic and biotic counterparts. We must remember to reassemble the elements and to see the ways in which they complement one another. We must also identify opportunities for enhancing the weave of all three. The case discussed in this chapter affords many possibilities, for fundamentally the history of the Central Ontario Lakelands has been one of symbiosis between human culture and the natural environment. We see emergent strengths and future potential rather than the polarity, disparity, struggle, and defiance that often dominate landscape analysis. Our discussion highlights three recurrent tendencies revealed by a scrutiny of continuity and change in the Central Ontario Lakelands.

The first point of discussion is the way in which this georegion is an area defined by transition. The Central Ontario Lakelands represent a wide "living edge" between the fertile plains of the St Lawrence Lowlands to the south and the rugged Pre-Cambrian rock and boreal forest of the Canadian Shield to the north. Historically, this georegion has usually been in flux, at least since the time of European exploration and colonization. According to our analysis, it did not settle into its contemporary roles until sometime in the early twentieth century as tourism came to dominate its activity patterns.

As a liminal space, the greatest contemporary importance of the Central Ontario Lakelands may well be in how its abiotic, biotic, and cultural characteristics distinguish it from the more urbanized parts of the province to the south (Luka 2008a,b). What have come to be all-important second-home areas of central Ontario are also characterized by an awareness of the landscape on the part of those users who are greatest in number: the urban sojourner. This is rare, and it represents a great opportunity for positive change in behaviour – an idea that we develop in a moment.

A second hallmark of the Central Ontario Lakelands can be seen as both cause and effect of the first. Long dominated by the Toronto-centred metropolitan region, this georegion is in many ways part of that very system. Its history as a second-home destination is historically telling in this respect. While many other cultural contexts have given rise to second homes, the catalyst for the rise of cottaging in the Central Ontario Lakelands is to be found in industrialization and the extensive mercantile capitalism that marked the late-nineteenth- and early-twentieth-century growth of cities in the Great Lakes basin. The georegion in which city folk eventually undertook extended sojourns was already a handy resource hinterland by the mid-nineteenth century for the St Lawrence Lowlands, as well as points throughout the Great Lakes and northeastern US states. Yet the resource hinterland yielded rather quickly to a longer-lasting role as a *leisure* hinterland.

Quite apparent in contemporary processes of transformation across the Central Ontario Lakelands are ways in which its waterfront residential settings are now more than ever pieces of the metropolitan puzzle. By many indications, this georegion is now a vast twentieth-century dormitory for the so-called "Greater Golden Horseshoe" of Toronto. Though many of its users might beg to differ, it has now become part of the "métapolis" or "hundred-mile city" to which François Ascher (1995) and Deyan Sudjič (1992) have respectively referred. This georegion is thus vitally part of the regional planning imperative that has inspired this volume.

The second-home patterns of this georegion have always been tightly bound up in the ebb and flow of the Toronto-centred metropolitan housing market, even if they do not clearly present themselves as extensions of the more spatially contiguous urban form of the Toronto region. This is true in more "objective" terms – as indicated by patterns of residential mobility, for instance (Luka 2008b) – and more "subjectively" in that growth is arguably being driven by representations, meanings, or ideologies that also motivate people to live in exurban settings (Luka 2008a).

The final point, and a sobering one, concerns the way in which the Central Ontario Lakelands are or have been a place of dwelling and encounter among aboriginal groups. At different moments, it has been the site of friendly exchange but also one of conflict exacerbated by European colonization, which has yielded, in effect, to the gradual obliteration of the aboriginal presence in its landscapes. The forced migration of the Huron nation is a case in point, but it is not the only example. Could the decline of the aboriginal presence in the Central Ontario Lakelands be substantively linked to its historical emergence as a cherished landscape in European eyes? A landscape is, of course, a cognitive-affective creation – it comes to exist by the accretion of meaning and identity that stems from its manipulation and inhabitation by humankind. But this has meant that this georegion was literally invisible to European eyes before it was captured in the paintings done by Tom Thomson and the Group of Seven, and with the development of the now stereotypical Ontario cultural practice of cottage life. Is it a coincidence that this vocation has been coupled with obliteration of aboriginal life? Such a question has been asked by Jonathan Bordo (1993), and we present it here both as a challenge for contemporary planning and for more general discussion of how so much of Ontario's history is a shameful one of colonization, conquest, and exploitation.

CONTEMPORARY CHALLENGES AND OPPORTUNITIES

Where does this leave us? We have in the Central Ontario Lakelands a liminal zone – an edge – that is part of the Toronto-centred metropolitan region and that was once a meeting place for several Aboriginal nations, but is no more. Without seeking to develop a full set of planning and design responses, we instead suggest several important considerations about where to go from here. Our discussion and comments on future directions have been distilled into five major points, each of which builds on its predecessor(s).

1. We have seen in the Central Ontario Lakelands a striking weave of abiotic, biotic, and cultural elements. This weave is strong and persistent through time for the georegion. The cultural patterns in particular cannot be understood nor would they have manifested themselves were it not for the particular combination of abiotic and biotic elements. This is in and of itself remarkable, but also suggests an important set of learning opportunities for planning and design.
2. Measured in its fullest sense – that is, in non-market terms – the cultural value of this georegion is fundamentally and irrevocably a function of the integrity of its abiotic and biotic systems. Users and non-users alike have a sense of this georegion as an important

and meaningful place (Luka 2008a). The Central Ontario Lakelands play notable roles in the everyday lives of tens of thousands of Canadians and occupy an intriguing niche in the annals of Ontario culture and folklore. Their ecosystems have both intrinsic and extrinsic value, in that they have demonstrably sustained interest well beyond their spatial limits. The settlement patterns that have resulted from the peculiar confluence of abiotic, biotic, and cultural factors are noteworthy and command both our attention and respect.

3 Even if measured solely in conventional market terms, the value placed on this georegion through the land market as a cultural instrument is once again fundamentally and irrevocably a function of the integrity of its abiotic and biotic systems. The fact that second-home use has persevered, or some might say triumphed, over other modes of place-use and resource exploitation is an indication of the deep importance that the *landscape* itself has as a unique confluence of nature and culture. It has perhaps problematically been assigned what in Marxian terms is called high exchange value as a commodity that can be bought and sold; it has cultural importance as an amenity, a notion discussed by Bunce (1994), Coppack (1988), Curry (1992), McIntyre and Pavlovich (2006), and Mueller-Wille (1990). This cultural value is nevertheless contingent on the integrity of the abiotic and biotic components that sustain the georegion's healthy, diverse, complex ecosystems. Users have demonstrated *care* in this respect through their direct investment, in monetary terms, in the cultural landscapes. They and other stewards of the local land market – notably the various levels of government – therefore have a direct sustained interest in maintaining this integrity, of which they are likely to be quite aware.

4 Evidence suggests that contemporary changes are upsetting a delicate balance both in the eyes of users and/or stewards and according to other measures of sustainability. Concern is now being expressed from several quarters that the fundamental qualities of cottage country are being jeopardized by current transformations, often described as the "urbanization of cottage country" (see, e.g., DeMont 2001; Fahner 2004; Lees 2001; MacGregor 2001). In one respect this is somewhat ironic given the innate links with urban growth already discussed. In another respect it is a classic reaction to various transformations of place – the same

sort of reaction that drives NIMBYism. Regardless, there is compelling evidence that concern is being expressed by users who are aware of the threat of undesirable change to the balance of abiotic, biotic, and cultural elements (Luka 2008a,b). The question of sustainability plays out in this context, not in abstract terms, but in very palpable ways: the lakewater must be clean and safe enough for swimming, but also for the aquatic species that are part of its ecosystem; similarly, the forests must flourish; new construction must respect the qualities of place that have been articulated more or less satisfactorily through the cultural interventions of the past. Perhaps most compelling, however, is the fact that many users spontaneously express what Relph (1981) described as environmental humility: the sense that we are all agents acting in and on the landscape and that we therefore have a direct responsibility for our actions. In the Central Ontario Lakelands, many of the feedback loops are adequately tight and self-evident, so that users are willing to temper their own behaviour.

5 We find ourselves presented with an exciting learning opportunity here. The challenge is twofold. First, we can nurture ways to make people more aware of the operational attributes linking nature and culture in this georegion – once we have, of course, ascertained what is "sustainable" in terms of biodiversity and carrying capacity, the latter being measured in both ecological and social terms. This project can build, for instance, on work done in the 1970s in Ontario by Jaakson et al. (1976, 1979) and in the 1980s by the provincial government (Teleki and Herskowitz 1986). The development and dissemination of waterfront homeowner's "manuals" (e.g., Kipp and Callaway 2003) is an example of a useful outreach strategy. The second and more daunting challenge is ensuring widespread recognition of how uncertainty and change are both inevitable in human settlements, as with healthy, functioning ecosystems of any sort (Lister 1998). Urban history in the Anglo-American world abounds with examples of similar second-home settings that have come to be usefully and happily integrated into the fabrics of the metropolitan region to which they are appended, often transforming themselves dramatically in the process (Aubin-Des Roches 2006; Borchert 1996; Luka 2006). It is more useful to recognize relative permanencies and to assess how changing

patterns can be taken as opportunities to ensure the ongoing transformation of this georegion in a manner that is sustainable and desirable.

Significant questions arise from the fact that tens of thousands of people are motivated to expand their life-spaces to cover an area hundreds of square kilometres in size. There is no denying the negative impacts of contemporary changes in this georegion. Pressures include the fragmentation of habitat with intensified settlement patterns; loss of biodiversity; the historic legacy of modified shorelines, where, for instance, artificial walls disrupt fish spawning zones; water pollution, including phosphates used in lawn fertilizers; and so on (cf. Hansen et al. 2005; Kipp and Callaway 2003). At the same time, there are exciting opportunities for helping people to learn about and act intelligently on the ways in which nature and culture intersect in this georegion and potentially in other settlement contexts. In acknowledging that the settlement patterns of the Central Ontario Lakelands are characterized by the well-known social practice of "cottaging" by residents of south-central Ontario's major urban centres, we must remember that these patterns are not timeless or unchangeable.

If useful change is to be encouraged in the Central Ontario Lakelands, there must be a collective recognition of how these cottage-country landscapes represent the complicated yet ultimately ambivalent Anglo-American relationship with natural processes – the workings of ecosystems in all their diverse, messy, and uncertain complexities (Lister 1998). The quest for the many charms and undeniably positive effects of the fresh air, water, and woods that characterize much of central Ontario seems to coincide almost invariably with the knowledge that this "wilderness" is unrelenting – an unforgiving deep, dark forest in which one could easily stray from the comforts of "civilization" and perish. In consequence, the settlement patterns of the Central Ontario Lakelands have been informed by a "conquest" mentality rather than one of knitting human wants and needs into ecosystem processes. This is a familiar condition in the Anglo-American context (Bixler and Floyd 1997; Bunce 1994; Castree and Braun 1998; Cronon 1999; Marx 1964, 2005; Smith 1990). It may be, however, that meeting this all-too-familiar challenge is less difficult in the waterfront settings of central Ontario than it is in other contexts.

It may be thanks to a quirk of history that the Central Ontario Lakelands were first opened up to European settlement during the great rise of tourism in the late nineteenth and early twentieth centuries. The "colonization" of this georegion nevertheless also came at a moment when attention was being directed to the increasingly strained interface between nature and culture, which was so romantically embodied in Thoreau's tales of his retreat to Walden. The ascendancy of central Ontario's cottage country involved developing effective and crowd-pleasing ways to reconcile a European sense of the tamed landscape with the menacing and untamed wilds of the Canadian "bush." The lodges and grand hotels of the Golden Era gave genteel city folk opportunities to see the "wilderness" without having to get too close for comfort. This continued apace with the surge in second-home use in the postwar years. With the contemporary renaissance, or feeding frenzy, of cottage life, we may indeed be at an all-important juncture where disaster looms but where opportunity also beckons. Will we rise to the challenge of ensuring that the Central Ontario Lakelands are more than cottage country?

NOTES

1 While reference is typically made to "cabins" or "camps" in parts of Ontario, the term "cottage" is predominantly used in the central part of the province (Green et al. 1997, 314).
2 Aboriginal history is discussed by Druke (1986), McMillan and Yellowhorn (2004), and Murray (1963).
3 See Dickason (2002), Druke (1986), and McMillan and Yellowhorn (2004).
4 The first forestry congress was convened in Montreal in the 1880s, soon followed by the establishment of Algonquin Provincial Park, the first in Ontario (Killan 1993).
5 Wolfe (1951) analyzed 1941 post office data and estimated that there were at least eight thousand cottage-owning households in the Central Ontario Lakelands, which corresponds to a working total of thirty-five thousand, based on the average household size of 4.3 as reported in the 1941 Census. This total of course excludes guests at commercially operated hotels and resorts.
6 For instance, paint and chemical works existed in and around Parry Sound, including Canadian Industries Limited (CIL), which established a dynamite plant in nearby Nobel.

7 Spectorsky (1955) coined the term "exurbia" to describe a dispersed form of quasi-rural suburban growth; these are usually linked with a highly constructed set of cultural narratives about the "countryside" and/or "rural" life (Blum et al. 2003; Bunce 1994; Cadieux 2005; Crump 2003; Jobes 2000; Nelson 1992). Toronto-area manifestations of the exurban phenomenon have been documented and discussed by Birnbaum et al. (2003), Bourne et al. (2003), and Walker (2000), and elsewhere in this volume.

8 Data from Statistics Canada (2004) indicate an increase of 73.1 percent in the number of second homes across Ontario from 1991 to 2003 (from 216,000 in 1991 to 365,000 in 1997, and 374,000 in 2003).

REFERENCES

Alley, P. 2004. "Could a Significant Natural System in Southern Ontario Be Overlooked?" In C.J. Lemieux, J.G. Nelson, T.J. Beechey, and M.J. Troughton, eds., *Protected Areas and Watershed Management*, 373–83. Waterloo, ON: Parks Research Forum of Ontario.

– 2006. "Designing an Initial GIS Analysis for The Land Between." In J.G. Nelson, T. Nudds, M. Beveridge, B. Dempster, L. Lefler, and E. Zajc, eds., *Protected Areas and Species and Ecosystems at Risk: Research and Planning Challenges*, 405–12. Waterloo, ON: Parks Research Forum of Ontario.

Angus, J.T. 1988. *A Respectable Ditch: A History of the Trent-Severn Waterway, 1833–1920*. Montreal and Kingston: McGill-Queen's University Press.

– 2000. *A Work Unfinished: The Making of the Trent-Severn Waterway*. Orillia: Severn Publications.

Ascher, F. 1995. *Métapolis ou l'avenir des villes*. Paris: Éditions O. Jacob.

Attfield, R. 2000. *Browning Island, Lake Muskoka: Cottagers Remember the Good Old Days*. Huntsville, ON: Fox Meadow Creations.

Aubin-Des Roches, C. 2006. "Retrouver la ville à la campagne: La villégiature à Montréal au tournant du XXe siècle." *Urban History Review/ Revue d'histoire urbaine* 34(2): 17–29.

Baldwin, D.J.B., J.R. Desloges, and L.E. Band. 2000. "Physical Geography of Ontario." In A. Perera, D. Euler, and I.D. Thompson, eds., *Ecology of a Managed Terrestrial Landscape: Patterns and Processes of Forest Landscapes in Ontario*, 12–29. Vancouver/Toronto: UBC Press/Ontario Ministry of Natural Resources.

Barry, J.P. 1995. *Georgian Bay The Sixth Great Lake.* 3d ed. Toronto: Stoddart.
Betcherman, L.R. 1975. *The Swastika and the Maple Leaf.* Toronto: Fitzhenry and Whiteside.
Birnbaum, L., L. Nicolet, and Z. Taylor. 2004. *Simcoe County: The New Growth Frontier.* Toronto: Neptis Foundation.
Bixler, R.D., and M.F. Floyd. 1997. "Nature Is Scary, Disgusting, and Uncomfortable." *Environment and Behavior* 29(4):443–67.
Blum, A., V. Cadieux, N. Luka, and L. Taylor. 2004. "'Deeply Connected' to the 'Natural Landscape': Exploring the Places and Cultural Landscapes of Exurbia." In D. Ramsey and C. Bryant, eds., *The Structure and Dynamics of Rural Territories: Geographical Perspectives,* 104–12. Brandon, MB: Brandon University.
Borchert, J. 1996. "Residential City Suburbs: The Emergence of a New Suburban Type, 1880–1930." *Journal of Urban History* 22(3): 283–307.
Bordo, J. 1993. "Jack Pine: Wilderness Sublime or the Erasure of the Aboriginal Presence from the Landscape." *Journal of Canadian Studies* 27(4): 98–128.
Bourne, L.S., M.F. Bunce, L. Taylor, N. Luka, and J. Maurer. 2003. "Contested Ground: The Dynamics of Peri-urban Growth in the Toronto Region." *Canadian Journal of Regional Science/Revue canadienne des sciences régionales* 26(2–3): 251–70.
Brown, R. 1983. *Ghost Towns of Ontario.* Vol. 2. Toronto: Cannonbooks.
Bunce, M.F. 1994. *The Countryside Ideal: Anglo-American Images of Landscape.* London and New York: Routledge.
Cadieux, K.V. 2005. "Engagement with the Land: Redemption of the Rural Residence Fantasy?" In S.J. Essex, A.W. Gilg, and R. Yarwood, eds., *Rural Change and Sustainability: Agriculture, the Environment and Communities,* 215–29. Cambridge: CABI.
Castree, N., and B. Braun. 1998. "The Construction of Nature and the Nature of Construction: Analytical and Political Tools for Building Survivable Futures." In B. Braun and N. Castree, eds., *Remaking Reality: Nature at the Millennium,* 3–42. London: Routledge.
Chapman, L.J., and D.F. Putnam. 1984. *The Physiography of Southern Ontario.* Toronto: Ontario Ministry of Natural Resources.
Coppack, P.M. 1988. "The Role of Amenity." In P.M. Coppack, L.H. Russwurm, and C.R. Bryant, eds., *Essays on Canadian Urban Process and Form.* Vol. 3, *The Urban Field,* 41–55. Waterloo, ON: Department of Geography, Faculty of Environmental Studies, University of Waterloo.

Cronon, W. 1996. "The Trouble with Wilderness; or, Getting Back to the Wrong Nature." In W. Cronon, ed., *Uncommon Ground: Rethinking the Human Place in Nature*, 69–90. New York: W.W. Norton.

Crump, J.R. 2003. "Finding a Place in the Country: Exurban and Suburban Development in Sonoma County, California." *Environment and Behavior* 35(2): 187–202.

Curry, N. 1992. "Recreation, Access, Amenity and Conservation: The Failure of Integration." In I. Bowler, C. Bryant, and M. Nellis, eds., *Contemporary Rural Systems in Transition*. Vol. 2. Wallingford, England: CAB International.

Dahms, F. 1996. "The Greying of South Georgian Bay." *Canadian Geographer/Géographe canadien*, 40(2): 148–63.

Dahms, F., and J. McComb. 1999. "'Counterurbanization,' Interaction and Functional Change in a Rural Amenity Area: A Canadian Example." *Journal of Rural Studies* 15(2): 129–46.

DeMont, J. 2001. "Paradise Lost." *Maclean's*, 28 May, 22–5.

Dickason, O.P. 2002. *Canada's First Nations: A History of Founding Peoples from Earliest Times*. 3d ed. Don Mills, ON: Oxford University Press.

Druke, M.A. 1986. "Iroquois and Iroquoian in Canada." In R.B. Morrison and C.R. Wilson, eds., *Native Peoples: The Canadian Experience*, 302–23. Toronto: Oxford University Press.

Epp, A.E. 2000. "Ontario Forests and Forest Policy before the Era of Sustainable Forestry." In A. Perera, D. Euler, and I.D. Thompson, eds., *Ecology of a Managed Terrestrial Landscape: Patterns and Processes of Forest Landscapes in Ontario*, 237–76. Vancouver/Toronto: UBC Press/Ontario Ministry of Natural Resources.

ESWG-Ecological Stratification Working Group. 1996. *A National Ecological Framework for Canada*. Ottawa: Centre for Land and Biological Resources Research, Research Branch, Agriculture and Agri-Food Canada.

Fahner, S. 2004. "Cottage Country: Protecting the Dream Demands Skill and Hard Work." *Ontario Planning Journal* 19(2): 13–14.

Gray, W. 1994. "Obajewanung: The First Peoples." In Muskoka Lakes Association, ed., *Summertimes: In Celebration of 100 Years of the Muskoka Lakes Association*, 25–36. Erin, ON: Boston Mills Press.

Green, S., J. Harkness, D. Friend, J. Keeler, D. Liebman, and F. Sutherland, eds. 1997. *Nelson Canadian Dictionary of the English Language*. Toronto: ITP Nelson.

Hague, J. 1893. "Aspects of Lake Ontario." *Canadian Magazine* 1:263.

Halseth, G. 1998. *Cottage Country in Transition: A Social Geography of Change and Contention in the Rural-Recreational Countryside.* Montreal and Kingston: McGill-Queen's University Press.

Halseth, G., and M.W. Rosenberg. 1995. "Cottagers in an Urban Field." *Professional Geographer* 47(2): 148–59.

Hansen, A.J., R.L. Knight, J.M. Marzluff, S. Powell, K. Brown, P. Gude, et al. 2005. "Effects of Exurban Development on Biodiversity: Patterns, Mechanisms, and Research Needs." *Ecological Applications* 15(6): 1893–1905.

Hodge, G. 1970. *Cottaging in the Toronto Urban Field: A Probe of Structure and Behaviour.* CUCS Report No. 29. Toronto: Centre for Urban and Community Studies, University of Toronto.

– 1974. "The City in the Periphery." In L.S. Bourne, R.D. MacKinnon, J. Siegel, and J.W. Simmons, eds., *Urban Futures for Central Canada: Perspectives on Forecasting Urban Growth and Form,* 281–301. Toronto: University of Toronto Press.

Jaakson, R. 1979. "A Spectrum Model of Lake Recreation Carrying Capacity Estimation." In J. Marsh, ed., *Water-Based Recreation Problems and Progress,* 63–84. Peterborough, ON: Department of Geography, Trent University.

Jaakson, R., M. Buszynski, and D. Botting. 1976. "Carrying Capacity and Lake Recreation Planning." *Town Planning Review* 47(4): 359–73.

Jobes, P.C. 2000. *Moving Nearer to Heaven: The Illusions and Disillusions of Migrants to Scenic Rural Places.* Westport, CT and London: Praeger.

Killan, G. 1993. *Protected Places: A History of Ontario's Provincial Parks System.* Toronto: Dundurn Press/Ontario Ministry of Natural Resources.

King, P. 1994. "The Promised Land: Early Settlers." In Muskoka Lakes Association, ed., *Summertimes: In Celebration of 100 Years of the Muskoka Lakes Association,* 59–74. Erin, ON: Boston Mills Press.

Kipp, S., and C. Callaway. 2003. *On the Living Edge: Your Handbook for Waterfront Living.* Ontario edition. Ottawa: Living by Water Project/Centre for Sustainable Watersheds.

Lees, D. 2001. "Subdivide and Conquer." *Cottage Life* 14 (March): 40–6, 108–13.

Levitt, C.H., and W. Shaffir. 1987. *The Riot at Christie Pits.* Toronto: Lester & Orpen Dennys.

Lister, N.M. 1998. "A Systems Approach to Biodiversity Conservation." *Environmental Monitoring and Assessment* 49:123–55.

Lower, A.R.M. 1938. *The North American Assault on the Canadian Forest: A History of the Lumber Trade between Canada and the United States.* Toronto: Ryerson Press.

Luka, N. 2006. "From Summer Cottage Colony to Metropolitan Suburb: Toronto's Beach District, 1889–1929." *Urban History Review/Revue d'histoire urbaine* 35(1): 18–31.

– 2008a. "Waterfront Second Homes in the Central Canada Woodlands: Images, Social Practice, and Attachment to Multiple Residency." *Ethnologia Europaea (Journal of European Ethnology)* 37(1/2): 71–87.

– 2008b. "Le 'cottage' comme pratique intergénérationnelle: Narrations de la vie familiale dans les résidences secondaires du centre de l'Ontario." *Enfances, Familles, Générations,* 8. http://www.erudit.org/revue/efg/2008/v/n2008/018493ar.html

MacGregor, R. 2001. "Forgive Me Ancestors, for What I Have Done ... and What I Am Planning to Do." *Cottage Life* 14 (March): 36–8, 101–4.

Marsh, J. 1983. "Cottaging and Land-Use Decision Making: A Case Study of the Kawarthas." *Recreation Research Review* 10(1): 5–13.

Marsh, J., and C. Moffatt. 1979. "The Historical Development of Resorts and Cottages in the Kawartha Lakes Area, Ontario." In J. Marsh, ed., *Water-Based Recreation Problems and Progress,* 27–50. Peterborough, ON: Department of Geography, Trent University.

Marsh, W.M. 2005. *Landscape Planning: Environmental Applications.* 4th ed. New York: Wiley.

Marx, L. 1964. *The Machine in the Garden: Technology and the Pastoral Ideal in America.* New York: Oxford University Press.

– 2005. "The Idea of Nature in America." In N. Tazi, ed., *Keywords: Nature,* 37–62. New York: Other Press.

McDonald, M. 2003. "Smog Sleuth: One of Canada's Leading Ecologists Investigates What Toronto's Bad Air Is Doing to Communities Downwind." *Canadian Geographic* 123(3): 70–81.

McIlwraith, T.F. 1997. *Looking for Old Ontario: Two Centuries of Landscape Change.* Toronto: University of Toronto Press.

McIntyre, N., and K. Pavlovich. 2006. "Changing Places: Amenity Coastal Communities in Transition." In N. McIntyre, D. Williams, and K. McHugh, eds., *Multiple Dwelling and Tourism: Negotiating Place, Home, and Identity,* 239–61. Cambridge, MA: CABI Publishing.

McMillan, A.D., and E. Yellowhorn. 2004. *First Peoples in Canada.* 3d updated and rev. ed. Vancouver: Douglas & McIntyre.

Mueller-Wille, C. 1990. *Natural Landscape Amenities and Suburban Growth: Metropolitan Chicago, 1970–1980.* Geography Research Paper no. 230. Chicago: University of Chicago.

Murray, F.B. 1963. *Muskoka and Haliburton, 1615–1875: A Collection of Documents.* Toronto: Champlain Society.

Nelles, H.V. 1974. *The Politics of Development: Forests, Mines, and Hydroelectric Power in Ontario, 1849–1941.* Toronto: Macmillan of Canada.

Nelson, A.C. 1992. "Characterizing Exurbia." *Journal of Planning Literature* 6(4): 350–68.

Øverlund, O. 1989. *Johan Schrøder's Travels in Canada, 1863.* Montreal: McGill-Queen's University Press.

Oxford English Dictionary. 2d ed. 1989. New York: Oxford University Press.

Parson, H.E. 1987. "The Colonization of the Southern Canadian Shield in Ontario: The Hastings Road." *Ontario History* 74(3): 263–73.

Relph, E. 1981. *Rational Landscapes and Humanistic Geography.* London: Croom Helm.

Royal LePage Real Estate Services. 2001. *Royal LePage Recreational Property Report.* Toronto: Royal LePage Real Estate Services.

– 2002. *Royal LePage Recreational Property Report.* Toronto: Royal LePage Real Estate Services.

– 2004. *Royal LePage Recreational Property Report.* Toronto: Royal LePage Real Estate Services.

Statistics Canada. 2004. *Survey of Household Spending: Household Spending on Shelter, by Province and Territory, Annual (1997–2003).* CANSIM Table 203–0003 (Updated November 2004, accessed June 2005).

Sanderson, M. 2004. *Weather and Climate in Southern Ontario.* Waterloo, ON: University of Waterloo Department of Geography.

Smith, A. 1990. "Farms, Forests and Cities: The Image of the Land and the Rise of the Metropolis in Ontario." In D. Keane and C. Reade, eds., *Old Ontario: Essays in Honour of J.M.S. Careless,* 71–94. Toronto: Dundurn Press.

Spectorsky, A.C. 1955. *The Exurbanites.* New York: J.B. Lippincott.

Stevenson, A. 1886. "Camping in the Muskoka Region." *The Week,* 13 May, 382.

Sudjič, D. 1992. *The 100 Mile City.* London: A. Deutsch.

Tatley, R. 1983. "The Steamboat Era in the Muskokas." Vol. 1, *To the Golden Years.* Erin, ON: Boston Mills Press.

- 1984. "The Steamboat Era in the Muskokas." Vol. 2, *The Golden Years to the Present*. Erin, ON: Boston Mills Press.
Teleki, G.C., and J. Herskowitz, eds. 1986. *Lakeshore Capacity Study*. Toronto: Research and Special Projects Branch, Ministry of Municipal Affairs, Government of Ontario.
Thompson, I.D. 2000a. "Forest Vegetation of Ontario: Factors Influencing Landscape Change." In A. Perera, D. Euler, and I. D. Thompson, eds., *Ecology of a Managed Terrestrial Landscape: Patterns and Processes of Forest Landscapes in Ontario*, 30–53. Vancouver and Toronto: UBC Press/Ontario Ministry of Natural Resources.
- 2000b. "Forest Vertebrates of Ontario: Patterns of Distribution." In A. Perera, D. Euler, and I.D. Thompson, eds., *Ecology of a Managed Terrestrial Landscape: Patterns and Processes of Forest Landscapes in Ontario*, 54–73. Vancouver and Toronto: UBC Press/Ontario Ministry of Natural Resources.
Varley, W.B. 1900. Tourist Attractions in Ontario. *Canadian Magazine* 15, 29.
Voigt, D.R., J.A. Baker, R.S. Rempel, and I.D. Thompson. 2000. "Forest Vertebrate Responses to Landscape-Level Changes in Ontario." In A. Perera, D. Euler, and I.D. Thompson, eds., *Ecology of a Managed Terrestrial Landscape: Patterns and Processes of Forest Landscapes in Ontario*, 198–233. Vancouver and Toronto: UBC Press/Ontario Ministry of Natural Resources.
Walker, G. 2000. "Urbanites Creating New Ruralities: Reflections on Social Action and Struggle in the Greater Toronto Area." *Great Lakes Geographer* 7(2): 106–18.
Wall, G. 1977. "Recreational Land Use in Muskoka." *Ontario Geography* 11:11–28.
Wolfe, R.I. 1951. "Summer Cottagers in Ontario." *Economic Geography* 27(1): 10–32.
- 1962. "The Summer Resorts of Ontario in the Nineteenth Century." *Ontario History* 54(3): 149–61.
- 1965. "About Cottages and Cottagers." *Landscape* 15(1): 6–8.
Wong, T. 2004. "Renovation Nation: Price of Cottage Serenity Keeps Rising." *Toronto Star*, 23 May.

9

The Kingston Georegion: Going "Glocal" In a Globalizing World

BRIAN S. OSBORNE

SETTING THE SCENE AND CONTEXTUALIZING THE PLACE

Toronto and its "Golden Horseshoe" megalopolis is Ontario's principal economic engine. However, there is considerable concern over the environmental, social, and economic impacts of its quintessential gigantism and exponentially sprawling land use. Perhaps more damaging than the growing material presence is the expansion of an associated universalizing mind-set and way of life. Taken together, they threaten the essential character and sense of place exhibited by Ontario's several regions and challenge their difference and uniqueness. Indeed, what is being assaulted is the diverse tapestry of cultural landscapes and living regions that have been produced by distinctive patterns of land use, ways of life, and natural and human histories over time.

This is the central concern of this chapter. It is because of these trends that it is posited here that a "georegional" approach to planning our futures will accommodate the needs and future destinies of Ontario's distinctive places. Such an approach identifies the natural and cultural character of specific regions and the significant contribution they make to the livelihood and way of life and, perhaps more importantly, the quality of life and sense of connectedness to place and time of their residents.

But this is not a uniquely Toronto-Ontario phenomenon. Toronto is but the local agent of that world-wide process known as globalization. Indeed, there has long been a concern over the oppressive dominance of metropolitanization and the mind-numbing expansion of sameness propagated by a monolithic "Westernizing" capitalism and global communications. Some have labelled this process an "Americanization" effected through such agencies as "MacDonaldization," "Coca-Colaism," or "Starbucki-city"! It is a process that David Harvey has charged must be challenged by a sense of "geographical situatedness" and a "liberatory process [that] can never take place outside of space and time, outside of place making, and without the engagement with the dialectics of socio-natural relations" (2009, 260).

While this is central to the neologism "georegion," it is not really a new idea. Regions of ecological and cultural distinctiveness have long been known to human geographers as "natural regions," "pays," or "landesgeshichte," places that have "personalities" (Gregory 1994; Wolfart 1997). That is, the cultural landscape is the visual and symbolic manifestation of underlying processes that are integral to local heritage, lifestyles, memories, and identities. That these have been constructed materially and emotionally over time is central to Yi Fu Tuan's concept of "topophilia."

The erosion of these meaningful places by the ubiquitous forces of global marketing has long attracted the attention of scholars, and the titles of their studies have provocatively captured the essence of the problem of a growing anomie: "place and placenessness," "no sense of place," "the view from nowhere," and "the geography of nowhere," (Relph 1976; Meyrowitz 1985; Nagel 1986; Kunstler 1994). At issue has been the threat to the powerful cultural universal of people's sense of belonging and identity anchored in particular places. However, such discrete connections to lived-in worlds are now being replaced by a predictable sameness, a "betweenness of place," that is produced by, and diagnostic of, the conditions of "modernity" (Entriken 1991).

But there is an emerging reaction to these universalizing trends as "global" forces provoke "local" reactions that prompt the commodification of difference. This marshalling of the "lure of the local" (Lippard 1998) is central to the concept of "glocalization," another clever neologism that captures that relationship between the nurturing of the local at a time of the assertion of the global, both in culture and in economy (Swyngedouw 1992, 2000; Robertson 1994, 1995). Thus, the "invention" of locality is reminiscent of the constructions flagged by Hobsbawm and Ranger's "invention of tradition" (1983) and Anderson's "imagined community" (1991 [1983]).

No mere exercise in metaphysics, the fundamental praxis of celebrating the local underpins the economic strategy of "micromarketing" in the new economic sector, the "experience economy" (Pine and Gilmore 1999); that is, the production of goods and services to fit the diverse demands of local markets, and the insatiable demand of global consumers to experience difference in places that retain their local identities. While globalization is engaged in producing sameness, the local is the key site for "place-marketeers" and "place-entrepreneurs" to gain advantage in an international competitive environment (Robins 1991). From this perspective, the global and the local are intertwined and mutually constituted in that global processes influence local actions and prompt reactions. At the same time, local coalitions of financial and political actors are assuming a lot more power to regulate their local economies, set objectives that are place-specific, and maximize unique histories and assets of local places (Shone and Memon 2008).

This is the context for this study of the East Lake Ontario–Upper St Lawrence Georegion (Osborne 2007). It will be considered as a "city-region" centred on Kingston, which is sensitive to the need for new economic strategies in a post-industrial world as it encounters considerable economic and social change (Figure 9.1).[1]

A CITY-REGION MARSHALS ITS PAST

If Kingston is to do the heavy lifting in advancing a "global" strategy for its "georegion," it has to marshal its resources for a forward-looking development strategy for marketing its economic and social assets. Four domains come to the fore: the water, the countryside, the forest, the city (figure 9.2).

Interestingly, these four domains were always integral to the subsistence and cultural practices of the georegion's First Nations. For the historical Iroquois and the subsequent Anishabe, Mississauga, and Algonquins, the Lake Ontario–St Lawrence waters and their tributary rivers were key to their identity (Osborne and Ripmeester 1995; Osborne, Huitema, and Ripmeester 2002, 2004; Osborne 2003). Indeed, the word "Mississauga" means "persons who inhabit the country where there are many mouths of rivers" (Blair 2008, 3). Their holistic view of the diverse resources of the forests and lakes of the back country generated a seasonally driven set of resource regimens. Further, their philosophies of *bimaadiziwin* and *menoyawin*

Figure 9.1 The Kingston Georegion. By the author, with permission

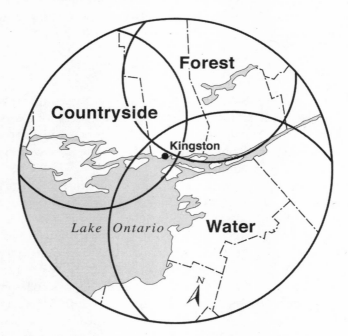

Figure 9.2 Kingston's Four Worlds. By the author, with permission

ensured that they pursued a lifestyle of peace and harmony with themselves, others, and, their lived-in world values. This is also central to the contemporary search for a sustainability that underpins the creative social planning of the georegional approach. It is no wonder that so many current initiatives in the regional marketing of new life strategies, local resources, and local tourism integrate the values and stories of the Kingston Georegion's First Nation's. Further, at all stages of the history of the development of the Kingston Georegion, the First Nations have, in varying degrees, been part of this story.

The Water: A Resource Base, Line, and Routeway

For the First Nations, the earlier settlers, and generations of commercial fishing communities, the St Lawrence and Lake Ontario was a valuable resource. Much history turns its back on the water and focuses on the resources of the soil and the forests. For many people, however, the variegated and seasonal yields of the water constituted a rich means of subsistence and, later, commerce (Osborne 1990).

That said, however, much of Kingston's character was influenced by its location at the eastern end of the Great Lakes where the junction with the headwaters of the St Lawrence produces an integrated waterway system that connects the continental interior with the Atlantic Ocean, albeit for but part of the year (figure 9.1). To the north-east, the Kingston Georegion includes the upper and most westerly section of the Upper St Lawrence corridor represented by the labyrinthine channels through the Thousand Islands (Nelson et al. 2005). To the north, much of its hinterland drains east to the Ottawa River. To the west, the sheltered Batteaux Channel and the Bay of Quinte have long integrated the "County" of Prince Edward into Kingston's commerce and imagination and from there links with the chain of the Great Lakes to the Canadian West. To the south, across the archipelago of islands at the eastern end of Lake Ontario, the Mohawk-Hudson corridor connects the Kingston region to the year-round port of New York.

This role as a regional (even continental!) nexus, has done much to influence Kingston's political, commercial, and cultural history. Indeed, transcending the local setting, Kingston's specific site and general situation have long ensured its centrality in geopolitics and nation-building, as well as contributing to the essence of the place.

These considerations were to the fore with the commencement of Kingston's Euro-Canadian history in 1673. This date marks the French assertion of their geopolitical control of their western frontier in general and negotiations with the Iroquois to protect their fur trade from Dutch and British interference in particular. Accordingly, the French established a fort at Cataraqui, later known as Fort Frontenac, that served as a functional and symbolic presence until its capture by the British in 1758. With the subsequent Treaty of Paris in 1763, this outpost of New France became integrated into British North America as part of the colony of Quebec. However, despite the formal political transfer and even though the French settlement there was never extensive, Cataraqui/Fort Frontenac became part of Kingston's remembered past and intangible cultural heritage.

In 1783, the arrival of British refugees from the new republic of the United States established a new political culture and a new military regimen that was to last for more than a century. Henceforth, Kingston was a western outpost of British imperial power situated on a "hot frontier" confronting an equally expansive United States. The current Kingston landscape is replete with reminders of its "citadel" role that lasted well into the nineteenth century: fortifications, garrisons, military lands. Even after the British military departed in the late nineteenth century, a new Canada turned to Kingston's facilities for a military college for the militia and barracks for its own forces. Indeed, the military presence continues to be a major component of the economy and ethos of Kingston.

That same water location was also crucial commercially. The key terms are "break-in-bulk" and transshipment. Simply put, lake transport could not negotiate the St Lawrence River, especially before the rapids were by-passed by canals. Initially, therefore, lake schooners transshipped their cargoes into bateaux or onto rafts to run the rapids to Montreal. Even after the introduction of canals along the river route at Cornwall, the Long Sault, and Gallop Rapids, lakers were obliged to off-load onto canallers. And this became a major element of Kingston's commercial economy, stimulating a host of marine businesses, employing hordes of wharf workers, and requiring a range of ship-repair and servicing operations. But, it was not without problems. In fact, there is a general law-of-traffic applied to Kingston: whenever navigation improvements "below" Kingston allowed larger vessels to negotiate the St Lawrence, Kingston's trade declined; whenever navigation improvements "above" Kingston allowed larger vessels to enter the lower lakes, Kingston's economy flourished. But all of this was rendered redundant by two factors. First, in 1856 the Grand Trunk Railway was completed and introduced the major competitor to water-borne trade. Nevertheless, with the promise of major navigational improvements above Kingston at the Welland Canal in the early twentieth century, for some two decades Kingston advanced its case for being the "Foot-of-the-Great Lakes-Terminus" (Osborne 1989). However, the second factor came into play with the completion of the St Lawrence–Great Lakes Seaway in the 1959. It by-passed Kingston completely and its shipping function disappeared (Osborne and Swainson 1988; Nelson et al. 2005; Parham 2008).

What was left, however, was much valuable waterfront available for development and as a tangible reminder of an important element of a crucial past function. These were subsequently transmuted into a valuable resource for high-density lakefront residential development and the marketing of Kingston's heritage. If past priorities looked on the river-lake corridor in terms of military-commercial imperatives, contemporary society evaluates waterfront through the lenses of the aesthetics of waterscapes and the economic opportunities of the new "quaternary" sector of the "experience" economy (Pine and Gilmore 1999).

The "Countryside": A Settled Place

"The countryside" connotes a settled agricultural landscape. To be sure, this landscape was constructed on the limestone and glacial clay plains that make up the islands and the lands fronting onto the Bay of Quinte to the west of Kingston. In 1783, the Crown negotiated with new First Nations occupants of the region, the Mississauga, and, by the "Crawford Purchase," acquired lands fronting onto the Bay of Quinte and the Upper St Lawrence to accommodate the Loyalists displaced north by the American Revolution (Osborne and Swainson 1988; Osborne and Ripmeester 1995). Like the rest of Ontario, these lands were surveyed into the geometric cadastre of concessions and lots that transformed the "wilderness" into a living landscape of agrarian production through several stages: first, a pioneer clearing of the original mixed deciduous forests; then the establishment of a commercial-crop economy focused on supplying the local Kingston civil and military market and the trans-Atlantic grain trade; finally, a shift to livestock with an emphasis on dairying for cheese and, later, bulk milk production. All this was followed by a gradual decline in farming with competition from elsewhere and rural-urban migration.

However, within the first decade of British settlement of the Kingston region, it was noted that the town's greatest deficiency was "the want of a fertile back-country." Where the Ordovician limestone was covered with glacial clays, the soils were often wet and cold and unsuited to wheat production. But elsewhere exposed extents of limestone pavement rejected the plough and even improved pasture. With the introduction of tile drainage in the late nineteenth century, the heavy clay lands were improved somewhat and supported a dairying economy. Of course, nothing could be done with the ubiquitous limestone pavements. Taken together, these factors determined that the Kingston Georegion was never a major region of agricultural production.

The "Forest": A Challenging Marginal Place

And then there were the granites and gneisses that define so much of Kingston's hinterland. The descriptor "forest" really refers to the lands of the Frontenac Axis, the extension of the Canadian Shield that runs from the Algonquin Arch in the northwest and across the St Lawrence at the Thousand Islands, reappearing in the southeast as the Adirondacks. Thus, within a short distance, the Kingston back country features the forests, lakes, rock outcrops, and disrupted drainage that were encountered to the east of Kingston in the first decades of settlement and were characteristic of the rear townships. They constituted the traditional hunting grounds of the Mississauga and Algonquin peoples. Some three million acres of this Ottawa-Huron tract were purchased by the Crown and later opened for settlement and development by the Public Lands and Colonization Act of 1853 and the Homestead Act of 1868 (Osborne 1976, 1977, 1982).

Of necessity, what agriculture was attempted was marginal, and often at a mere subsistence scale, using the patches of clay lands between rock outcrops, and ubiquitous beaver-marshes for hay-lands. But, there were supplementary income opportunities. The model was as simple as the predictions were grand. Government boosters claimed that the Ottawa-Huron Tract was capable of accommodating millions of pioneers to be engaged in producing goods and supplies for the lumbering economy, which, in turn, would provide supplementary income for farmers-cum-lumbermen. For half a century, this putative symbiosis of a marginal agriculture and commercial lumbering did provide a rationale for some hardy settlers.

And for others, there was another opportunity: mining. If deficient in agricultural lands, many of the Shield farms benefited from the opportunities for small-scale, part-time mining of mica, phosphate, feldspar, and other minerals and ores. Again, as with lumbering, several commercial mining operations offered employment in iron and lead mines throughout the region, while, for others, small-scale mines on their properties afforded the only commercial crop their Precambrian fields afforded them!

Despite these marginal strategies, Shield settlement was always precarious, and it peaked in the region in the 1880s. Ironically, the exodus was facilitated by a development that had been intended to support the region's economic development: colonization railways. Lines such as the Napanee-Tamworth ("Nip and Tuck") and the Kingston-Pembroke ("Kick and Push") were designed as tap-lines for produce – particularly lumber and minerals – to travel to Lake Ontario communications. Soon, they became connectors with the growing national railway system that transported frontier settlers out of the region to the better lands of the United States and Canadian Prairies.

But if the permanent population of Kingston's forest-backcountry declined in some parts, it was replaced by a growing number of seasonal residents as recreational fishing, hotel-tourism, and, later, cottaging, were attracted to the area by the new culture of nature worship and the search for bucolic leisure. In particular, the Algonquin-Rideau-Thousand Islands axis emerged as a preferred destination for the affluent of the Toronto-Ottawa-Montreal triangle, as well as for those from cities of the adjacent United States, which were linked to the area by rail and steamer communications.

The City: Kingston, "A Drowsy Place"!

Given these trends, it may be argued that the key to the prevailing ambience and, some would say, even the pervasive culture of Kingston, is failure! Certainly the Toronto *Globe and Mail* considered it to be a somewhat less-than-dynamic place on 14 June 1881:

"[Kingston is] a drowsy antiquated place, the relic of a past when everybody was not all the time on a jump, but took life easily, a land where it is always afternoon, and people would sooner sit on the wharf and fish than engage in the eternal chase after the dollar." To be sure, there was much to support this observation in 1881, and the next century was to reinforce certain grim realities.

The decision to locate the national capital in Ottawa shattered Kingston's aspirations for that political role, albeit it was left with penitentiaries, a major university, and an array of provincial and federal institutions. The cooling-off of British-Canadian and American political tensions rendered the considerable investment in fortifications redundant, transforming them into historical edifices and public open spaces. With the advent of rail, larger vessels, and the Seaway, Kingston's role in water-based commerce had ended by the mid-twentieth century. And, finally, despite energetic attempts to do so, heavy industry was not attracted to bolster the local economy, and even the war-stimulated industries (Alcan, Dupont, Kingston Shipyards, the Locomotive Works) gradually declined and dropped out of the economic picture.

Accordingly, Kingston cascaded down through the urban hierarchy. The rise of Toronto – and Hamilton, London, Ottawa – was matched by the decline of Kingston (Osborne 1975). But its failure to grow is now an essential dimension of its potential strategy for success in a world seeking distinctiveness and quality of life. Its distinctive ambience is generated by echoes of former pre-modern functions, streets, and urban spaces, together with the visual prompts afforded by a rich legacy of nineteenth-century domestic, commercial, and institutional architecture that has survived because of the overall low replacement rate of its original architectural stock and streets and spaces.

More than thirty years ago, I commented that these "vestiges" of Kingston's past "might well constitute one of Kingston's most valuable resources" and that the city had "a unique opportunity, and even a responsibility, to protect these elements which have survived": "They are aesthetically pleasing and contribute significantly to the quality of the inhabitants' urban environment. Furthermore, Kingston's accessibility to the major urban centres of North America makes its unique character an important economic resource and the development of tourism in 'Historic Kingston' could make a significant contribution to the future economy of the city" (Osborne 1973).

These thoughts are more appropriate than ever. Indeed, Kingston's essential task is to preserve and enhance the character of its georegion by celebrating the "local" in an ever-standardized globalized world.

HOW DOES THE KINGSTON GEOREGION SEE ITS FUTURE?

This is why I have presented this reprise of the region's history here. No exercise in antiquarian reflection, it is central to new theoretical perspectives on how regions such as Kingston can approach their future. The point is that communities everywhere are seeking strategies to sustain economically viable communities and enhance the "quality of life" and "quality of place" for their residents. It's an old problem, but it's now being worked out in a new context. The disruptions and opportunities associated with a globalizing and cosmopolitanizing world are eroding formerly well-established economic functions in many communities. These losses are accompanied by decreasing municipal tax bases at a time when cities are encumbered with ageing infrastructures and escalating demands for social services and amenities (Throsby 2001). It is in this environment that some cities and regions in the twenty-first century are turning to new paradigms for development: "smart" growth; "creative" cities; and the marketing of the "glocal."

A Growth That's "Smart"?

For long, urban planning relied on the apparently proven assumptions of the "accumulation model" of economic growth: cities are economic entities; they have costs; costs have to be covered by tax revenues; the yields of tax revenues depend on a viable urban economy; this economy employs resident workers who also contribute to the municipal tax base; employed persons contribute to the local consumer economy. From this perspective, the attraction of new economic opportunities increased employment, stimulated the consumption of local goods and services, and thus ensured both growth in the economy and population growth. Exponential growth rendered itself as a model to be pursued rather than as an irrational strategy. However, a combination of demographic shifts, a strengthening environmental ethic, global economic dislocations, the pragmatics of long-distance/time commuting, burgeoning gas prices, and the economic realities of the cost of urban infrastructure are prompting new priorities.

All these trends are calling for more nuanced views of growth and alternative paradigms for economic development and social enhancement. Rather than the constant quest for additional economic multipliers and an exponential burgeoning of population and the associated tax base, "new urbanism" seeks community sustainability through rational environmental, economic, and social practices (Bullard 2007; CMHC 2005; Neal 2003; Portney 2003; Szold and Carbonell 2002; Bohl 2002; Grant 2006; Kelbough 1999; Katz 1994; Lorinc 2006; Wagner 2005; Wyman 2004). In particular, both ecological and economic realities call for the end of the hitherto implacable, ever-outward, and ever-standardizing urban sprawl and a new focus on inner-city revitalization and growth. Central to this "new urbanism" is a renaissance in transport options incorporating public transit systems with pedestrian and bicycle mobility.

To this end, a decrease in the individual use of automobiles would be facilitated by increased downtown residential and commercial densities. Rather than additional "green-field" development on the urban fringe, "infill development" in existing neighbourhoods is favoured by a greater awareness of the consequent fiscal, environmental, and social benefits. Not only does more compact vertical building design reduce the construction footprint, it also allows more open space, supports wider transportation choices, and accommodates economies in the provision of water, sewer, electricity, and other utilities. Moreover, the higher densities of residents contribute much to the diversity and economic viability of downtown enterprises. Finally, the melding of the public good and private enterprise works for cost-effective smart growth through private-public sector partnerships.

If the logistics and economics of "smart growth" are to the fore, so also are the more qualitative dimensions of life in the "new city." A wide range of quality housing for people of all income levels is integral to the precepts of "smart growth." A diversity of accommodation options and people in well-designed, higher-density urban environments leads to pedestrian-friendly communities that come to be seen as people places because of the opportunities for enhanced social interaction. And when such developments are also sensitive to a community's architectural distinctiveness, unique heritage, and preferred patterns of urban living, they foster social cohesion and a rich and inclusive community fabric. Thus, the provision of new transport options, diversity in housing, and increased downtown vitality are all indicators and products of a smart growth.

Certainly, this is being recognized in Kingston. In 2004, the article "Urban Revitalization and Renewal" looked at initiatives that seek "new uses for neighbourhood facilities and underused or abandoned spaces" by "blending history and architecture to enliven the retail core of a city" (Anon. 2004). It went on to laud Kingston for launching an urban renewal of its downtown with culture, building towards a heritage district to anchor the cultural strength of downtown Kingston. Local businesses, which eagerly support the process, recognize that it is economically attractive for businesses, will help social and cultural cohesion, and will raise the profile of the community through its culture.

Other initiatives are also part of this strategy: the refurbishing of Kingston's downtown Grand Theatre; the construction of a large Community Entertainment Centre (the K-Rock Centre) adjacent to the principal commercial district; and the construction of several downtown apartment and condominium complexes, especially along the Ontario Street waterfront.

While these initiatives are indicative of an appreciation for some of the tenets of the "new urbanism," some observers are concerned that issues of social justice and diversity have been neglected. It is this concern that recently prompted a new city council to propose a "principled" approach to establishing priorities for a "green, sustainable city": such "environmentally friendly" projects as public transit, pedestrian- and cycle-friendly initiatives, energy-efficient public buildings and vehicles, household organic waste collection; enhancement of neighbourhoods with parks, pools, and rinks; promotion of economic sustainability by increasing the commercial tax base, more middle-income jobs, more public sector employment; strategies for more affordable housing and reducing homelessness and poverty; improving heritage preservation and cultural assets (Hutchins 2007). This mission is moving ahead with the pursuit of a plan to "adopt a culture of sustainable lifestyles that place personal, organizational and community initiatives in a global context" based on the four pillars of environmental, economic, social, and cultural sustainability (City Kingston 2009: 1).

A "Creative" City?

The pragmatics of a more qualitative approach to the social engineering of urban environments are to the fore in the work of the advocates of "creative cities" (Landry 2000; D'Auria 2001). For these urban theorists, communities will benefit from policies that focus on the quality of lived experience for permanent residents and transient visitors alike. Such "creative cities" embrace alternative lifestyles and invest in their creative communities: "They are nurturing cultural activities that serve local residents, create and celebrate community identity, facilitate exchange and understanding among socially, economically, and culturally diverse groups, and attract visitors and new businesses" (*Creative City News* 2006).

Clearly, the arts, culture, and heritage are dimensions of life that promote beauty, creativity, and individual and social health and cohesion. They have long been indicative of social progress and sensitivity. But from the pragmatic perspective of the creative city, their role transcends the usual rationale of aesthetics and urbanity, as they are seen to prompt economic and social benefits: urban revitalization and renewal, cultural tourism, economic investment, community identity and pride, and the personal and social development of youth.

Building on the philosophy of Jane Jacobs (1961, 1996), Richard Florida (2002) has advanced the proposition that people want to live in places that offer rich lifestyles and opportunities for social interaction in interesting meeting points; they are attracted to demographic diversity; they seek out what is authentic not generic; and they want to nurture the need for identity and security in an ever-changing postmodern world. For Florida, enhancing a community's quality of life and thus generating its spatial equivalent, a "quality of place," produces a vibrant urban society defined by its unique and attractive characteristics. Accordingly, communities should invest in defining their unique and most attractive assets and focus on creating a rich living experience for permanent residents, as well as for transient visitors.

It is in this context that Florida speaks of the role of the "creative class" in nurturing human capital, attracting people, and, thus, stimulating appropriate growth in urban places. This correlation between culture, quality of life, and economic development has been referred to as the "artistic dividend" (Markusen and King 2003).

And it is in this vein that Florida, while accommodating the tenets of the economic "growth model," focuses on a new component: a "human capital" theory of regional development in which the cultural sector is a key part of a community's core "business" aimed at attracting new residents, visitors, and businesses, and thus increasing the municipal tax base (Florida 2002, 215–34). The essential premise is that if cities are to attract new investment and employment, "creativity" has to be an essential component of their economic planning. That is, creativity is the engine of the new economy and new urbanism, and creativity is fostered by investments in the arts, culture, heritage, and environments conducive to innovation (Florida and Martin 2009).

But if some find Florida's prescriptions for a new urbanism to be elitist and classist, others keep their eyes on a more comprehensive and inclusive view of optimal development. For architect Bing Thom, "one of the things we've forgotten is that cities are communities of people. But what is the glue that holds them together? Common values. Common ideas. In recent years, it's also the growing importance of culture. Our economic well-being is based on our ability to celebrate culture, and celebrate the true identity of our cities" (quoted in *Creative City News* 2006). Moreover, cultural vitality plays a major role in supporting social and economic health. One of the central tenets of the Creative City Network is that the calculus of economic-cultural development sees community vitality, identity, and pride fostered by festivals, libraries, museums, public art, local performance groups, societies, and heritage buildings. Taken together, these dimensions of a progressive urban life animate the downtown core, reverse the move of businesses to peripheral malls, and create lived-in downtowns.

Further, an additional benefit in this age of the "experience economy" is that such authentic community attractions attract tourists (Richards 2007). Simply put, cultural tourism is big business. Canadian cultural tourists spent $3 billion domestically in 2001: 6.5 million trips to museums/art galleries, 7.8 million trips to historic sites, 5.7 million trips to cultural events, 6.2 million trips to festivals or fairs (Canadian Tourism Commission 2004). However, tourism also has to be planned if it is to be competitive and sustainable. If festivals, events, and promotional publications are not put forward as legitimate expressions of a city's local character and history, then they become just generic, a sign of decline, not development. "The city without its own story to tell has nothing distinctive to promote" (Garret-Petts and Dubinsky 2005).

Certainly, there is no reason for Kingston to fall into this trap. It has an authentic story that links its history to that of the rest of Canada and the United States. Its list of sites of national and provincial significance is impressive: 5 National Historic Sites; 33 Provincial Heritage Sites; 247 Local Heritage Sites; 181 Architectural Heritage Sites; 25 Interpretive Shipwreck sites. Perhaps the stellar site is Fort Henry. Its military trappings and performances have elevated it to one of Ontario's leading attractions; in 2006 it drew some 120,000 visitors. This has been further strengthened by the designation of the Rideau Canal and Kingston Fortifications as a World Heritage Site in 2007. Other didactic tourist attractions are the long list of houses, offices, and bars associated with Canada's first and most bibulous prime minister, Sir John A. Macdonald, as well as an impressive array of museums and exhibitions displaying facets of the city's past involvement with shipping, penitentiaries, and medicine. For the more venturesome, the hyper-real attractions of Baudrilliard's simulacrum are marketed in the very popular excursions into Kingston's spectoral landscape of ghost-tours and the recent transmogrification – at least for Halloween – of Fort Henry into Fort Fright!

Nor are the attractions of the lakes and forests and islands neglected. In 2002, a section of the Frontenac Arch running from Westport to the Thousand islands was designated as one of Canada's thirteen UNESCO Biosphere Reserves. It attracts local, national, and international visitors to enjoy its distinctive ecosystem and its long history of peoples' diverse engagements with nature here. On a more commercial tack, the ecological beauties of the Thousand Islands and Lake Ontario shore continue to be available through the long-standing culture of boat-tourism (Osborne and Osborne 1999; Osborne 2001, 2009; Nelson et al. 2005).

Increasingly, therefore, the marketing of the Kingston region's ecological, historical, and cultural assets is big business, and the promise of increased volumes because of the UNESCO/WHO designations is encouraging further development of the culture-tourism industry. However, the essence of Kingston's marketing of its identity is the overall ambiance of place, as manifest in its streetscapes, public open spaces, and diverse opportunities to experience the sense of this place in a passive, undirected way. Indeed, the activities of overzealous marketers is a concern for some converts to "new urbanism" and "creative cities." Are the crowds off-loaded by tour buses and cruise ships conducive to the quality of place being nurtured for long-time residents and discerning visitors? Will they threaten the sustainability of the character of this place and its tourist economy (Osborne 2007)?

Going "Glocal"?

These questions are central to the concerns of an Italian invention, the Citta Slow and Slow Food movements that advance the ideological precept that "Region is Reason" (Anon. 2002, Arnott 2002; Leitch 2003; Knox 2005; Parkins 2005; Petrini 2001; Petrini and Padovani 2006; Morgan 2000). That is, they aim to preserve local traditions and sustain the local economy. Initiated in 1986 in reaction to the McDonaldization of local cuisine, the Slow Food strategy has been to forge an equilibrium between the modern and the traditional, the city and the country, and to improve the quality of urban life by nurturing a distinct landscape, distinct cities, and distinct people. The League of Slow Cities started in Tuscany, and now, a decade later, the movement has spread throughout Europe as communities seek different strategies to combat the universalizing effects of globalization. The League's manifesto speaks of the need for an escape from frenetic modernity by seeking a more tranquil and reflective way of living.

Targeting "discerning tourists," they are marketing themselves as places that are ecologically sensitive, technologically non-intrusive, and regionally focused in terms of production and diet. Simply put, it's a celebration of the local. To this end, the Slow City movement challenges the increasingly frenetic pace of living, cherishes local traditions and diversity, and resists the insidious globalization of culture (Pink 2008). While criticized as being too nostalgic and unrealistic, the *Slow City Manifesto* is very pragmatic given the problems of this post-Kyoto world and favours displacing cars with pedestrian areas, cutting noise pollution, and increasing green spaces. But apart from ecological rationality and aesthetics, enhancing the quality of life is to the fore. The movement encourages and supports local farmers and shopkeepers who produce and sell local products and delicacies. In this way, it preserves and engenders local traditions and forges links between the modern and the traditional in a manner that promotes and enhances the sustainability and economic viability of distinctive places (Hinrichs 2003; Swynegedouw 2000).

In particular, by encouraging the production of local foodstuffs by natural, eco-compatible techniques that exclude transgenic products, the Slow Food/Slow City movement vitalizes the local rural economy at a time when food producers are being challenged by global producers and industrialized mass production and distribution. By creating occasions and spaces for direct contacts between consumers and quality producers in towns, they create profitable niche-marketing for restauranteurs and ensure the supply of interesting foodstuffs for discerning diners sensitive to the threats posed by distant producers (Urry 1995).

In this way, the "eating locally" initiative is situated at the interface between urban planning, regional development, agricultural economics, tourist studies, and culture. Indeed, it constitutes a viable model for one dimension of an economic development strategy based on several interdependencies: economic development of local resources, environmental protection, sustainability, quality of life, and social justice (Campbell 2004; Rast 2005; Imbroscio 1997; Stone 2004). But if it is to succeed, it is essential that the four constituencies involved (local farmers, local distributors, local restaurants, local consumers) be integrated into this strategy if "the goals of environmental protection, economic localism, and sustainability are to satisfy grassroots environmental groups and the locally rooted business community" (Mayer and Knox 2006, 331).

More than being simply about eating, the protection of the cultivated countryside and associated communities of the rural world also attracts potential tourists. Recognizing the opportunities to host visitors wishing to escape urban crowds and pressure, enjoy rural scenery and society, and participate in local attractions and activities, some ruralities have embraced agro-tourism. Agro-tourism can be defined as a segment of the cultural-heritage tourist industry. Agricultural entrepreneurs and their families pursue an economic strategy that allows them to continue to be connected to the land in ways that are complementary to their traditional farming activities. It promises to be a means of preserving the countryside and historic rural environments as well as promoting income generation. In so doing, it serves as a model of sustainable development in the face of changing economic circumstances that threaten to disrupt rural society (Butler, Hall, and Jenkins 1998; Roberts and Hall 2001; Morgan 2006; Lapalme et al. 2009).

As for the agro-tourists, it makes it possible for tourists to enjoy an "authentic" rural world by staying at real working farms that supplement their income through charging for room and board, as well as through the sale of souvenirs, local wine, and produce. Of course, apart from benefiting the farming community and the transient tourist, agrotourism promotes the conservation of the rural environment and a "living countryside." By encouraging farmers to remain on the land, agrotourism serves as a small-farm survival tactic for those threatened by agribusiness and metropolitanism and ensures the continued presence of stewards of the rural heritage in a living countryside.

These initiatives are being melded into Kingston's strategy for a georegional development of a sustainable and economically and socially viable community. At the same time that it is nurturing a vital downtown core and celebrating the authentic expressions of its rich heritage it is also seeking links with its outlying areas and communities. One initiative sponsored by the local National Farmers Union is Food Down the Road, a development that seeks to nurture a "sustainable local food system for Kingston and countryside." While meeting demands of discerning local consumers, participants are engaged in organizing the production, distribution, marketing, and consumption of local foodstuffs.

At the street level, members of the Downtown Kingston Business Improvement organization are playing a leading role in implementing a strategy of "culinary tourism" based on connecting local food producers and restaurateurs and filling a niche in demand by discerning diners, both local and visiting. In so doing, they contribute to the attractions of this lived-in-place, as well as supporting the local rural economy (Kingston BIA 2009).

LOOKING TO THE NEXT CENTURY

In these various ways, the Kingston georegion is striving to develop a sustainable economy and society in the face of an ever-standardizing, globalizing, metropolitanizing, and cosmopolitanizing world. And, ultimately, it's all about sustainability, that is, about communities seeking optimum states of environmental, cultural, spiritual, and economic well-being. But while sustainable communities need economic wealth and employment, even the new opportunities afforded by the tenets of Smart Growth, Creative Cities, and Slow Cities must also be environmentally, economically, and socially compatible with the values of the local communities. In particular, the "devil's bargain" of tourism needs particularly to be monitored to ensure the carrying capacity and sensitivity of the host environment.

In summary, it is argued here that the central premise of a georegional approach is a "glocalizing" strategy that embraces the "lure of the local." As Hans Schmidt, Denmark's then minister of the environment put it: cities and regions face great challenges because of globalization. Cities are often the driving force of the global economy. The challenge is to determine how cities can pull an entire region forward without compromising its identity; that is to say, we must "remain locally anchored in a global world" (Quoted in Hague 2005).

NOTE

1 This "city-region" essay is a reflexive rendering of some four decades of my research based on Kingston's Georegion.

REFERENCES

Anderson, Benedict. 1991 [1983]. *Imagined Communities*. London: Verso.

Anon. 2002. "Slow Cities: Neo-Humanism of the Twenty-first Century." *New Perspectives Quarterly*. 17(4): 20–1.

Arnot, Chris. 2002. "Italian Towns Embrace the Idea That Slow Is Beautiful." *Guardian Weekly*, 6 February, 28.

Blair, Peggy, J. 2008. *Lament for a First Nation: The Williams Treaties of Southern Ontario*. Vancouver: University of British Columbia Press.

Blay-Palmer, A., M. Dwyer, J. Miller. 2006. *Sustainable Communities: Building Local Foodshed Capacity in Frontenac and Lennox-Addington Counties through Improved Farm to Fork Links*. Harrowsmith: Frontenac Community Futures Development Corporation; Picton: Prince Edward-Lennox and Addington Community Futures Development Corporations.

Bohl, C.C. 2002. *Place Making: Developing Town Centers, Main Streets, and Urban Villages*. Washington, DC: Urban Land Institute.

Bullard, Robert D., ed. 2007. *Growing Smarter: Achieving Livable Communities, Environmental Justice, and Regional Equity*. Cambridge: MIT Press.
Butler, R.W., C.M. Hall, and J. Jenkins. 1998. *Tourism and Recreation in Rural Areas*. Chichester: John Wiley.
Campbell, M.C. 2004. "Building a Common Table: The Role for Planning in Community Food Systems." *Journal of Planning, Education, and Research* 23(4): 341–55.
Canadian Tourism Commission. 2004. *Sharing Manitoba's Culture with the World*. Winnipeg, MB [Ottawa]: Manitoba Culture, Heritage and Tourism.
City of Kingston. 2009. *Kingston Integrated Community Sustainability Plan (ISCP): Sustainability Summit Results, 25–29 May 2009*. Kingston: Municipality of Kingston.
CMHC. 2005. *Smart Growth in Canada: A Report Card*. Ottawa: Canada Mortgage & Housing Corporation.
Creative City News. 2004a. "Urban Revitalization and Renewal: The Arts Revitalize City Neighbourhoods Including Downtown Cores." *Creative City News*, Special Edition, 1.
– 2004b. "Creative Economics and Social Benefits for Communities." *Creative City News*, Special Edition, 1.
D'Auria, A.J. 2001. "City Networks and Sustainability: The Role of Knowledge and of Cultural Heritage in Globalization." *International Journal of Sustainability in Higher Education*, 2(1): 38–47.
Entriken, J.N. 1991. *The Betweenness of Place: Towards a Geography of Modernity*. Baltimore: Johns Hopkins.
Florida, Richard. 2002. *The Rise of the Creative Class: And How It's Transforming Work, Leisure, Community, and Everyday Life*. New York: Basic Books.
Florida, Richard, and Roger L. Martin. 2009. *Ontario in the Creative Age*. Toronto: University of Toronto, Rotman School of Management, Martin Prosperity Institute.
Friedmann, H. 2003. "Eating in the Gardens of Gaia: Envisioning Polycultural Communities." In J. Adams, ed., *Fighting for the Farm*, 252–73. Philadelphia: University of Pennsylvania Press.
Garret-Petts, W.F., and Lon Dubinsky. 2005. "Working Well Together: An Introduction to the Cultural Future of Small Cities." *The Small Cities Book*. Vancouver: New Star Books.
Grant, Jill. 2006. *Planning the Good Community: New Urbanism in Theory and Practice*. London: Routledge.

Gregory, Derek. 1994. *Geographical Imaginations*. Oxford: Blackwell.
Hague, Cliff. 2005. "Planning and Place Identity." In Cliff Hague and Paul Jenkins, eds., *Place Identity, Participation and Planning*. New York: Routledge, 3–17.
Harvey, David. 2009. *Cosmopolitanism and the Geographies of Freedom*. New York: Columbia University Press.
Hinrichs, C. 2003. "The Practice and Politics of Food System Localization." *Journal of Rural Studies* 19:33–45.
Hobsbawm, Eric, and Terence Ranger. 1983. *The Invention of Tradition*. Cambridge: University of Cambridge Press.
Hutchins, Bill. 2007. "Council Takes a 'Principled' Approach to Civic Priorities." *Kingston Heritage* 24(45): 1–2.
Imbroscio, D. 1997. *Reconstructing City Politics: Alternative Economic Development and Urban Regimes*. Thousand Oaks, CA: Sage.
Jacobs, Jane. 1961. *The Death and Life of Great American Cities*. New York: Random House.
– 1996. *Edge of Empire: Postcolonialism and the City*. New York: Routledge.
Katz, Peter., ed. 1994. *The New Urbanism: Toward an Architecture of Community*. New York: McGraw-Hill.
Kelbaugh, D. 1999. *Common Place: Toward Neighborhood and Regional Design*. Seattle: University of Washington Press.
Kingston BIA. 2009. "Local Foods Local Chefs Initiative." Application to Ontario Market Investment Fund.
Knox, P. 2005. "Creating Ordinary Places: Slow Cities in a Fast World." *Journal of Urban Design* 10(1): 1–11.
Kunstler, J. 1994. *The Geography of Nowhere*. New York: Touchstone.
Landry, C. 2000. *The Creative City: A Toolkit for Urban Innovation*. London: Earthscan Publications.
Leitch, Alison. 2003. "Slow Food and the Politics of Pork Fat: Italian Food and European Identity." *Ethnos* 68(4): 437–62.
Lippard, Lucy R. 1998. *The Lure of the Local: Senses of Place in a Multicentered Society*. New York: The New Press.
Lorinc, John. 2006. *The New City: How the Crisis in Canada's Urban Centres is Reshaping the Nation*. Toronto: Penguin Canada.
Markusen, Ann, and David King. 2003. *The Artistic Dividend: The Art's Hidden Contributions to Regional Development*. Minneapolis: University of Minnesota.
Mayer, Heike, and Paul L. Knox. 2006. "Slow Cities: Sustainable Places in a Fast World." *Journal of Urban Affairs* 28(4): 321–34.

Meyrowitz, J. 1985. *No Sense of Place.* Oxford: Oxford University Press.

Morgan, Kevin. 2006. *Worlds of Food: Place, Power, and Provenance in the Food Chain.* Oxford: Oxford University Press.

Nagel, T. 1986. *The View from Nowhere.* New York: Oxford University Press.

Neal, Peter, ed. 2003. *Urban Villages and the Making of Communities.* New York: Spon Press.

Nelson, Gordon, et al. 2005. *The Great River: A Heritage Landscape Guide to the Upper St. Lawrence Corridor.* Heritage Landscape Series, no. 4. Waterloo, ON: Environments Publications, University of Waterloo.

Osborne, Brian S. 1973. "The Site of Kingston." In G. Betts, ed., *Kingston 300: A Social Snapshot,* 8–12. Kingston: Hanson and Edgar.

– 1975. "Kingston in the Nineteenth Century: A Study in Urban Decline." In J.D. Wood, ed., *Perspectives on Landscape and Settlement in 19th Century Ontario,* 159–82. Ottawa: Carleton University Press.

– 1976. "The Settlement of Kingston's Hinterland." In G. Tulchinsky, ed., *To Preserve & Defend: Essays on Kingston in the Nineteenth Century,* 63–79, 351–3. Montreal and Kingston: McGill-Queen's University Press.

– 1977. "Frontier Settlement in Eastern Ontario in the 19th Century: A Study in Changing Perceptions of Land and Opportunity." In David Harry Miller and Jerome O. Steffen, eds., *The Frontier: Comparative Studies,* 201–26. Norman, OK: University of Oklahoma Press.

– 1982. "The Farmer and the Land." In Bryan Rollason, ed., *County of a Thousand Lakes: The History of the County of Frontenac,* 81–94. Kingston: Frontenac County Council.

– (with Donald Swainson). 1988. *Kingston: Building on the Past.* Westport: Butternut Press.

– 1989. "The Long and the Short of It: Kingston as the Foot-of-the-Great-Lakes Terminus." *Freshwater* 4:1–19.

– 1990. "Organizing the East Lake Ontario Fishery." *Historic Kingston* 38:81–93.

– (with M. Ripmeester). 1995. "Kingston, Bedford, Grape Island, Alnwick: The Odyssey of the Kingston Mississauga." *Historic Kingston* 43:84–112.

– (with Geraint B. Osborne). 1999. *Oral History, Front of Leeds and Lansdowne: Interviews.* Maitland, ON: Parks Canada, Thousand Islands.

– 2001. "The St. Lawrence in the Canadian National Imagination." *Études Canadiennes* 50 (June): 257–75.

– (with M. Huitema and M. Ripmeester). 2002. "Imagined Spaces, Constructed Boundaries, Conflicting Claims: A Legacy of Colonial Conflict in Eastern Ontario." *International Journal of Canadian Studies*, no. 25, 87–112.
– 2003. "Barter, Bible, Bush: Strategies of Survival and Resistance among the Kingston–Bay of Quinte Mississauga, 1783–1836." In Bruce Hodgins, Ute Lischke, and David McNab, eds., *Blockades & Resistance: Studies in Actions of Peace and the Temagami Blockades of 1988–89*, 85–104. Waterloo: Wilfrid Laurier University Press.
– (with M. Huitema and M. Ripmeester). 2004. "Shared Places, Shifting Spaces: The First Nations of the Thousand Islands." In K. Burtch, ed., *Life on the Edge: The Cultural Landscape of the Thousand Islands Area*, 13–26. Maitland, ON: Canadian Thousand Islands Heritage Conservancy.
– 2007. "Heritage, Tourism, Quality of Life: Constructing New Urban Places." In Geoffrey Wall, ed., *Approaching Tourism*, 155–82. Geography Occasional Paper No. 21. Waterloo: University of Waterloo,
– 2009. "Remembering and Constructing Intangible Heritage along Canada's Upper St. Lawrence." In Sergio Lira, R. Amoeda, C. Pinheiro, J. Pinheiro, and F. Oliveira, eds., *Sharing Cultures 2009: International Conference on Intangible Heritage*, 231–8. Barcelos, Portugal: Green Lines Institute for Sustainable Development.

Parham, Claire Puccia. 2008. *The St. Lawrence Seaway and Power Project: An Oral History of the Greatest Show on Earth*. Syracuse: Syracuse University Press.

Parkins, Wendy. 2004. "Out of Time: Fast Subjects and Slow Living." *Time and Society* 13, no. 2/3, 363–82.

Portney, K.E. 2003. *Taking Sustainable Cities Seriously: Economic Development, the Environment, and Quality of Life in American Cities*. Cambridge, MA: MIT Press.

Roberts, L., and D. Hall. 2001. *Rural Tourism and Recreation: Principles to Practice*. Wallingford, UK: CABI Publishing.

Petrini, Carlo, ed. 2001. *Slow Food: Collected Thoughts on Taste, Tradition, and the Honest Pleasures of Food*. White River Junction, VT: Chelsea Green.

Petrini, Carlo, and Giggi Padovani. 2006. *Slow Food Revolution: A New Culture for Eating and Living*. New York: Rizzoli.

Pietrykowski, Bruce. 2004. "You Are What You Eat: The Social Economy of the Slow Food Movement." *Review of Social Economy* 62(3): 308–21.

Pine II, Joseph, and James H. Gilmore. 1999. *The Experience Economy: Work is Theatre and Every Business a Stage*. Boston: Harvard Business School Press.

Pink, Sarah. 2008. "Sense and Sustainability: The Case of the Slow City Movement." *Local Environment* 13(2): 95–106.

Rast, J. 2005. "The Politics of Alternative Economic Development: Revisiting the Stone-Imbroscio Debate." *Journal of Urban Affairs* 27(1): 53–69.

Relph, E. 1976. *Place and Placenessness*. London: Pion.

Richards, Greg. 2007. *Cultural Tourism: Global and Local Perspectives*. London: THHP.

Robertson, R. 1994. *Globalization: Social Theory and Global Culture*. London: Sage.

– 1995. "Glocalization: Time-Space and Homogeneity-Heterogeneity." In Mike Featherstone, Scott Lash, and Roland Robertson, eds., *Global Modernities*, 25–44. Sage: London.

Robins, K. 1991. "Tradition and Translations: National Culture in a Global Context." In J. Corner and S. Harvey, eds., *Enterprise and Heritage*. London: Routledge.

Sack, R. 1997. *Homo Geographicus*. Baltimore: Johns Hopkins University Press.

Shone, Michael C., and P. Ali Memon. 2008. "Tourism, Public Policy and Regional Development: A Turn from Neo-Liberalism to the New Regionalism." *Local Economy* 23(4): 290–304.

Stone, C.N. 2004. "It's More than the Economy after All: Continuing the Debate about Urban Regimes." *Journal of Urban Affairs* 26(1): 1–19.

Swyngedouw, Erik. 1992. "The Mammon Quest: 'Glocalisation,' Interspatial Competition and the Monetary Order: The Construction of New Scales." In M. Dunford and G. Kafkalas, eds., *Cities and Regions in the New Europe: The Global-Local Interplay and Spatial Development Strategies*. London: Belhaven Press.

– 2000. "Elite Power, Global Forces, and the Political Economy of 'Glocal' Development." In G.L. Clark, M.P. Feldman, and M.S. Gertler, eds., *The Oxford Handbook of Economic Geography*. Toronto: Oxford University Press.

Szold, T.S., and A. Carbonell, eds. 2002. *Smart Growth: Form and Consequences*. Cambridge, MA: Land Institute of Land Policy.

Throsby, David. 2001. *Economics and Culture*. Cambridge: University of Cambridge Press.

Tuan, Yi-Fu. 1974. *Topophilia: A Study of Environmental Perception, Attitudes, and Values*. Englewood Cliffs: Prentice-Hall.

Urry, John. 1995. *Consuming Places*. London: Routledge.
Wagner, F.W. 2005. *Revitalizing the City: Strategies to Contain Sprawl and Revive the Core*. Armonk, NY; London: M.E. Sharpe.
Wolfart, Philip. 1997. "The 'Region' in German Historiography." In Serge Courville and Brian S. Osborne, eds., *Histoire Mythique et Paysage Symbolique/Mythic History and Symbolic Landscape: Actes du Projet d'Echange Laval-Queen's*, 88–95. St Foy, QC: Cheminements de CIEQ.
Wyman, Max. 2004. *The Defiant Imagination: Why Culture Matters*. Vancouver: Douglas & MacIntyre.

10

The Ottawa Valley Georegion

MARK SEASONS

SUMMARY

Here we explore one of Ontario's, and Canada's, most diverse georegions – the Ottawa Valley. The chapter provides basic facts and figures on the Valley, then offers some of the nuances and subtleties that characterize life in this georegion. The Ottawa Valley is the city of Ottawa and the National Capital, certainly, and the Ottawa River watershed, but it is so much more. The Valley plays a highly symbolic role in our nation as the point of intersection between Canada's two founding European cultures and their expression in the political constructs of the provinces of Ontario and Quebec.

We seek to describe the physical geography of this georegion, as well as its political, economic, and social geographies. The Valley's historical evolution has been very ably explained in chapters 1 and 2 by Nelson and Troughton. We know that the Valley was inhabited for hundreds of years by the Algonquin peoples in pre-European times. Unlike the more agrarian culture of their Iroquoian contemporaries in southern Ontario, the Algonquin of the Ottawa Valley were hunters, fishers, and gatherers. They used the Ottawa River and its tributaries as indigenous trading routes with other First Nations groups via connections east to the St Lawrence, and north and west to Georgian Bay and on to the Great Lakes. French explorers and voyageurs arrived in the seventeenth century, followed in the eighteenth and nineteenth centuries by successive waves of European settlers.

The pace and extent of European settlement accelerated following the 1812–14 war with the establishment of settlements along the Ottawa Valley watershed taking advantage of water power to drive mills. The great forests of white pine were cut and floated down the Ottawa River to the burgeoning mill towns a practice that continued until the 1960s. By the mid-nineteenth century, an extensive network of rail lines and the Rideau Canal connected the Ottawa Valley with the rest of Canada and the northern United States. Ottawa became the capital of the fledgling nation and its growth, and influence, has proceeded apace since.

In this chapter, recent trends, current realities, and future prospects are considered in terms of the Valley's economic, socio-cultural, political, and environmental context. It is important to note that there is no single, homogenous Ottawa Valley. Rather, we can identify three sub-regions, each with its own identity, yet related and complementary: the urban, multi-cultural and multi-lingual Ottawa; the suburban/exurban communities within the Ottawa commuter-shed (Lanark and the western edge of Prescott-Russell); and the smaller urban centres, rural and traditional towns and villages elsewhere in the Valley (Nipissing and Renfrew). In keeping with a major theme of this book, this chapter concludes with implications for urban and regional planning, as well as for economic-development planning in this part of Ontario.

THE SPATIAL CONTEXT (BOUNDARIES)

One of the first challenges is to establish the spatial extent and character of the Ottawa Valley (box 10.1). For the purposes of this chapter, the Ottawa Valley is not, as Raffan (2005, 52) notes, simply "the Ottawa River watershed, which is twice the size of New Brunswick and sprawls across western Quebec and eastern Ontario." Instead, we focus on those lands and communities that are adjacent to the Ottawa River, specifically on the Ontario side.

BOX 10.1. SUMMARY OF MAJOR CHARACTERISTICS OF THE OTTAWA VALLEY

Size	7,645km^2
Length	1,271km^2
Population (2006)	1,036,050
Economic structure	Government, high technology, tourism, resource development

Landscape character	Mixedwood Plains Ecozone: St Lawrence Lowlands ecoregion
	Boreal Shield Ecozone: Algonquin-Lake Nipissing ecoregion
	Boreal Shield Ecozone: Southern Laurentians ecoregion
Socio-cultural composition	Ottawa: highly multicultural
	Upper Ottawa Valley: British Isles origins
	Lower Ottawa Valley: mix of British Isles and French cultural origins

By the author, with permission

The convention is to delineate the Upper Valley and its Lower Valley counterpart as separate and quite identifiable geographical, economic, and cultural sub-regions. The Upper Valley sub-region comprises the communities west of Ottawa, extending about 220 kilometres up the Ottawa River. The Lower Valley sub-region begins on Ottawa's eastern border and ends approximately 80 kilometres to the east at the Ontario-Quebec provincial border (Canadian Geographic 2005a).

The Ottawa Valley covers over 7,645 square kilometres (Canadian Geographic 2005a,b) (figure 10.1). Geologically, the Valley was formed as a result of ancient tectonic and glacial processes that produced the Canadian Shield. Consequently, the north side of the river in Quebec, which is hilly and features countless lakes, reflects uplift and folding. To the south, on the Ontario side, we see a less hilly region that comprises flatlands and gentler, rolling hills. The Ottawa River is clearly the dominant geographical and defining feature of this region (Ottawa River Institute 2007). The river is 1,271 kilometres long; its origins are considered to be in the Temiskamingue region of north-west Quebec and its terminus where it joins the St Lawrence River at Montreal (Canadian Geographic 2005a).

The Valley straddles three ecoregions: the St Lawrence Lowlands ecoregion, the Algonquin-Lake Nipissing ecoregion, and the Southern Laurentians ecoregion on the north side of the Ottawa River (Environment Canada 2007a, b, c, d, e). An ecoregion is characterized by distinctive regional ecological factors, including climatic factors, physiography, vegetation, soil, water, fauna, economic factors,

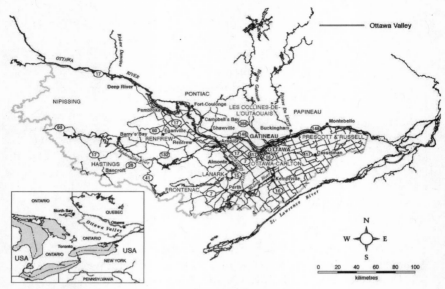

Figure 10.1   Ottawa Georegion. By the author, with permission

and land use (Environment Canada 2007a). Small to major differences in these factors differentiate the three ecoregions, which were originally heavily forested with white pine and other conifers, as well as maple and other deciduous trees.

POPULATION

In political-administrative terms, the Ontario side of the Valley comprises three counties adjacent to the Ottawa River, the United Counties of Prescott and Russell and the Counties of Lanark and Renfrew; one district, the Nipissing District; and the City of Ottawa. The population of the Ontario portion of the Valley is 1,138,311 (Statistics Canada 2007b). The dominant urban centre, accounting for approximately 77 percent of the Ottawa Valley population in Ontario, is the city of Ottawa (2006 population 812,129). Ottawa's urban counterpart, across the river in the Outaouais region of Quebec, is the city of Gatineau (2006 population 283,959). Together, the Ottawa-Gatineau CMA (census metropolitan area, 2006 population 1,130,761) is the fifth largest urban area in Canada. Table 10.1 depicts population trends from 1996 to 2006 in this area.

Table 10.1
Population by County, 1996–2006

County	1996	2001	Percentage Change 1996–2001	2006	Percentage Change 2001–6	Percentage Change 1996–2006
Lanark	59,845	62,495	4.4	63,785	2.1	6.2
Nipissing	84,832	82,910	–2.3	84,688	2.1	0.0
Ottawa	721,136	774,072	7.3	812,129	4.9	11.3
Prescott and Russell	74,013	76,446	3.3	80,184	4.9	7.7
Renfrew	96,224	95,138	–1.1	97,545	2.5	1.4
Ottawa Valley (ON)						
Ontario	10,753,573	11,410,046	6.1	12,160,282	6.6	11.6

Source: Data from Community Profiles 2006 (Statistics Canada 2007a). By the author, with permission.

Clearly, the city of Ottawa is the major urban centre for the Valley, a role that it has played for over 160 years. Ottawa has experienced significant population growth, with a rate of 7.3 percent from 1996 to 2001. This was followed by slower, yet still impressive, growth of 4.9 percent from 2001 to 2006 (Statistics Canada 2007a,b). The Counties of Ottawa-Lanark and Prescott-Russell, adjacent to the city, have also experienced steady rates of population growth, attributable in part to their location within the commutershed for the Ottawa-based regional economy. In contrast, the jurisdictions located distant from Ottawa, to the northwest of the Valley – Renfrew County and the district of Nipissing – have experienced slower rates of population growth. Indeed, both Renfrew and Nipissing lost population from 1996 to 2001. It is interesting to note that in 2006 none of the study-area municipalities matched the province-wide growth figures. During 1996–2001, only the city of Ottawa exceeded the provincial numbers, and none of these municipalities matched the provincial trend from 2001 to 2006. With the exception of the city of Ottawa, Ottawa Valley communities, while not stagnant in population growth, have lagged behind such dynamic growth generators as the Greater Toronto Area.

Other demographic data reveal few differences from the provincial median figures. The median age of residents of Ottawa Valley municipalities is generally slightly older than the provincial pattern. Table 10.2 indicates that the populations in these municipalities

Table 10.2
Median Age by County, 2001–6

County	2001	2006	Percentage Change 2001–6
Lanark	39.9	43.1	7.5
Nipissing	39.2	42.2	7.1
Ottawa	36.7	38.4	5.4
Prescott and Russell	37.7	40.5	7.0
Renfrew	39.2	42.1	6.9
Ontario	37.2	39.0	4.6

Source: Data from Community Profiles 2006 (Statistics Canada 2007a). By the author, with permission

Table 10.3
Percentage 15+ Years, by County, 2001–6

County	2001	2006	Percentage Change 2001–6
Lanark	80.3	82.8	3.0
Nipissing	81.4	83.7	2.7
Ottawa	81.1	82.4	1.6
Prescott and Russell	78.7	81.3	3.2
Renfrew	80.6	82.7	2.5
Ontario	80.4	81.8	1.7

Source: Data from Community Profiles 2006 (Statistics Canada 2007a). By the author, with permission.

can be up to three to four years older than the provincial average, although Ottawa is slightly lower than the provincial average.

This can be attributed to its role as an economic magnet and thus as an attractor of a younger demographic profile. It is also apparent that the populations in these communities are aging at a faster rate than the provincial median. In Lanark County, this could be attributed to an increase in retirees from the Ottawa area and other urban centres, but in more remote places that lie beyond commuting range from Ottawa, such as Nipissing and Renfrew, we see populations aging in place (Service Canada 2007).

These trends are confirmed by table 10.3, which indicates that each Ottawa Valley municipality, including Ottawa, has more residents older than fifteen years of age than the provincial average. Several factors may be at play. Many of these communities, especially those within driving range of Ottawa, and including the city of

Table 10.4
Immigration Profile, by County, 2001

County	Total Population	Canadian-born (% of total)	Foreign-born (% of total)
Lanark	60,995	57,015 (93.5)	3,905 (3.9)
Nipissing	81,595	77,910 (95.4)	3,570 (4.3)
Ottawa	763,795	589,015 (77.1)	166,745 (21.8)
Prescott and Russell	76,446	71,765 (93.9)	3,160 (4.1)
Renfrew	93,760	88,330 (94.2)	5,340 (6.1)
Ontario	11,285,545	8,164,860 (68.9)	3,030,075 (26.9)

Source: Data from Community Profiles 2001 (Statistics Canada 2007). By the author, with permission.

Ottawa, are attractive to retirees. Data from the 2001 census showed that one census tract in Ottawa, in the near west end of the city, was second only to Victoria in numbers and concentration of senior citizens. It may be that Ottawa Valley counties have lost younger populations to employment opportunities elsewhere. With the exception of Ottawa, these places have not attracted young families.

Immigration has a direct bearing on population size and composition. As national fertility rates decline, immigration plays an increasingly important role in domestic population and economic growth. Communities that attract significant numbers of immigrants generally thrive (see Simmons and Bourne 2007). The Ottawa Valley data are especially interesting in this respect. In 2001, approximately 69 percent of Ontario residents were born in Canada. However, the profile in Ottawa Valley municipalities presents a considerably different situation. Table 10.4 indicates that the vast majority of Ottawa Valley residents, with the exception of those in the city of Ottawa, were born in Canada. The numbers are quite remarkable: each of the counties and Nipissing registered over 90 percent Canadian-born residents.

On a national, large-city scale, Ottawa is considered a second-tier settlement-location for immigrants; immigration tends to account for 15–20 percent of total population in cities like this. Setting

Table 10.5
Immigration Stages, by County, 2001

County	Foreign-born (% of total)	Immigrated before 1991 (% of total)	Immigrated 1991–2001 (% of total)	Non-permanent residents (% of total)
Lanark	3,905 (2.2)	3,540 (90.7)	365 (9.3)	35 (0.9)
Nipissing	3,570 (2.0)	3,260 (91.3)	315 (8.8)	105 (2.3)
Ottawa	166,745 (92.8)	103,045 (61.8)	63,705 (38.2)	8,030 (4.8)
Prescott and Russell	3,160 (1.8)	2,435 (77.1)	720 (22.8)	60 (1.9)
Renfrew	5,340 (3.0)	4,550 (85.2)	790 (14.8)	90 (1.7)
Ontario	3,030,075	2,007,705 (66.3)	1,022,370 (33.7)	90,615 (0.3)

Source: Data from Community Profiles 2001, 2006 (Statistics Canada 2007a). By the author, with permission

Ottawa aside as an anomaly, the overall picture is of a region that has not been significantly affected by inter-regional domestic migration or by extensive immigration.

Table 10.5 indicates that the majority of immigrants who reside in municipalities outside Ottawa arrived before 1991; they are not part of the large wave of post-1991 immigration that the GTA and much of the rest of the province have experienced. While the data do not demonstrate this clearly, they suggest that immigrants probably arrived in the Valley in waves that started in the late nineteenth century and continued post–World War Two. We also see that Ottawa attracted a greater share of post–1991 immigrants than the provincial average. This reflects greater economic opportunity in the Ottawa economy as the tendency for cultural groups to migrate where family, resources, and support groups exist to ease the immigrant transition experience. Similarly, the data confirm a common pattern experienced across the country: immigrants are less attracted to smaller, rural communities where there are fewer adjustment resources or social networks.

The data presented in table 10.6 tell us that most immigrants to Ottawa Valley communities come from "traditional" sources such as Western Europe, while in Ottawa the immigrant profile is considerably more diverse. Visible minorities comprise a miniscule

Table 10.6
Visible Minorities, by County, 2001

County	Total Population	Visible minorities (% of total)	All others (% of total)
Lanark	60,955	470 (0.8)	60,480 (99.2)
Nipissing	81,595	1,015 (1.2)	80,580 (98.8)
Ottawa	763,790	137,245 (18)	626,545 (82)
Prescott and Russell	74,980	965 (1.3)	74,010 (98.7)
Renfrew	93,760	1,495 (1.6)	92,265 (98.4)
Ottawa Valley (ON)	1,075,080	141,190 (2.6)	933,880 (95.4)
Ontario	11,285,545	2,153,045 (19.1)	9,132,500 (80.9)

Source: Data from Community Profiles 2001 (Statistics Canada 2007). By the author, with permission.

proportion of the populations in Ottawa Valley municipalities, with the exception of the city of Ottawa. Indeed, these data reflect the three Ottawa Valley experiences: the urban, multicultural, and multilingual Ottawa; the suburban/exurban communities within the Ottawa commuter-shed (Lanark and the western edge of Prescott-Russell); and the smaller urban centres, rural and traditional towns and villages elsewhere in the Upper and Lower Ottawa Valley (Nipissing and Renfrew).

Similarly, there is considerable polarity in linguistic profile across the Valley. Table 10.7 indicates that only Ottawa can claim a multilingual community that is similar to Ontario as a whole. Two counties are firmly anglophone (Lanark and Renfrew), and two are significantly francophone (Prescott and Russell, and Nipissing). These figures reflect the historical and apparently well-entrenched settlement patterns of the Valley, specifically the predominance of Irish, English, and Scottish settlers in Lanark and Renfrew counties, and the descendants of Quebec migrants in the Upper and Lower Ottawa Valley. In a sense, Ottawa sits at the geographic centre and cultural fulcrum of these traditional Ottawa Valley groups (Environment Canada 2007e). It is also very much the exception to a well-established pattern in the Ottawa Valley of Western European cultural origins dominated by the two official languages.

HERITAGE AND CULTURAL LANDSCAPES

The area's cultural and heritage landscapes have also received considerable attention. The federal government, through the National

Table 10.7
Linguistic Profile, by County, 2001

County	Total population	English only (% of total)	French only (% of total)	Both English and French (% of total)	Other languages (% of total)
Lanark	60,955	57,345 (94.1)	1,805 (3.0)	205 (0.3)	1,600 (2.6)
Nipissing	81,595	56,635 (69.4)	20,885 (25.6)	1,375 (1.7)	2,695 (3.3)
Ottawa	763,790	485,825 (63.6)	115,220 (15.1)	7,445 (1.0)	155,295 (20.3)
Prescott and Russell	74,980	21,210 (28.3)	50,120 (66.8)	1,200 (1.6)	2,455 (3.3)
Renfrew	93,760	83,800 (89.4)	4,270 (4.6)	595 (0.06)	5,090 (5.4)
Ottawa Valley (ON)	1,075,080	704,815 (69)	192,300 (23)	10,820 (4)	167,715 (16)
Ontario	11,285,545	7,965,225 (70.6)	485,630 (4.3)	37,135 (0.3)	2,797,555 (24.8)

Source: Data from Community Profiles 2001 (Statistics Canada 2007). By the author, with permission.

Capital Commission (NCC), has long been instrumental in the preservation of locally and nationally significant cultural landscapes in the National Capital Region. This has involved the long-range planning of the capital and its surrounding region; the acquisition and restoration of heritage properties of local and national significance; and strategic investments in infrastructure, amenities, and land uses of national importance. While the NCC and its forebears have been planning the capital since 1899, the seminal capital planning vision was articulated in the Gréber Plan of 1950. That plan laid the foundation for much of the current capital's structure and regional-scale land use.

Gréber's legacy included the designation of twenty thousand hectares of land on the Ottawa side as the Greenbelt and the creation of Gatineau Park on the Outauoais side. These lands were set aside to control growth and act as an "urban fence," to provide low-intensity recreational opportunities, and to preserve environmentally sensitive lands. The Gréber Plan also led to the purchase and control of most of the lands adjacent to both sides of the Ottawa River, the relocation of rail lines, and the creation of federal employment nodes outside the core area (National Capital Commission 2007a). The result is an impressive green capital.

During the 1950s and 1960s, the NCC was also the regional leader in the acquisition and restoration of nineteenth-century heritage properties, principally along Sussex Drive in Ottawa. The facades of these buildings were restored, and interpretive plaques were installed to tell each building's role in the capital and in the historic ByWard Market area, now a key tourist draw. The buildings were then modernized and have since been used for retail, commercial, and office purposes. Over the years, the NCC has also been active in towns and villages elsewhere in the National Capital Region through its purchase and restoration of mills and heritage structures. More recently, the NCC has acted to research and preserve ancient sites of value to the Ottawa Valley's First Nations, such as the Lac Leamy area along the Ottawa River in Gatineau (National Capital Commission 2007b).

The Ottawa georegion's cultural landscape has benefited from Parks Canada's ownership and enlightened management of the Rideau Canal, built in the 1830s and recently designated a world heritage site by UNESCO (Parks Canada 2007). At the municipal level, the City of Ottawa has a well-established and respected expertise in heritage preservation. Its land use planning and development review processes are well informed by consideration for heritage landscapes (City of Ottawa 2007a, 2007b).

ECONOMIC DEVELOPMENT

As noted previously, there is no single Ottawa Valley in terms of demographic characteristics. The same is true for economic structure and economic development. The potentials to be realized in economic development seem largely a function of proximity to the dominant Ottawa-based regional economy.

As might be expected, planning for population and economic growth is a preoccupation of the City of Ottawa, which is both the national capital and a major national urban centre. It has experienced considerable structural economic change over the past twenty-five years as the city-region's economy has evolved and matured. However, the federal government, while still a major regional economic factor and a stabilizing influence, is no longer the dominant economic force in the Ottawa area.

Ottawa is no longer only a federal "company town." The size of the federal civil service was reduced significantly through the 1980s and 1990s because of downsizing efforts, the privatization or downloading of federal government services, and the relocation of functions and staff to other Canadian centres and regions. At the same time, the region's services base, its tourism, and its information and technology sectors have become the region's economic-growth engines. The federal government's post–World War Two investment in research and development, carried out under the auspices of agencies such as the National Research Council (NRC), has generated multiple small and medium communications and biotechnology enterprises and fostered the dramatic growth of large high-tech corporations (City of Ottawa 2007d). The business sector arguably comprises the most powerful and influential group of stakeholders in the Ottawa area. Private sector employment accounted for 66 percent of all employment in Ottawa in 2006 (City of Ottawa 2007c).

Quality of life is a critical factor in the attraction and retention of businesses and their employees. There is very much a vested interest on the part of business to maintain a green city that has a vibrant urban and social life. The residents of Ottawa are well educated, active, and affluent, and they demand a high quality of urban life. There is an increasing market for high-end retail businesses, sophisticated cultural amenities, and recreational facilities (City of Ottawa 2007c). The combined influence of economic stability and demography has generated a significant demand for higher-density, condominium-style housing in the inner suburbs. The dramatic and rather startling transformation of Westboro Village, in Ottawa's western inner suburbs, from a sleepy neighbourhood centre to a thriving, "hip," and trendy retail, dining, and residential district is a case in point.

For communities within the Ottawa commuter-shed – within a radius of about a forty-five minute drive – Ottawa's burgeoning economic power has created a high demand for affordable, accessible homes in outlying communities that are linked to the city by regional public transit and by improved provincial road networks. Small towns in this zone have grown in size through the addition of new subdivisions. In these places we see big-box retail outlets that service the immediate community and also act as second-order service centres for retail goods.

With the arrival of full-size, big-box retail outlets such as the Canadian Tire and Staples stores in Carleton Place (Lanark County), there is less need to travel to Ottawa for shopping. These communities have become slightly more self-contained in terms of urban-type amenities and services. However, an increasing proportion of employment is still generated by the Ottawa economy. While retail and service sector activities have increased in importance, we are also aware of the waning influence or complete loss of the traditional manufacturing and resource-sector employment base in many of these towns. Indeed, these communities' adjacency to Ottawa has eased and often solved local economic-development issues.

It is a different story in the outlying counties of Renfrew, Nipissing district, and the eastern edge of Prescott-Russell in the Upper and Lower Ottawa Valley, respectively. These communities are located beyond Ottawa's economic shadow, outside the city's commuter-shed. Here, there is comparatively little change. The rate of population growth is comparatively much slower, as is economic growth. In fact, decline has been part of their historical reality. There is still considerable dependence on the more traditional economic foundations of agricultural, resource extraction, and manufacturing. Small towns in this sub-region continue to fulfill an important service centre role; they are not high-tech hubs, nor is there extensive and high-volume subdivision development. However, tourism related to natural features such as the Ottawa River has gained in importance in these places, for example, river-rafting expeditions, as has cultural tourism.

GOVERNANCE

The Ottawa Valley lies at the intersection between the provinces of Quebec and Ontario, separated by the Ottawa River. The legal, linguistic, and political systems on each side of the river are quite different. Too often they exist in "splendid isolation." These systems, their agencies, and their participants are brought together over issues of common interest that coalesce around the river, for example, river crossings, water-level control, and hydro-electric generating facilities, but there is no interprovincial regional-development planning bureau. Interprovincial transportation planning and coordination takes place in the National Capital Region and elsewhere in the Upper and Lower Valley around specific bridge crossings.

Local economic development, education, health care, transportation, land use planning – all these and more are planned and carried out on either side of the river as though they exist in parallel universes. In both provinces, we have witnessed recent efforts to consolidate and amalgamate municipal governments, specifically in the Ottawa-Gatineau region. There have been promises made and great expectations of improved efficiencies and savings in service delivery, economic development, and health care. The new City of Ottawa incorporates what was, until 2000, ten separate local municipalities and one regional government. On the Quebec side, in the Outaouais region, we have seen the recent consolidation of the former municipalities of Gatineau, Hull, and Aylmer into the new Ville de Gatineau.

The federal government is a major actor in the Ottawa Valley. Through its Crown corporation, the National Capital Commission (NCC), the federal government plans and manages about five hundred square kilometres of land in a national capital region (NCR) of approximately forty-seven hundred square kilometres. In essence, the federal government plays a key broker, facilitator, and negotiator role between the two provinces and their respective agencies. Its ownership and planning control of river edges in the NCR and of key transportation and river-crossing corridors gives the federal government considerable strategic influence on at least the central section of the Ottawa Valley.

IMPLICATIONS FOR PLANNING

This chapter demonstrates that the Ottawa Valley is not a homogenous social or economic entity. It comprises three sub-regions: the City of Ottawa, the smaller towns and villages located within the Ottawa commuter-shed, and outlying communities that are beyond the main economic reach of the city of Ottawa. These communities may also be classified as belonging to the Upper or the Lower Ottawa Valley. It is apparent that the City of Ottawa is, by far, the dominant economic force in this ecoregion. Its economy has become highly diversified, and it is a destination for immigrants and inter-regional migrants. In contrast, most other Valley municipalities seem quite unaffected by immigration; they remain much as they have for almost two hundred years of settlement, at least in socio-cultural terms.

The planning challenges in the Ottawa area concern the management of current and future growth pressures in a sustainable manner. We can anticipate continued efforts to set physical boundaries on urban growth and increased support for the intensification of development and redevelopment of urban sites for residential uses. We can also predict continued and stronger support for the protection and effective management of natural resources – the rivers, waterbodies, and ecosystems that contribute to the area's high quality of life.

It should be noted that the City of Ottawa, and its predecessor the Regional Municipality of Ottawa-Carleton (RMOC), are considered Canadian leaders in environmental planning and management. The former City of Ottawa was one of the first municipalities in Canada to establish a municipal-level environmental assessment process – one that complemented and exceeded the Ontario standard for environmental assessments (City of Ottawa 2007b). Clearly, the city's well-educated populace has high expectations of appropriate resource management and environmentally based decision making.

Affordability, especially of housing, will remain a high-profile policy issue for Ottawa, as will efforts to diversify the regional economy, reducing its dependence on the federal government and a technology sector that experiences cyclical behaviour. The replacement of obsolescent and inefficient infrastructure, for example water and sewage systems and public transit, will be a major planning and municipal finance issue for years to come.

In the commuter-shed municipalities, there will be continued pressure to provide urban-standard services and amenities. There will be intentions to diversify these economies, but it is likely that these communities will remain bedroom suburbs for the Ottawa employment market. With time, these communities will start to resemble the more diverse socio-cultural make-up of Ottawa. Immigrants will be drawn by more affordable housing and the quality of life offered by a smaller, urban-edge community. In response, these communities will need to offer urban-style services and facilities.

As for the third group – the municipalities located at a distance from Ottawa – we can expect few changes in their socio-cultural composition. These places are less likely to attract inter-regional migrants or immigrants; residents will age in place, and younger people will migrate to larger urban centres, such as Ottawa. The economy in these communities will continue to be based on agriculture, resource management, and tourism. Their economic role as agricultural service centres will likely continue. Growth will probably not be a concern. Instead, some of these places will need to consider planning for place decline or contraction. Difficult decisions will need to be made concerning the continued viability of smaller communities that seem to be aging in place, especially communities in the Upper and Lower Ottawa Valley.

The key for the future viability and sustainability of the Ottawa Valley Georegion will be to build on, and respect, a well-defined base of regional cultures, natural amenities, and resources, while positioning these communities for future economic development possibilities and challenges. There is a need for a regional planning and development strategy, such as an Eastern Ontario version of the province's well-known *Places to Grow* initiative that links investments in strategic infrastructure such as health services, the regional road network, and transit to the Valley's economic development potential, all within the framework of sustainability and the bioregion.

It would be appropriate to establish clear roles for Ottawa Valley communities within the context of a regional economic-development strategy. The designation of growth centres or growth poles would make infrastructure investment decisions, such as inter-urban transportation and health care, much clearer. There is also a need to better coordinate interprovincial planning for economic development, infrastructure, and resource management along the extent of the Ottawa Valley.

In summary, the Ottawa Valley is a georegion whose distinctiveness has been shaped by its geography: the mighty and dominant Ottawa River, the Canadian Shield, and the impacts of European settlement. Its location (and national role) at the point of intersection between the traditional European founding cultures make this georegion unique in the country. It is an economic and cultural crossroad that continues to play an important role as a major conduit for ideas, trade, and migration between Eastern and Western Canada, and its ecosystem is delicately balanced. For these reasons, there is a need for a planning and development strategy that acknowledges the variety of urban and rural experiences across the Valley.

REFERENCES

Canadian Geographic. 2005a. Retrieved from the World Wide Web on 1 August 2007. http://www.canadiangeographic.ca/magazine/so05/indepth/justthefacts.asp
- 2005b. Retrieved from the World Wide Web on 1 August 2007. http://www.canadiangeographic.ca/Magazine/so05/feature.asp.

City of Ottawa. 2007a. *Official Plan*. Vol. 1, s.4.6 *Cultural Heritage Resources*. Retrieved from the World Wide Web on 28 October 2007. http://www.ottawa.ca/city_hall/ottawa2020/official_plan/vol_1/review_dvlpmnt_apps/cultrl_heritge_rsrces/index_en.html
- 2007b. *Official Plan*. Volume 1, s.4.7.1. *Integrated Environmental Review to Assess Development Applications*. Retrieved from the World Wide Web on 28 October 2007. http://www.ottawa.ca/city_hall/ottawa2020/official_plan/vol_1/review_dvlpmnt_apps/enviro_protectn/integratd_review_en.html.
- 2007c. *Annual Development Report: Highlights*. Retrieved from the World Wide Web on 28 October 2007. http://www.ottawa.ca/city_services/statistics/dev_report_2006/pdf/highlights_en.pdf.
- 2007d. *Key Competitive Sectors*. Retrieved from the World Wide Web on 28 October 2007. http://www.ottawa.ca/business/economic_profile/key_competitive_en.html.

Environment Canada. 2007a. "Narrative Descriptions of Terrestrial Ecozones and Ecoregions of Canada." Retrieved from the World Wide Web on 1 August 2007: http://www.ec.gc.ca/soerree/English/Framework/Nardesc/ExNote.cfm.
- 2007b. "State of the Environment Database: Boreal Shield." Retrieved from the World Wide Web on 1 August 2007: http://www.ec.gc.ca/soer-ree/English/Framework/Nardesc/borshd_e.cfm.
- 2007c. "State of the Environment Database: Southern Laurentians." Retrieved from the World Wide Web on 1 August 2007. http://www.ec.gc.ca/soer]free/English/Framework/Nardesc/Region.cfm?region=99.
- 2007d. "State of the Environment Database: Algonquin-Lake Nipissing." Retrieved from the World Wide Web on 1 August 2007: http://www.ec.gc.ca/soerree/English/Framework/Nardesc/Region.cfm?region=98.
- 2007e. "State of the Environment Database. Great Canadian Rivers, 2007." Retrieved from the World Wide Web on 1 August 2007. http://www.greatcanadianrivers.com/rivers/ottawa/culture-home.html.

National Capital Commission. 2007a. "Planning Chronology." Retrieved from the World Wide Web on 28 October 2007. http://www.capcan.ca/bins/ncc_web_content_page.asp?cid=16300-20443-29365-23552&lang=1.
- 2007b. *Heritage and Capital Treasues*. Retrieved from the World Wide Web on 28 October 2007. http://www.capcan.ca/bins/ncc_web_content_page.asp?cid=16300-20450&lang=1.
Ottawa River Institute. 2007. Retrieved from the World Wide Web on 1 August 2007. http://www.ottawariverinstitute.ca/WatershedWayso4/wwWaters hed-map.htm.
Parks Canada. 2007. *Rideau Canal National Historic Site of Canada*. Retrieved from the World Wide Web on 28 October 2007. http://www.pc.gc.ca/lhnnhs/on/rideau/plan/plan3_e.asp.
Raffan, J. 2005. "The Valley." *Canadian Geographic* 125(5): 44–57.
Service Canada. 2007. Ontario Census Division Age Pyramids, Census 1996 and 2001. Retrieved from the World Wide Web on 1 August 2007. http://www1.servicecanada.gc.ca/en/on/lmi/eaid/ore/ceno1/cd/cd_population_census_1996_2001.shtml#3548.
Simmons, J., and L. Bourne. 2007. "Living with Population Growth and Decline." *Plan Canada* 47(2).
Statistics Canada. 2007. "Community Profiles 2001." Retrieved from the World Wide Web on 1 August 2007. http://www12.statcan.ca/english/Profilo1/CPO1/Index.cfm?Lang=E.
- 2007b. "Community Profiles 2006." Retrieved from the World Wide Web on 1 August 2007. http://www12.statcan.ca/english/censuso6/data/profiles/community/Index.cfm?Lang=E.

11

Learning from the Past, Assessing the Future: A Geo-historical Overview of Northeastern Ontario

RAYNALD HARVEY LEMELIN AND RHONDA KOSTER

SUMMARY

The development of Northeastern Ontario is often associated with extractive activities such as forestry and mining. Various other activities include railroad and hydroelectric initiatives, highway and other transportation projects, and agricultural settlements. Tourism and recreational developments have a rich history in this region, yet relatively little research has focused on these activities. Although various researchers have examined the role of First Nations and Euro-Canadian settlement in Northeastern Ontario, there is a limited understanding of the importance of this history in the new economy. The goal of this chapter is to examine how various economic ebbs and flows have affected Northeastern Ontario throughout the past three centuries. We also discuss how these activities have been related to various socio-cultural characteristics and economic opportunities in this region of Ontario.

IDENTIFYING THE GEOREGION

Often defined under the rubric of "New Ontario" and "Northern Ontario," in this chapter Northeastern Ontario includes the area north of the French River–Lake Nipissing and east of Lakes Superior and Huron (figure 11.1, table 11.1). It includes that part of the Kenora district north of 50°, Cochrane, Sudbury, the single-tier municipality of Greater Sudbury, Nipissing, and Timiskaming. Although

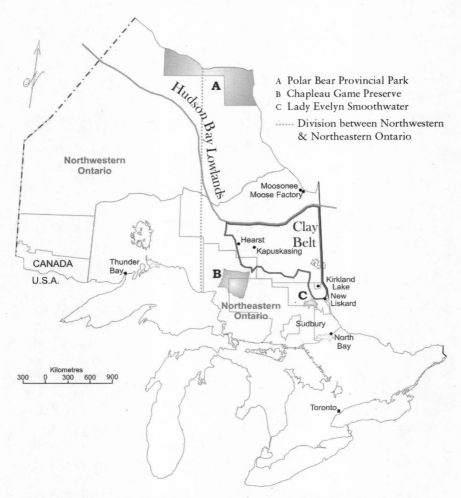

Figure 11.1 Map delimiting Northeastern Ontario. By the authors, with permission

they were considered part of Ontario's northern frontier during the nineteenth century, we do not consider the Algoma-Manitoulin, Muskoka, and Parry Sound Districts to be part of Northeastern Ontario. While some of the communities located along the Hudson and James Bay shores are geographically situated in the "western" part of Northern Ontario, these communities for historical and political reasons are considered as part of Northeastern Ontario in this chapter.

The term "New Ontario," which often has been used by historians to define all of late-nineteenth- and twentieth-century Northern Ontario, is somewhat misleading, since it fails to recognize the historic role of First Nations. It also downplays the exploration and early development of this region by the French and the English. Indeed, French explorers had been involved in the fur trade in Northeastern Ontario since 1610 (Choquette 1980). The English and the Hudson Bay Company established Moose Fort, or Moose Factory, on the shores of Moose River, close to James Bay in 1672–73. This fort was the first English settlement in what would later become Upper Canada and then Ontario (Marsh 2007).

There are similarities between Northeastern and Northwestern Ontario, including overlapping political jurisdictions, remoteness from major markets in southern Ontario, and overwhelming historical and current dependencies on primary-resource extraction industries (Bray and Epp 1984; Dunk 2003). However, there are characteristics of Northeastern Ontario that distinguish it from the Northwest: there is a distinct farming legacy in the region's Clay Belt (McDermott 1961), and proximity to the Quebec border has contributed to its high concentration of francophones, roughly one person out of four. In contrast, the francophone population of Northwestern Ontario is less than 2 percent of the population as a whole (Ontario Office of Francophone Affairs n.d.).

The Lands for Life process, later renamed the Living Legacy, which was initiated in the late 1990s by the government of Ontario, was in large part dedicated to increasing the number of protected areas and forest allocations in Northwestern and Northeastern Ontario. While the Northwest received a number of new protected areas and substantial increases in the existing ones, the Northeast acquired only modest additions to its protected-areas network, while receiving more forestry allocations. Nevertheless, Northeastern Ontario is renowned for its canoeing and other outdoor recreation opportunities.

Northeastern Ontario can be best described as a sparsely populated area with large tracts of wildlands. Most of the population is concentrated in major urban centres such as Sudbury and North Bay. The remainder is located in First Nations and smaller Euro-Canadian communities, the latter mostly dependent on primary-resource extractive industries such as mining and forestry (Dunk 1994, 2003; Southcott 2000, 2005).

Table 11.1
A Profile of Northeastern Ontario

Size	Descriptor	Total 360,000km²	Descriptive Features
Main landscape features	Boreal Shield	210,000 km²	
	Hudson Bay Lowlands	15,000 km²	
Population	Total	228,689	
	Urban	181,811	
	Rural	46,878	
	First Nation	2,704	
Age (average)	Euro-Canadian	39.6	
	First Nation	22.6	
Economic activities (numbers employed)	Agriculture and resource-based Industries	9,385	
	Manufacturing and construction	18,215	
	Wholesale and retail	16,875	
	Finance and real estate	3,940	
	Health and education	20,270	
	Business services	15,540	
	Other services	21,870	
Protected Areas	Polar Bear Provincial Park	24,087 km²	Largest provincial park in Ontario, includes a Ramsar Site (2,408,700 ha), established in 1987
	Chapleau Game Preserve	7,000 km²	Large game preserve in the world, established 1925
	Far North Planning Initiative (proposed)	225,000 km²	Interconnected series of protected areas proposed for 2009

Source: By the authors, with permission.

Although much has been written about the development of Northeastern Ontario and the associated trapping, forestry, mining, and other extractive activities, as well as transport and hydroelectric development, agricultural settlement, and the new economy based on knowledge and technology (Bray and Epp 1984; Southcott 2000, 2005), very little has been written about tourism and recreational initiatives. Existing research often highlights the barriers and obstacles to tourism (Johnston and Payne 2004), even though numerous opportunities for recreation and tourism exist in the region, including historic canoe routes and waterways such as the Albany and the Moose and Canadian Heritage Rivers such as the French River, the Mattawa River Provincial Park, and the Missinaibi Waterway Park.

Insufficient recognition is given to provincial parks such as Killarney, Lady-Evelyn Smoothwater, and Polar Bear, as well as other large protected areas, including the world's largest game preserve, Chapleau Game Preserve, in terms of their importance for recreational and tourism activities. Some parks and protected areas support tourism initiatives, but very little monitoring or research is conducted on the socio-economic returns. Some First Nations may benefit from proximity to these protected areas (for example, the Weenusk and Peawanuck First Nations offer locally guided tours into Polar Bear Provincial Park), but few First Nations have developed recreational and tourism strategies with Ontario Parks. Indeed, apprehension about the establishment of protected areas in Northeastern Ontario has resulted in suggestions that some provincial parks be removed from contested territories, notably where unresolved land claims exist.

More recent initiatives aimed at promoting tourism include snowmobile and all-terrain vehicle (ATV) trails, cultural tours, and the Champlain Canoe Circuit. We provide further details on some of these initiatives later. This chapter begins with a description of the physical landscape, which sets the stage for both extractive industries and tourism and recreation. The human development of this landscape reflects the interesting history and current economic realities of present-day Northeastern Ontario.

THE NATURE OF NORTHEASTERN ONTARIO

The varied physical landscape of Northeastern Ontario includes the "granitic" Canadian Shield and the Hudson Bay Lowlands. The

Shield area is dominated by a glaciated landscape of bare Precambrian rocks and numerous lakes. Although the landscape is rugged, the altitude is not great, varying between 150m and 300m above sea level. In contrast, the Hudson Bay Lowlands form a 150–300km-wide belt of flat, low-lying land adjacent to the coast. At the end of the last Ice Age these lowlands were flooded as the bay advanced over land depressed by the weight of the ice. Subsequent isostatic rebound exposed the previously deposited marine sediments, forming the extensive mudflats and marshes characteristic of the lowlands today.

The process of glaciation effectively removed much of the historical soil, leaving some relatively fertile deposits such as glacial till and gravel. The exceptions include soils that developed on thicker, more fertile deposits laid down in past large postglacial lakes created as the ice retreated. These soils comprise the clay lands that make up much of Southeastern and Northeastern Ontario (see box 11.1).

BOX 11.1   SETTLEMENT AND AGRICULTURE IN NORTHEASTERN ONTARIO'S GREAT CLAY BELT

The Great Clay Belt spans a region in Northeastern Ontario and Northwestern Quebec. On the Ontario side, it comprises almost the entire municipal region of Cochrane. The gray clay, laid down by a temporary glacial lake, is estimated to cover 6.47497 million hectares in Northeastern Ontario. The first settlements, located along the Timiskaming and Northern Ontario Railroad near Cochrane, were opened in 1910. The farming population peaked in 1941 and then continued to decline, in terms both of land size and of the total number of farms. McDermott (1961) assessed the reasons for the decline of farming to include the following:

- The settlers' unfamiliarity with the conditions of the Clay Belt, as well as the techniques of their newly adopted means of livelihood. These problems caused discouragement and ultimate abandonment. Many of the settlers to the region were former tradesmen from southern Ontario cities.
- Low financial returns from farming in comparison to other occupations. For example, the estimated average farm income in Cochrane for 1955 was $1,160, compared with an estimated average of $4,000 for forestry workers (McDermott 1961, 270)

> Despite the decline in farming, the Cochrane District continues to have the largest number of people engaged in agriculturally related industries in Northeastern Ontario. (Statistics Canada, 2002).
>
> By the authors, with permission.

Another important challenge is presented by permafrost. The southern limit of discontinuous permafrost in Northeastern Ontario follows a line from roughly west of the southern end of James Bay to the Manitoba border and comprises much of the region under study. At least a portion of the soils in this region remain frozen year round. Drainage is poor, and there is little potential for forest or agriculture.

Vegetation varies according to local conditions of climate, soil, and drainage, and species diversity decreases from south to north (figure 11.2). The boreal forest, which is the predominant vegetation type in the more southerly part of Northeastern Ontario, is characterized by a fairly dense cover of spruce, pine, and other needleleaf conifers mixed with birch, poplar, and other broadleaf deciduous species. Toward the north and in the angle between James Bay and Hudson Bay, the forest is more open, with stands of black spruce and tamarack separated by large expanses of muskeg or peat bogs. In the far northern reaches, a short growing season, poor drainage, and permafrost result in tundra vegetation of shrubby birch, willow, sedge, and low-growing plants such as blueberries.

Northeastern Ontario has a continental climate characterized by hot summers, cold winters, and limited precipitation, which falls mostly during the summer months. This weather pattern is relatively consistent throughout the region, since both Lakes Superior and Huron and James and Hudson Bays cause limited modifications locally.

Ungulates such as white-tailed deer, moose, and woodland caribou are distributed according to their habitat preferences; the caribou are mostly found in open woodland and lowland tundra, where lichens, their main food source, are abundant. In contrast, moose and white-tailed deer are found in the more southerly parts of the region, in part because commercial logging has provided an opportunity for

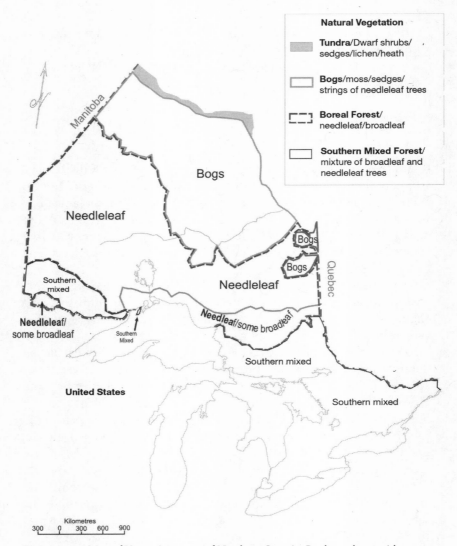

Figure 11.2 Map of Vegetation zones of Northern Ontario. By the authors, with permission

browse such as birch, willow, and poplar to flourish. Black bears are found throughout central, southern, and Northeastern Ontario. The world's most southerly population of polar bears is located within southern Hudson and James Bays. The population periodically uses the strip of tundra along the coast of Hudson Bay for staging areas, and females go further inland to den.

Beaver, otter, fox, and other fur-bearing mammals are of continued economic significance for First Nation hunters and trappers. However, it is the annual spring and fall hunt of Canada geese and other migratory waterfowl species that is of particular socio-cultural as well as subsistence importance to First Nations, especially Cree communities located along the James and Hudson Bay shores. During the spring hunt entire family groups gather in camps to participate in the harvest and benefit from this important food source.

According to *Ontario Nature* (2007), "the boreal forest is the single most important breeding ground for birds in North America. Approximately three hundred species and two billion individual birds breed there before migrating south. Lying at the heart of Canada's boreal region, Ontario's boreal forests form an enormous part of this songbird nursery." Changes brought about through human activities such as mining, forestry, hydroelectric development, and recreational projects can affect these animals, and great effects may arise from climate change.

Northeastern Ontario has been said to be especially vulnerable to climate change. As Thompson et al. (1999) explain, climate change could bring two broad effects to the forest landscapes of this region: changes in the fire regime and in the diversity and extent of forests. Thompson et al. (1999) suggest that the Northwestern and Northeastern regions of Ontario may experience minimal increases in fires or even remain the same or decline. The resulting forest landscape could therefore increase in diversity with a slow invasion from southern deciduous species. This is especially troubling for forest-management strategies in Northeastern Ontario, where forest fires traditionally have longer return cycles than in forests in Northwestern Ontario, and the implications for caribou habitat are not clear. The lack of guidelines and policies for woodland caribou management in Northeastern Ontario is discussed further in Brown et al. (2007) and Vors et al. (2007).

In their climate-change analysis of parks and protected areas in Ontario, Lemieux et al. (2004) found that the greatest increases in temperature and precipitation are predicted to occur in Tidewater Provincial Park near Moosonee and Moose Factory on the Moose River. In addition, Polar Bear Provincial Park on the south shore of Hudson Bay is expected to change biome type and therefore to no longer provide suitable denning habitat for polar bears (Lemieux et al. 2004).

HUMAN SETTLEMENT

The story of Northeastern Ontario "involves a complex and continuing interaction between the broader political and economic forces that have influenced development throughout Ontario, and distinctive local circumstances, both cultural and geographic" (Hodgins and Benidickson 1989, 6). Various Amerindian cultures put their early stamp on some of the most notable landscapes in the area, including the Algonquin name of *Temikami* and *Temikaming*, meaning "deep waters" for Lake Temiscaming, and the Eastern Cree name of *Winipekw* for the muddy waters of the Hudson and James Bays.

The earliest evidence of human habitation in Northeastern Ontario, notably in the Temagami country, dates from about 3000 BC (Hodgins and Benidickson 1989). The archaeological findings differ substantially from oral history, which indicates that the Amerindians were in Northeastern Ontario from nine thousand to fifteen thousand years ago. These early people relied on the hunting and fishing of deer, caribou, and other animals and the gathering of edible plants. The seasonal availability of these resources contributed to the nomadic existence of the bands. They travelled to their traditional lakes and rivers to fish in the warmer months of the year and inland to hunt for game in winter (Manore 1999). These nomadic patterns laid the way for numerous First Nations people later to pursue careers as guides for early explorers and as voyageurs for the fur companies.

In the seventeenth and eighteenth centuries, Amerindians, Europeans, Métis, and Euro-Canadians seeking fishing and hunting grounds and fur trade routes established a number of villages, forts, and transportation routes along northern waterways. Most of these routes, first traversed by canoes and sloops, provided the only mode of transportation until the nineteenth century, when some were replaced by steamships. These "waterways" were later used to move logs, while also transporting early geologists, tourists, and in some cases early colonists, for example, at Timiskaming Shores. As stated in chapter 2, the quest for various ore deposits such as nickel in Sudbury (1883), silver in Cobalt (1904), and gold in Timmins/South Porcupine (1909) led to the development of rail lines and other infrastructure for mining and the development of new settlements in Northern Ontario.

Growing North American demands for newsprint and pulp in the early to mid-twentieth century encouraged the development of manufacturing and wood-processing plants in Northeastern Ontario, while also providing the opportunity for the construction of dams. Some of the most impressive hydroelectric developments were the Moose River basin projects, completed in the late 1960s. Ontario Hydro and private developers built fifteen stations with a generating capacity greater than two megawatts on the Mattagami and Abitibi rivers (Anon. 1992).

From a First Nation's perspective, activities such as mining, lumbering, hydroelectric development, and the quest for agricultural land disrupted traditional lifestyles much more than the post-contact fur trade (Manore 1999). In fact, complaints by representatives of the Cree and Anishabee nations resulted in the negotiation and subsequent surrender of aboriginal land rights in various treaties including the Robinson-Superior Treaty (1854) and the Robinson-Huron Treaty (1850–54), the former covering the lands fronting on Lake Superior and the latter the lands fronting on Lake Huron, including northern and eastern Georgian Bay. Treaty 9, initially signed in 1905–6 and again in 1929–30, encompasses one of Ontario's largest treaty areas extending east to west from Quebec to the Manitoba border and from the coastline of James and Hudson Bays south beyond the watershed draining into the Great Lakes.

Assembled chiefs and headmen were apparently assured that lands involved in mining and forestry, the resources of greater concern, would remain in the hands of the Crown, that government interest would focus on lands suitable for European-style agriculture, and that Treaty No. 9 would allow traditional activities such as hunting and fishing to continue unimpeded. Restrictions on these activities would apply only on lands sold or leased to individuals or companies (Hodgins and Benidickson 1989). Omitted from these discussions was the value of these areas for conservation, recreation, and tourism. These "omissions" would later prove contentious, especially in the establishment of protected areas. Today, Treaty No. 9 is administered by the Nishnawbe Aski Nation (NAN), while First Nations located along the Ontario coastline of James Bay are represented by the Mushkegowuk Council of Chiefs (MCC). First Nations in the Robinson-Huron and Robinson-Superior Treaty areas are administered by the Union of Ontario Indians, while others are considered independent. More information about First Nations is provided later in this chapter.

Euro-Canadians of British descent tended to disperse throughout Northeastern Ontario, while ethnic groups such as Finns, Italians, and Germans tended to congregate in certain parts of the region. For French Canadians, it was the timber industry that first provided the impetus to pursue new employment opportunities and lucrative salaries in Northeastern Ontario (Choquette 1980). The construction of the Canadian Pacific Railway and other lines in the 1880s provided access to the forests and later the minerals of the Sudbury basin as well as agricultural areas in the Clay Belt. To this day, Northeastern Ontario continues to have a strong francophone presence.

The economy of Northeastern Ontario has centred on the primary resource-extraction industries of mining and forestry. As a result, the region has been subject to the "boom and bust" cycles associated with such industries. The 1950s to 1980s brought boom times with population growth, high employment rates, high wages, and expanding communities. By the mid–1980s, commodity prices and global competition began to have a negative impact, first on mining and later on forestry-based communities. Since the 1990s, declines in resource extraction have resulted in the transformation of certain communities. Some cities, such as North Bay and Sudbury, have benefited from economic diversification, while other smaller communities such as Smooth Rock Falls continue to struggle. In the early years of the twenty-first century the region appeared to be experiencing another "boom" phase.

THE NORTHEASTERN ONTARIO POPULATION

As noted earlier, despite its geographic extent, Northeastern Ontario is sparsely populated. Cochrane (85,247), Nippissing (82,910), Sudbury (22,894), Greater Sudbury (155,268), and Timiskaming (34,442) comprise the census districts of the region (Statistics Canada 2002). Within this large area, only three cities, North Bay, Sudbury, and Timmins have a population of more than 40,000 people. All the other towns and cities have individual populations of less than 12,000. The region contains several First Nations communities, including, from west to east, Fort Severn, Weenusk, and Moose Factory; and from north to south, Muskrat Dam, Webequie, Temagami, and Nipissing. Some communities, like Moose Factory, are

Table 11.2
Population and Population Change for Selected Communities in Northeastern Ontario, 2001

	Population in 2001	Population Change, 1996–2001 (%)
Province of Ontario	11,410,046	6.1
DISTRICT		
Cochrane District	85,247	–8.6
Nipissing District	82,910	–2.3
Sudbury District	22,894	–3.9
Timiskaming District	34,442	–8.9
CITY/TOWN		
Chapleau	2,832	–3.5
Cochrane	5,690	–4.5
Elliot Lake	11,956	–12.0
Kapuskasing	9,238	–8.0
Kirkland Lake	8,616	–13.0
Moosonee	936	–51.7
New Liskeard	4,906	–4.0
North Bay	52,771	–2.9
Subury	85,354	–7.3
Timmins	43,686	–8.0
FIRST NATIONS RESERVE		
Attawapiskat 91A	1,293	2.8
Fort Severn 89	401	10.8
Peawanuck	193	–19.2
Muskrat Dam	217	N/A
Webequie	600	35.4

Source: By the authors, with permission. Data from Statistics Canada.

larger, with 2,458 in total population. Others, like Peawanuck, are quite small, with a population of only 193 (see table 11.2 for further details). However, a note of caution is offered regarding census figures and statistics for First Nations communities, since they tend to be less reliable than those for Euro-Canadian communities.

While the province of Ontario experienced a population increase of 6.1 percent between 1996 and 2001, all districts and cities of Northeastern Ontario, with the exception of those of the First Nations, experienced population declines ranging from –2.3 percent for the Nippissing District to –51.7 percent for the town of Moosonee (table 11.2). The dwindling population rates are related to the economic challenges the region has experienced in the last decade, a topic addressed later in the chapter.

Two unique features of Northeastern Ontario are the predominance of Anishnawbee, Cree, Oji-Cree, and Francophones. In 2001 Ontario had 88,315 Aboriginal people accounting for less than 2 percent of the total population. Northeastern Ontario is unique in terms of its comparatively large Aboriginal population, which makes up almost 8 percent of the population of the region (Southcott 2002). Indeed, the population in the area north of the 50th parallel is almost entirely made up of First Nations (Southcott 2002).

As Tables 11.3 and 11.4 illustrate, the population, education and economy north of 50° differ considerably from that of both the province and the Northeastern region. The age structure reflects a much younger and growing population – importantly, First Nations have less out-migration of youth than does the rest of the province or the Northeastern region. Education levels remain much lower compared with the region and the province. Whereas the provincial average for French as a first language is approximately 4 percent, the Northeastern region has some districts where over half the population speaks French, for example, the Cochrane district. This is not surprising given the geographic proximity to Quebec and the settlement history previously discussed.

As Southcott's (2006) examination of the region has indicated, income levels for the First Nations population in the region are approximately 10 percent lower than the provincial average (see table 11.3). The labour force participation rates and employment rates for First Nations are considerably lower than for the province as a whole. In some instances the rate of unemployment exceeds that of employment (e.g., Webequie). In every case the rate of unemployment is at least 20 percent higher than of the provincial average. In addition, the percentage of income made up by government transfers is extremely high, ranging from 20 to 40 percent. Much of the employment centres on health and education and service industries. However, recent agreements with mining companies may provide employment opportunities previously unavailable to the First Nations. These are discussed in the next section.

NORTHEASTERN ONTARIO ECONOMY

Northern Ontario, including the Northeastern and Northwestern regions, was an important resource hinterland for Ontario, initially through the fur trade and later, through mining and forestry. Indeed

Table 11.3
Selected Demographic Characteristics for Ontario and First Nations in Northeastern Ontario, 2001

	Province of Ontario	Attawapiskat 91A	Fort Severn 89	Peawanuck	Webequie
AGE STRUCTURE					
Median age of population	37.2	19.5	21.1	26.6	23.3
Percentage of population aged 15+	80.4	59.8	62.5	73.7	69.2
ABORIGINAL POPULATION					
Aboriginal identity population	188,315	1,265	400	195	600
Non-Aboriginal population	11,097,235	25	0	0	0
EDUCATION CHARACTERISTICS					
Total population 15 years and over attending school full-time	1,060,115	15	10	10	35
Highest level of schooling – total population aged 20–34	2,263,910	285	90	45.0	160
Percentage with less than high school certificate	13.2	71.9	55.6	55.6	71.9
Percentage with post-secondary certificate (includes college and university)	86.8	26.3	38.9	22.2	0
Highest level of schooling – total population aged 35–44	1,949,840	120	40.0	30.0	70
Percentage with less than high school certificate	17.3	70.8	50.0	50.0	85.7
Percentage with post secondary certificate (includes college and university)	82.6	25.0	50.0	66.6	0
Highest level of schooling – total population aged 45–64	2,684,705	165	60.0	25.0	80
Percentage with less than high school certificate	27.5	90.9	50.0	100.0	93.8
Percentage with post secondary certificate (includes college and university)	72.6	12.2	25.0	0	0

Note: Information is not available from Statistics Canada for Moose Factory or Muskrat Dam.
Source: By the authors, with permission.

Table 11.4
Selected Labour Statistics for First Nations in Northeastern Ontario, 2001

Earnings and Income	Province of Ontario	Attawapiskat 91A	Fort Severn 89	Peawanuck	Webequie
Average earnings, working full time, full year	39,141	28,983	29,122	N/A	27,737
Earnings as percentage of income	68.9	62.3	78.2	N/A	59.3
Government transfers (%)	18.1	36.5	20.8	N/A	40.0
Other money (%)	13.0	1.2	1.0	N/A	0.4
LABOUR FORCE INDICATORS – PLACE OF WORK					
Home	406,230	10	0	0	10
No fixed workplace address	466,950	25	10	10	10
Worked at usual place	44,806,790	225	105	40	95
LABOUR FORCE INDICATORS – RATES OF EMPLOYMENT					
Employment rate (%)	63.2	34.0	44.0	35.7	28.9
Unemployment rate (%)	6.1	31.1	25.0	33.3	35.1
LABOUR FORCE INDICATORS – INDUSTRY					
Agriculture and resource industry	191,020	25	10	0	0
Manufacturing and construction	1,319,580	30	10	0	20
Wholesale and retail	950,730	45	15	0	25
Finance and real estate	401,445	0	0	0	0
Health and Education	902,990	80	25	15	35
Business services	1,145,910	20	10	10	20
Other services	1,084,090	120	70	25	65

Source: By the authors, with permission. Data from Statistics Canada.

as Southcott (2006) points out, the residents of Northeastern Ontario saw themselves as contributing to the economic growth of the province and were concerned that the wealth generated in this region more significantly benefited southern Ontario.

These concerns were mitigated somewhat by "boom cycles" and the associated high economic prosperity. But by the late 1980s a "bust cycle" began, and residents became concerned about loss of jobs in the primary sectors. Between 1990 and 2004 Ontario communities in the Northeast experienced an average job loss rate of about 1 percent (Di Matteo 2005). What this average belies, however, is the variation experienced throughout the region. For example, while employment rates declined by almost 7 percent in Sault Ste Marie during this period, North Bay and Sudbury experienced a growth range of between 6 and 10 percent. This growth is explained in part by increases in mineral exploration and economic diversification.

Both the mining and the forestry industries have increasingly used labour-saving technologies to reduce production costs. In some cases entire mills, such as Smooth Rock Falls and Cochrane, were closed. Mines such as Kirkland Lake also were shut down as resources were expended. Randall and Ironside (1996) point out that Northeastern Ontario is highly vulnerable to resource depletion, as well as to changes in commodity prices, corporate and government policies, and Canadian exchange rates. Indeed, most of the head offices for the corporations operating within the north are located elsewhere in the country or internationally. Two perceived repercussions are that corporations have little regard for the implications of their decisions on communities and that reliance on large corporations results in relatively low levels of locally owned businesses and a limited entrepreneurial culture. Another problematic feature of Northeastern Ontario is that it is made up of districts rather than regional governments and therefore that unlike other regions of Ontario it lacks intermediaries between the provincial government and local municipalities (Southcott 2006).

As Southcott (2006) indicates, studies done before 2001 for the whole of northern Ontario are limited, but statistical data provide some indication of socio-economic trends for the region as a whole. Broadly these are

- Slow population growth for the region (less than 5 percent between 1991 and 1996) and increased youth out-migration

- A population that is aging more rapidly than that of the rest of the province.
- A significant increase in the number of self-employed workers since the 1980s, although the amount is still lower than the average for the province as a whole.
- Traditionally lower levels of education, a trend that has continued.

An evaluation of the 2001 census data indicates that approximately 40 percent of the jobs in Northeastern Ontario are within the primary-resource economy, while the remainder are in service-based industries (table 11.5). Despite growth in knowledge-based industry as a whole for the province, Northeastern Ontario remains largely dependent on resource-extractive industries, and consequently a large portion of the workforce is blue-collar. Advances in technology have streamlined these industries and resulted in significant job loss. It would appear that although most communities are looking for diversification, for example in the form of a more varied knowledge base that enhances the primary industries of forestry and mining, some communities are also considering tourism and value-added forestry opportunities for economic diversification, success, and longevity.

Tables 11.5 and 11.6 indicate the state of income, employment, industrial structure, and education, based on the 2001 census, for the Northeastern region, which continues to have a higher number of primary resource industry occupations, and a lower number of manufacturing positions (table 11.5). As table 11.6 illustrates, relative to the province as a whole, the region has similar or higher earnings for full-time workers, with less reliance on government transfers, with the exception of Kirkland Lake and Elliot Lake. Northeastern Ontario has higher rates of unemployment than the provincial average, as well as that of Northern Ontario as a whole. Education levels in Northeastern Ontario remain lower than the provincial average, with lower percentages of university or college degrees and a higher percentage of trades' certification.

FORESTRY

Many factors, including the rising Canadian dollar, the softwood lumber dispute with the United States, high energy prices, overcapacity, and wood supply issues, have been cited as contributing to

Table 11.5
Employment Indicators for Selected Industries for Selected Communities in Northeastern Ontario, 2001

	Agriculture and Other Resource-Based Industries	Manufacturing and Construction Industries	Wholesale and Retail Trade	Finance and Real Estate	Health and Education	Business Services	Other Services
Province of Ontario	191,020	1,316,580	950,730	401,445	902,990	1,145,910	1,084,090
DISTRICT							
Cochrane District	5,205	7,440	6,430	1,325	7,590	5,340	7,350
Nipissing District	1,525	5,845	6,560	1,815	8,025	6,155	8,920
Sudbury District	905	2,305	1,430	295	1,460	1,545	2,400
Timiskaming District	1,715	2,585	2,395	505	3,090	2,460	2,985
CITY/TOWN							
Chapleau	140	310	215	20	225	275	260
Cochrane	175	555	455	65	485	510	535
Elliot Lake	270	305	660	145	835	590	870
Kapuskasing	340	1,115	780	125	880	470	700
Kirkland Lake	415	345	590	185	920	385	860
Moosonee	x	x	x	x	x	x	x
New Liskeard	130	325	510	70	485	390	460
North Bay	570	3,025	4,675	1,350	5,525	4,310	5,905
Subury	1,975	4,555	6,870	1,805	8,680	7,630	9,725
Timmins	3,315	2,715	3,675	730	3,870	2,995	3,900

Source: By the authors, with permission. Data from Statistics Canada.
Note: Figures for Moosonee are not available owing to Statistics Canada's guidelines for suppressing data for communities with a small population, to ensure anonymity.

Table 11.6
Rates of Income and Employment for Selected Communities in Northeastern Ontario, 2001

	Income in 2000			Rates of Employment	
	Average Earnings (worked full year, full time) ($)	Earnings as a Percentage of Income	Government Transfers as a Percentage of Income	Employment Rate	Unemployment Rate
Province of Ontario	39,141	68.9	18.1	63.2	6.1
DISTRICT					
Cochrane District	43,990	76.3	14.2	55.2	11.5
Nipissing District	39,445	70.1	15.9	54.5	9.1
Sudbury District	40,533	71.2	16.1	49.8	12.5
Timiskaming District	39,141	68.9	18.1	52.7	10.0
CITY/TOWN					
Chapleau	43,317	79.7	11.4	61.6	8.8
Cochrane	46,993	75.5	13.1	56.8	9.4
Elliot Lake	37,638	48.4	29.2	33.1	13.0
Kapuskasing	44,755	74.9	13.9	54.3	9.9
Kirkland Lake	41,552	66.7	20.5	48.5	13.5
Moosonee	x	x	x	x	x
New Liskeard	42,179	71.8	13.3	56.9	6.0
North Bay	40,700	70.5	14.6	56.3	8.3
Subury	43,749	70.1	14.5	54.7	9.5
Timmins	44,079	76.9	14.0	56.7	11.2

Source: By the authors, with permission. Data from Statistics Canada.
Note: Figures for Moosonee are not available owing to Statistics Canada's guidelines for suppressing data for communities with a small population to ensure anonymity.

Canada's and Northeastern Ontario's ailing forestry industry. Other factors, including the consolidation of the forest companies through globalization, mechanization, and clear-cutting practices are also cited (Forest Ethics 2007).

Government aid packages have provided temporary, short-term relief, but what is required according to many leading industry experts and advocacy groups is a long-term vision and diversification of the forest industry (see the Northern Ontario Sustainable Communities Partnership 2007 for further details). This vision would include better management of existing forest resources, as can be seen in the Lac Abitibi Model Forest, near Cochrane; investment in bio-technologies, for example, biomass and waste-heat recovery systems in Hearst; value-added products, including the use of black ash and birch for timber; the further development of non-timber forest products such as tree sap, mushrooms, berries, nuts, and medicines; and economic reimbursement for the value of the boreal forest as a mitigator of climate change, or as a carbon sink.

MINING

Because of low commodity prices during the 1980s and early 1990s, mining did not fare well, and limited mining and exploration occurred. Since the mid–1990s increasing commodity prices and improved technologies have until recently fuelled exploration and land acquisition by smaller mining companies, especially in the Sudbury region (Louiseize 2006a, 2007; Ross 2007; Ulrichse 2009); this despite new environmental laws and regulations such as the modifications to the Mine Development and Closure Act, part VII, in 2003, modifications requiring mining companies to rehabilitate mining sites.

Part of the diversification associated with the mining industry includes an increasing connection between post-secondary training, research, and business. For example, a partnership has been created between Laurentian University and Materials and Manufacturing Ontario to research and build mobile mining equipment (Ross 2002). In addition, Laurentian University is in the process of developing a Centre of Excellence in Mining Innovation, the focus of which will be mineral exploration, mining telerobotics, mine process engineering, deep mining, and environmental reclamation (Ross 2007). This initiative has been financially supported by Xstrata

Nickel and CVRD-Inco, the province of Ontario, and the City of Greater Sudbury.

Mining companies are also now required to consult with First Nations on whose traditional lands they are planning to explore. The Webequie First Nation has developed a First Nations Land Policy that sets out guidelines for development and requests that mining companies use town amenities like hotels and stores (Louiseize 2006b). This policy aids the First Nation in protecting fishing, hunting, and harvesting areas, as well as important cultural sites such as sacred burial grounds. First Nations are also able to benefit more fully from the development of mines within their region. As part of the Benefit Impact Agreement (which includes revenue sharing and training partnerships) signed between De Beers and the five First Nations in the James and Hudson Bay area, the James Bay Employment Training Committee was developed and will help to administer revenue sharing and training partnerships (Larmour 2006). More recently there have been changes to the mining act in Ontario. However, it is too early to determine the repercussions of these legislative changes for mining practices in Ontario.

Challenges to mining in the province include the regulatory environment and the recent Diamond Royalty Tax (see box 11.2). In addition, it is expected that there will be a shortage of skilled workers in mining and associated fields. As a result, First Nations communities are gaining training and skills experience through donations, for example, through a $3 million grant from the Aboriginal Human Resources Development Council of Canada (Louiseize 2007).

BOX 11.2 DIAMOND MINING IN NORTHEASTERN ONTARIO

The diamond mining potential of Northeastern Ontario is significant, even though the initial exploration of the region determined it to be low. The discovery of kimberlites (a host mineral) near Timiskaming with 15 carats per 100 tonnes illustrated that the potential was far greater than initially thought (Louiseize 2006a). The kimberlites form in small clusters and follow an imaginary line from Attawapiskat, down through Kirkland Lake and Cobalt in Northeastern Ontario. Since 1987 De Beers has spent $130 million in exploration and evaluation at the Victor Mine site, near Attawapiskat. The company committed.

> $982 million and began construction in 2006. Construction has created approximately 600 jobs, while production will provide 400 additional jobs in the open pit, processing plant, workshops, and warehouses (Ross 2007). A variety of joint ventures are planned with First Nations communities and businesses in Northeastern Ontario (Ulrichsen 2007).
>
> By the authors, with permission

TOURISM

The Ontario Ministry of Tourism divides the province into numerous destination areas, several of which are located in Northeastern Ontario: Sudbury, Algoma, Cochrane, Marten Falls, Fort Albany, Peawanuck, Attawapiskat, and Polar Bear Provincial Park. Statistics for visitation are difficult to come by at a regional level. However, the Ministry of Tourism (2007, 2006a,b,c,d,e) has compiled a number of reports on patterns of visitation to Ontario that have been further analyzed at regional levels. The most recent available data illustrate that a total of 3.7 million Canadians visited northern Ontario as a whole in 2004–5 and that these travellers consisted of 35 percent of all travellers to Ontario. However, the Northeastern region of the province receives only approximately 15 percent of Ontario's total visitation. American visitors make up the bulk of total visitation for outdoor and pleasure tourism purposes, while Canadian and Ontario visitors comprise the bulk of those visiting for cultural, business, and convention purposes.

Traditional activities such as local and non-local fishing and hunting on- and off-road continue to be popular activities in Northeastern Ontario. Many of these traditional activities, along with new tourism and recreational initiatives, are often sought as a source of economic diversification for rural and small urban centres when downturns occur in their economic cycle. This has certainly been the case in Northeastern Ontario for Elliot Lake, a former uranium mining town between Sudbury and Sault St Marie, which transformed itself into a retirement community, perhaps best exemplifying this strategy.

Other notable examples of successful tourism attractions include

- The Polar Bear Conservation and Education Habitat in Cochrane, which was awarded the 2005 Innovation Award from the Tourism Federation of Ontario and the 2006 Tourism Industry Association of Canada award for Business of the Year
- Science North in Sudbury, which has won numerous awards, including the 2002 award by Attractions Canada for Best Indoor Site.

Other, more specific projects include viewing the rare and recently protected white moose in the Foleyet area and bear-viewing in the Chapleau Game Preserve. Thus it appears that the tourism industry, supported by various local entrepreneurs and various levels of government, is gaining momentum in the Northeastern region (Ross 2006).

Given its long winter season, Northeastern Ontario is one of the most popular destinations for recreational snowmobiling. An estimated 364,200 snowmobiles in the province of Ontario travel thousands of kilometres of groomed trails and established trail loops. An example is the Arctic Edge Snowmobile Tour to James Bay (Wawa to Hearst to Moosonee, five-day tours, spanning over a thousand kilometres). The Ontario Federation of Snowmobile Clubs (OFSC) further estimates that the annual economic activity generated by the snowmobile trails is valued at approximately one billion dollars in Ontario (DeMarco et al. 2002).

However, there growing concerns over the vulnerability of recreation to climate change, which has resulted in declining precipitation and deteriorating trail quality (Scott and Jones 2006; Scott et al. 2006). Studies have determined that both downhill and cross-country skiing and the snowmobile industry are particularly at risk, snowmobiling even more so because it relies on naturally occurring snowfall. In the 2005–6 season, snowmobile trails remained closed until late January in most of Northeastern Ontario because of warm weather and lack of snow. Northeastern Ontario currently has the longest number of snowmobiling days in the province (one hundred days). Projections indicate that under climate change, the number of days could be reduced by twenty-five to forty-five days by 2020 and by as much as eighty-seven days by 2050 and that more southerly regions will likely experience the greater loss of days more quickly.

So snowmobile tourism is vulnerable and may not be sustainable or economically viable in the long term. An alternative may be the use of all terrain vehicles (ATVs) instead of snowmobiles, since they are suited to the snowmobile trails and the terrain and are not limited by the absence of snow. Guided tours for ATV enthusiasts are now available in Cochrane, Dubreuilville, and Elliot Lake. While ATV tours are somewhat modest compared to snowmobile tours, the potential growth of this activity in Northeastern Ontario has to be considered. However, the potential repercussions of the use of ATVs on Crown land, will require further research and possibly significant changes to existing laws and management plans, since these vehicles are currently not allowed to use snowmobile trails, nor are they permitted in most protected areas in Ontario.

ABORIGINAL TOURISM

Tourism initiatives such as the Eagles Earth near Hearst, Ontario, the Eco-Lodge in Moose Factory, the Temagami Anishnabai Tipi Camp on Bear Island near Temaugamee, and guided polar bear expeditions in Polar Bear Provincial Park illustrate a cultural renaissance of sorts for First Nations in Northeastern Ontario. What is even more important is that these activities promote cultural and natural heritage and are owned and operated by local entrepreneurs.

A great example highlighting the adage "that the old is new again" is the Mattawa-Bonfield Economic Development Corporation and its Voyageur Days. Other events include the Circuit Champlain, which retraces Samuel de Champlain's exploration of the French and Mattawa Rivers, Georgian Bay, and Lake Nipissing in Northeastern Ontario. Another is Le grand portage du nord, which promotes canoe expeditions throughout Northeastern Ontario, with a particular focus on Moosonee and Moose Factory. Retracing the routes of early Amerindian and European explorers, these tourism initiatives are aimed at untapped tourism markets in Quebec and France, while capitalizing on the Francophone heritage of Northeastern Ontario.

DISCUSSION

From the fur trade in Moose Factory to the mining industry in Sudbury and Timmins; from forestry in Kapuskasing to the agricultural legacy on the Timiskaming shores, Northeastern Ontario is an area rich in natural and cultural heritage. It is also home to well-known rivers such as the Albany and the Missinaibi, numerous protected areas, including Chapleau, the world's largest game preserve, and Ontario's largest protected area, Polar Bear Provincial Park. Northeastern Ontario also offers habitat for unique wildlife including the white Moose of Folyet, woodland caribou, and the world's most southerly population of polar bears. Northeastern Ontario is also one of the areas in Canada that is likely to be significantly affected by climate changes. Indeed, climate change presents a threat to all northern ecosystems, the magnitude of which is not yet well understood (Scott et al. 2006).

World energy shortages are increasing pressure for hydroelectric generation across the Northeast. Mineral exploration and development are a priority in all regions, and timber harvesting may expand significantly beyond the 51st parallel, where no logging is currently permitted. Compounding these changes are the potential linkages between them. For example, how will climate change affect current forest fire succession rates in Northeastern Ontario? Will the melting permafrost in the Hudson Bay Lowlands affect the breeding patterns of geese in the area or the staging grounds of polar bears? How will agricultural possibilities increase in the Clay Belt through warming?

While these changes are complex and uncertain, they do not leave the people of the Northeast without recourse. Indeed, if any lesson can be gleaned from this chapter, it is that Northerners are adaptable and resilient. Perhaps we can learn from the original settlers of Northeastern Ontario, the First Nations, who cautioned and opposed unrestrained development projects such as hydroelectric developments along the Mattagami River and the Moose River. Their perseverance coalesced with changing court decisions and legislation, empowering them to negotiate a moratorium with Ontario Hydro on further development on the Mattagami River (Manore 1995, 1999). They also received compensation for hydroelectric projects on the Moose River. Other projects like the DeBeers diamond mine near the Attawapiskat First Nation are supported because of the opportunity to cooperate in gaining employment and generating revenue. Time will tell if more recent initiatives like the changes to the mining act in Ontario and the Far North Initiative, which calls for the establishment of Canada's largest protected area (50 percent of the boreal forest in Northern Ontario will be set aside as protected), will have positive impacts in First Nation communities.

Northeastern Ontario faces a number of challenges from globalization, emigration, lower levels of education, and a continued dependence on primary extractive activities such as mining and forestry. The downturn in the latter illustrates what can transpire without proper planning or diversification. While protests about these woes are explained as "northern alienation," the challenge for Northeastern Ontario is not strictly a north-south discourse in which resources are sent "southward" never to be seen again. In fact, many communities and individuals have benefited from the boom in resource extraction. The result has been a mindset that has left some residents with an apparent sense of entitlement. Challenges currently facing Northeastern Ontario are also sometimes seen as someone else's responsibility, thereby creating a cycle of dependency.

To overcome this cycle, which is perpetuated by the bust cycle of forestry or mining, a new narrative based on adaptability, responsibility, and empowerment is required. It is these values that will help to inform decision makers in southern Ontario about the challenges of doing business in Northeastern Ontario; it is this approach that will transform policy from a reactive process to an adaptive course of action. An example of such positive discourse has already occurred with the provincial government's shift away from the proposed 15 percent royalty tax on diamonds. Another example is the Ontario Stewardship program, which provides opportunities for various stakeholders – First Nations, landowners, management agencies, NGOs, and interested individuals – to come together and seek solutions (OMNR 2006). Other examples include the Sustainable Forest Management Network, the Ontario Forestry Coalition, the Blueprint for Forestry Transformation, the Federation of Northern Ontario Municipalities (FONOM), and the Northern Ontario School of Medicine.

Even as some communities continue to mourn the most recent layoffs in forestry, with two hundred jobs lost in Kapuskasing and Cochrane in October 2007, others, like Elliot Lake, Hearst, and Cobalt, are attempting to diversify their economies. If Northeastern Ontario is going to break its cycle of dependency on primary extractive resources, it will require new approaches, including the use of resources for multiple tasks such as wood for panelling, waste for heating, and greenhouse nurseries. And Crown land and protected areas could be managed for value added, instead of value deficits. Individuals would be empowered with a *sense d'appartenance* and responsibility. Examples of new thinking are "new economy hubs" in Northeastern Ontario, including the greater City of Sudbury and the City of North Bay, which recently received $1.375 million from FedNor to enhance development of their waterfront, providing for an immediate increase in local tourism and spurring public (Dale 2011). And the Moose Cree First Nation are indicative of smaller "new economy hubs." It thus appears that many of the solutions to the challenges facing Northeastern Ontario will be local.

CONCLUSION

By the time Northern Ontario was labelled New Ontario several centuries of exploration, settlement, and development had passed. So there was nothing very new about New Ontario. Apart from this label being somewhat Eurocentric in disregarding the rich heritage of First Nations in the area, it would have been more appropriate to call the area "Rediscovering Old Ontario," since so many of Ontario's oldest settlements, such as Moose Factory and Fort Albany, are located in the North.

Provided with strong cultural links and a *sense d'appartenance*, many of the communities in the northern part of Northeastern Ontario are examining and developing various strategies to cope with new economic forces and climate change, including renewable-energy technologies, tourism, mining, health, and education. These strategies should move people beyond the cycle of survival to one of adaptability. They are, according to many First Nation leaders, the tools that will generate equity while promoting empowerment. Perhaps the rest of Ontario could learn something if we all simply gazed northward and listened to the voices that have always adapted.

REFERENCES

Bray, M., and E. Epp, eds. 1984. *A Vast and Magnificent Land*. Sudbury, ON: Laurentian University.

Brown, G.S., W.J. Rettie, R.J. Brooks, and F.F. Mallory. 2007. "Predicting the Impacts of Forest Management on Woodland Caribou Habitat

Suitability in Black Spruce Boreal Forest." *Forest Ecology and Management* 245:137–47.

Choquette, R. 1980. *L'Ontario francais, historique*. Montreal, QC: Editions Etudes vivantes.

Dale, D. 2011. "FedNor Whistle Stop in North Bay." *Sudbury Star*. Accessed 2 November 2011 at http://www.thesudburystar.com/Article Display.aspx?archive=true&e=732145.

DeMarco, W., J. Hohenadel, and N. Giesbrech. 2002. "Snowmobilers in Northeastern Ontario: A Report on Characteristics, Experiences and Strategies for increasing Snowmobile Safety." Online document available at http://www.porcupinehu.on.ca/Epidemiology/documents/SnowmobileReport.pdf (accessed 14 April 2005).

Di Matteo, L. 2005. "Sudbury Development Growing More Quickly than Thunder Bay." *Chronicle-Journal*, 30 November, C4.

Dunk, T.W. 2003. *It's A Working Man's Town: Male Working-Class Culture*. 2d ed. Montreal: McGill-Queen's University Press.

– 1994. "Talking about Trees: Images of the Environment and Society in Forest Workers' Discourse." *Canadian Review of Sociology and Anthropology* 31:14–34.

Forest Ethics. 2007. "Overview of Ontario Forests and Forestry." Online article available at http://forestethics.org/downloads/Ontario%20Report.pdf (accessed 16 July 2007).

Hodgins, B.W., and J. Benidickson. 1989. *The Temagami Experience: Recreation, Resources, and Aboriginal Rights in the Northern Ontario*. Toronto, ON: University of Toronto Press.

Johnston, M., and R.J. Payne. 2004. "Ecotourism and Regional Transformation in Northwestern Ontario." In C.M. Hall and S. Boyd, eds., *Nature-based Tourism in Peripheral Areas: Development or Disaster?* 21–35. Toronto, ON: Channel View Publications.

Kelly, Peter E., and Douglas W. Larson. 2002. "A Review of the Niagara Escarpment Ancient Tree Atlas Project: 1998–2001." In Shannon McFayden and Richard Murzin, eds., *Leading Edge '01 Conference Proceedings*. 2002. Georgetown, ON: Niagara Escarpment Commission.

Larmour, A. 2006. "Impact of Victor Mine Starts to Trickle Down." *Northern Ontario Business*, 26 June (8): 26

Lemieux, C., D. Scott, and R. Davis. 2004. "Climate Change and Ontario's Provincial Parks: A Preliminary Analysis of Potential Impacts and Implications for Policy, Planning and Management." In C. Rehbein, G. Nelson, T. Beechey, and R. Payne, eds., *Parks and Protected Areas Research in Ontario: Planning Northern Parks and Protected Areas*.

Proceedings of the Parks and Research Forum of Ontario (PFRO) Annual General Meeting, 4–6 May 2004, Lakehead University, 83–104. University of Waterloo: Parks Research Forum of Ontario.

Louiseize, K. 2007. "Stellar Results for 2006 Mining Sector." *Northern Ontario Business*, 27 June (8): B2.

– 2006a. "Contact Hunts for Viable Mine." *Northern Ontario Business*, 26 June (8): 25.

– 2006b. "First Nations Must Be Involved Early On: Expert." *Northern Ontario Business*. 26 May (7): 5.

Manore, J.L. 1999. *Cross-Currents: Hydroelectricty and the Engineering of Northern Ontario*. Waterloo, ON: Wilfrid Laurier University Press.

– 1995. "Nature's Power and Native Persistence: The Influence of First Nations and the Environment in the Development of the Mattagami Hydro-Electric System during the Twentieth Century." *Journal of the Canadian Historical Association* 6:157–77.

Marsh, J. 2007. "Moose Factory." *The Canadian Encyclopedia*. On-line article available at http://thecanadianencyclopedia.com/index.cfm?PgNm=TCE&Params=A1ARTA0005424 (accessed 15 July 2007)

Matthes, U., J.A. Gerrath, and D.W. Larson. 2003. "Experimental Restoration of Disturbed Cliff-Edge Forests in Bruce Peninsula National Park, Ontario, Canada." *Restoration Ecology* 11 (2): 174–84.

Ministry of Tourism. 2007. *Canadian Travellers Who Visited Northern Ontario: A Profile for Marketing Implications*. Tourism Research Unit, Ministry of Tourism. Queen's Printer.

– 2006a. "Ontario's Overnight Pleasure Travel Market, 2004." *The Tourism Monographs*, no. 55, March. Queen's Printer for Ontario.

– 2006b. "Ontario's Overnight Visitors with an Outdoors Interest, 2004." *The Tourism Monographs*, no. 54, March. Queen's Printer for Ontario.

– 2006c. "Ontario's Overnight Cultural Tourist Market and Its Economic Impact, 2004." *The Tourism Monographs*, no. 58, April. Queen's Printer for Ontario.

– 2006d. "Ontario's Overnight Visiting Friends and/or Relatives Travel Market, 2004: Who's Coming to Ontario for Visiting Friends and/or Relatives?" *The Tourism Monographs*, no. 57, March. Queen's Printer for Ontario.

– 2006e. "Ontario's Business and Business Convention Market, 2004." *The Tourism Monographs*, no. 59, April. Queen's Printer for Ontario.

McDermott, G. 1961. "Frontiers of Settlement in the Great Clay Belt, Ontario and Quebec." *Annals of the Association of American Geographers* 51(3): 261–73.

Nelson, Gordon, Patrick Lawrence, and Catherine Beck. 2000. "Building the Great Arc in the Great Lakes Region." In Stephen Carty et al. eds., *Leading Edge '99: Making Connections*. Conference Proceedings. Georgetown, ON: Niagara Escarpment Commission.

Niagara Escarpment Commission and Ontario Ministry of Natural Resources. 1976. *Landscape Evaluation Study, Niagara Escarpment Planning Area*. Unpublished report. Niagara Escarpment Commission Library, Georgetown, Ontario.

Niagara Escarpment Commission. 2004. *About the Niagara Escarpment*. http://www.escarpment.org/About/facts.htm, with permission.

– 2011. *ONE Monitoring Program*. http://www.escarpment.org/education/monitoring/index.php

No Author. 1992. "Hydroelectric Development in the Hudson Bay Bioregion." *Northern Perspectives* 20:2.

Northern Ontario Sustainable Communities Partnership. 2007. "Community-Based Forest Management for Northern Ontario: A Discussion and Background Paper." http://noscp.greenstone.ca/Portals/5/NOSCP_Letterhead_Background.pdf (accessed November 2 2011)

Ontario Nature. 2007. "The Time Is Now to Protect the Songbirds of Ontario's Boreal Forest." Webpage: http://www.ontarionature.org/news/template.php3?n_code=380 (accessed July 15 2007).

Ontario Office of Francophone Affairs. n.d. "Regional Characteristics–Francophones in Ontario." Webpage: http://www.ofa.gov.on.ca/english/stats/genprof-charac.html (accessed July 15 2007).

Ontario Ministry of Natural Resources. 2006. "Northern Ontario Has First Official Stewardship Council: Group Will Have Say in Local Resource Management Decisions." http://www.mnr.gov.on.ca/MNR/csb/news/2006/apr13nr_06.html (accessed April 13 2006)

Randall, J.E., and R.G. Ironside. 1996. "Communities on the Edge: An Economic Geography of Resource-Dependent Communities in Canada." *Canadian Geographer* 40(1): 17–36.

Ross, I. 2007. "Xstrata Gives $5M for Mines Research Centre." *Northern Ontario Business* 27 January (3): 6.

– 2006. "EDC Looks to Boost Sagging Tourism Industry." *Northern Ontario Business* April 2006 32:26.

– 2002. "Connecting Researchers Industry." *Northern Ontario Business* 1 February (22)4: B2.

Scott, D., G. McBoyle, and A. Minogue. 2006. "Climate Change and the Sustainability of Ski-based Tourism in Eastern North America: A Reassessment." *Journal of Sustainable Tourism* 14(4): 376–98.

Southcott, C. 2006. *The North in Numbers: A Demographic Analysis of Social and Economic Change in Northern Ontario*. Lakehead University: Centre for Northern Studies.
– 2005. "Old Economy/New Economy Transitions and Shifts in Demographic and Industrial Patterns in Northern Ontario." Paper presented to the conference Old Economy Regions in the New Economy, Thunder Bay, Ontario, 24–25 March 2006.
– 2002. *Population Change in Northern Ontario: 1996 to 2001. 2001 Census Research Paper Series*: Report no. 1.
2000. "Single Industry Towns in a Post-Industrial Era: Northwestern Ontario as a Case Study." *Research Reports, Centre for Northern Studies*. Lakehead University.
Statistics Canada. 2002. *2001 Community Profiles*. Statistics Canada Catalogue no. 93F0053XIE.
Thompson, I.D., M. Flannigan, B.M. Wotton, and R. Suffling. 1999. "The Effects of Climate Change on Landscape Diversity: An Example in Ontario Forests." *Environmental Monitoring and Assessment* 49:213–33.
Ulrichsen, H. 2007. "Local Mining Company Signs $45M Deal." *Northern Ontario Business* 27 February (4): 38.
Vors, S.L, J.A. Schaefer, B.A. Pond, A.R. Rodgers, and B.R. Patterson. 2007. "Woodland Caribou Extirpation and Anthropogenic Landscape Disturbance in Ontario." *Journal of Wildlife Management* 71(4): 1249–56.

12

Northwestern Ontario and the Lakehead Georegion: A Region Apart?

MARGARET E. JOHNSTON AND R.J. PAYNE

SUMMARY

The vast Northwestern Ontario Georegion is characterized by key geographical features – the boreal forest, mineral deposits, abundant wildlife, and numerous lakes and rivers – as well as by a small human population, which is spread across many small communities and one large one. The region consists of just over half a million square kilometres and is bounded by an Arctic shoreline to the north, Lake Superior and the international boundary to the south, and Manitoba to the west. To the east is Northeastern Ontario, with some cultural and geographical similarities to the northwest, yet with its own distinct development through the centuries. The boundary used for this chapter is a political one: the region includes the Districts of Kenora, Rainy River, and Thunder Bay (figure 12.1). This is a practical approach to defining the region that coincides with delineations used by ministries and municipal leaders when working on a regional basis.

The resource-extraction history of this region has endured from the fur trade of the nineteenth century, through the establishment of productive mines for copper, zinc, iron, nickel, silver, and gold, and the harvesting and secondary processing of forest products. The region has experienced a gradual transition over the past three decades as the dominance of resource extraction is being overshadowed by the depopulation of small communities, including those that are single-sector towns. There is also a very general increase in the service and education sectors. Further and accelerated transformation

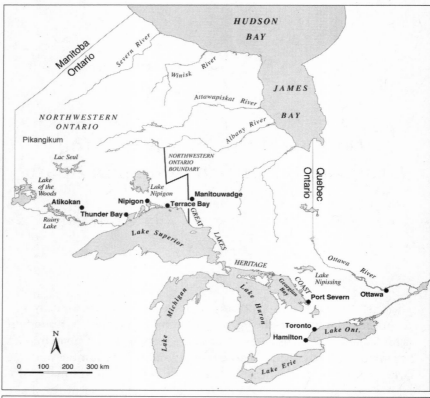

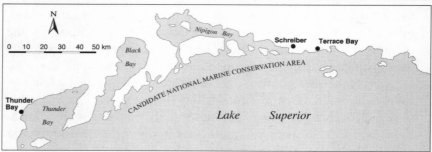

Figure 12.1   Location map of Northeastern Ontario. By the authors, with permission

has occurred in the last several years as forestry continues to decline and other industries have shown some promise. This region participated in the recent international mining boom, though not to the same extent as Northeastern Ontario.

As a region of transformation located many hundreds of kilometres from the provincial capital and the seat of decision making, it should not be surprising that periodically in Northwestern Ontario the idea is raised that the region should separate from Ontario and join Manitoba. This idea arises in response to a feeling of alienation from the urban south and from the expectation of obtaining greater recognition at the provincial level of government with a shift to Manitoba. Further, questions of local control over resources and land bring a uniquely Northwestern Ontario flavour to land management. This chapter uses several examples to explore the mix between regional desires and the policy and planning directed from afar. We hope that through this chapter readers will begin to understand the interplay between the economy, geography, and cultures of Northwestern Ontario that forms the foundation for a sense of separation from the rest of the province.

THE SETTING

The region covers a huge geographical area that extends from the transitional forest at the international boundary at 49 degrees north to the tundra of the Hudson Bay coast at 56 degrees north, just west-northwest of East Pen Island. Most of the region lies within the Hudson Bay watershed and is drained by four large rivers, the Severn, the Winisk, and the Attawapiskat, all of which are fully within in the region, and the Albany, which forms the boundary with Northeastern Ontario. Just over half the total area of Ontario comprises Arctic drainage (Rouse 1994).

Climate and biome regions do not overlap perfectly. In Northwestern Ontario the arctic biome extends along the shoreline of Hudson and James Bays and inland about one hundred kilometres, while the arctic climate zone along this coast is somewhat smaller (see Bone 2003). This arctic area is underlain by continuous permafrost, while less than half of the rest of the region is underlain by sporadic or discontinuous permafrost (Rouse 1994). The vast majority of Northwestern Ontario is defined as having a subarctic climate and biome; one small area along the border with Minnesota is not included in the subarctic (Bone 2003). This climate is characterized by short, warm summers and long, cold winters, with just under half the annual precipitation falling in winter as snow. The large continental climate zone is moderated to a small degree by the presence of Lake Superior and Hudson Bay (Kemp 1994). The southern boundary of the subarctic biome corresponds with the southern limit of the boreal forest.

A large portion of the region is on the ragged granite rocks of the Canadian Shield, while the northernmost portion is sedimentary Hudson Bay Lowland. Rivers flow quickly on the Shield and are broader and slower in the lowland (Rouse 1994). The exposed Shield is characterized by thin soils, rock outcrops, and boreal forest, close-crowned in the south with a boreal-tundra transition zone to the north. It is this geologic foundation that has provided the rich natural-resource base for economic development. Forest species include balsam fir, black spruce, birch, ash, jack pine, larch/tamarack, red pine, white pine, and eastern white cedar. Mineral deposits include gold, copper, zinc, palladium, and diamonds. The agricultural land in the region is confined to clay plains around Dryden and Rainy River, an area along the Kaministiqua River, and some river valleys off Lake Superior, totalling about 120,000 hectares (Dilley 1994).

The vast size and varied nature of the region and its relatively sparse human population have enabled a diversity of animal populations to thrive. Moose, deer, and caribou exist through much of the region, though in unequal concentrations and quantities. Black bear are common throughout, while polar bear inhabit the northernmost portion of the Arctic coast. Fox, lynx, marten, wolf, and their prey inhabit the forests and tundra, while otter and beaver take advantage of the bounty of suitable water habitat in the region. Various fish species abound in the lakes and rivers; edible plants are found throughout. Climate change, along with other stressors, is having an effect on animal and plant population, health, and distribution (see, for example, Meadows 2007a; *Chronicle Journal* 2007).

The social and cultural history of this region is well covered in various texts by authors such as Bone (1993), Coates and Morrison (1992), Dickason (1992), Wightman and Wightman (1992) and Zaslow (1971, 1988). Bray and Epp (1984) describe some of the history of northern Ontario. For a detailed description of the pre-contact, early trade, and resource exploration periods, readers are encouraged to explore these sources; this chapter provides only a brief background and then focuses on recent developments that help explain policy and planning outcomes in the region.

Northwestern Ontario has been populated since the last retreat of the glaciers, first around nine thousand to seven thousand years ago by Paleo-Indians, who were primarily hunters. Gradually, large-game hunting was replaced by a mix of foraging and hunting, alongside the further development of tools. In the few centuries before contact with Europeans, life followed a seasonal-mobility pattern in response to the availability of food. This First Nations foundation remains a defining cultural feature in most of the region. Seasonal migration has been affected by the development of treaties and by settlement in communities with new forms of economic and social activity, but reliance on wildlife availability continues in many places.

The advent of Europeans meant the development of infrastructure such as railroads and settlements and the emergence of mining and forest industries (box 12.1). Economic development in the region has been founded on resource extraction, but there has been an overreliance on natural-resource exploitation, which has left the region vulnerable to international forces and boom-and-bust economic cycles. Although this is a regional concern, it has particular implications for the small communities created to serve infrastructural and industrial developments (Johnston 1995). Southcott (2007) categorizes settlements in northern Ontario into three types: the small and medium-sized cities that tend to be diversified service centres, the resource-dependent communities that are smaller and less diversified, and the aboriginal communities that dominate north of the 50th parallel. In Northwestern Ontario, Thunder Bay exemplifies the first category, Manitouwadge and Atikokan the second, and Kitchenuhmaykoosib and Webequie the third.

BOX 12.1   TRANSPORTATION IN NORTHWESTERN ONTARIO

In such a large area, transportation is an important issue for social and economic development.

Both the Canadian National (CN) and Canadian Pacific (CP) railways traverse the region, although the CN main line approximates the 50th parallel, well away from the major population centres. Passenger rail traffic (Via Rail) has used the CN line since 1990, leaving the southern route, on CP tracks through more populated and scenic areas, for freight. These arrangements are not popular with residents: they point to the abandonment of the southern route for passenger service as yet another example of government neglect.

The Trans-Canada Highway – Highway 17 in Ontario – and, to a lesser extent, Highway 11 provide the main road access to and from the region. The completed Trans-Canada Highway dates from 1971 but remains variable in the number of lanes and the quality of the road surface. Suggestions that Highway 17 be twinned through the northwest in order to facilitate pleasure travel are regularly heard from the region's tourism industry.

Air service is centred on Thunder Bay, with scheduled flights to and from Toronto, as well as to and from the west, by Air Canada and Westjet. A second tier of services, operated by companies such as Bearskin Airlines, connects smaller communities to Thunder Bay. Wasaya Airways, owned by a group of remote First Nations communities, has become an important player in regional air transportation, servicing communities that otherwise would be without air links to the southern parts.

By the authors, with permission.

Nearly all the population resides in the southern half of the region; there are only a few scattered communities in the north, where economic development is limited to mineral exploration, temporary mining camps, and fly-in tourism. In the southern half, which is accessible by good highways and all-weather roads, economic activity is more varied, but it still bears the imprint of natural-resource development.

Southcott (2007) describes the social and economic developments that have caused people in northern Ontario to move from worries about regionally produced wealth leaving the region to worries about jobs leaving the region. Slow population growth since 1971 in Northwestern Ontario reflects the resource dependency of the region and is related to a decrease in the number of workers needed to produce resource commodities. Further, the region has attracted a lower proportion of immigrants than the rest of the province, in direct contrast to its draw for immigrants in the decades before the 1970s.

Both Rainy River and Thunder Bay Districts have had declining populations since 1996, yet Kenora District has seen an increase (table 12.1). Much of this increase seems to be fuelled by small but steady increases in less populated and more remote communities that have primarily aboriginal residents. Kenora District has

Table 12.1
Population Counts for Northwestern Ontario Districts and Selected Municipalities, from 2006 and 1996 census data

	Population 2006	Population 1996
Thunder Bay District	149,063	157,619
Rainy River District	21,564	23,138
Kenora District	64,419	63,360
Atikokan	3,293	4,043
Attiwapiskat	1,293	1,258
Fort Frances	8,103	8,790
Kenora (city)	5,177	16,365
Kitchenuhmaykoosib	916	435
Manitouwadge	2,300	3,395
Marathon	4,791	3,863
Pikangikum	2,100	1,170
Schreiber	901	1,788
Webequie	614	443

2001 Census figures. By the authors, with permission

the highest proportion of aboriginal people in its population, 34.8 percent, while the figures for Rainy River District and Thunder Bay District are, respectively, 16.6 percent and 8.8 percent (see Statistics Canada 2002) The aboriginal populations across Northwestern Ontario are growing in both the larger centres and in these smaller, remote communities. Table 12.1 demonstrates population growth in primarily native communities such as Attawapiskat, Kitchenuhmaykoosib, Pikangikum, and Webequie over the last decade, while the other communities have lost population. These other communities are largely non-aboriginal, but even as they have experienced decline in the last ten years, the aboriginal component of the population appears to have risen.

Resource-dependent communities have faired poorly over the decade. Manitouwadge, for example, has seen a major decline since 1996 with the closure of its main mining employer, Geco, and subsequent employment losses in the Hemlo goldfields. Schreiber has seen the most dramatic declines over the decade, concomitant with employment declines in the transportation sector. Incidentally, it was concern over the fate of single-sector communities that brought about the policy of the Ontario Ministry of Municipal Affairs that does not allow new townsites to be constructed for resource-extraction enterprises; instead workers are to be accommodated in temporary and portable structures or in existing townsites (Johnston 1995). This policy resulted in the creation of a portable camp at the Musselwhite Mine, near Pickle Lake, that remains in use more than ten years after the mine opened and is now experiencing capacity problems given the recent mining boom.

Southcott (2007) discusses the related aspects of youth outmigration and an aging population as examples of change, although he notes that the aboriginal population has a younger age structure because of high birth rates and shorter life expectancy. In Kenora District, for example, 45.8 percent of the Aboriginal population is nineteen years of age or younger, well above the provincial figure of 38.4 percent in this group (see Statistics Canada 2002). Though industrial jobs in Northwestern Ontario continue to be more prevalent than in the province as a whole, even these jobs have been declining because of technological change and shifts in industry structure. At the same time, service sector jobs and self-employment have increased, though not to the same degree as at the provincial level (Southcott 2007).

Bone (2003) notes that population growth in the Canadian north, i.e., the provincial norths, as well as the Northwest Territories, the Yukon, and Nunavut, mirrors the cyclical pattern of international economic growth. When the world's economy is strong, demand for Canada's natural resources increases and so population in these northern areas increases as people move for jobs. Yet recent population growth in the northern parts of the western provinces and the far north is in stark contrast to declines in Northwestern Ontario, indicating that barriers are preventing the region from enjoying the boom experienced elsewhere.[1]

A REGION APART

A feature of life and, occasionally, politics in Northwestern Ontario has been a feeling of alienation from the centres of power and decision making in Toronto and Ottawa (Weller 1977; Martelli 1993; DiMatteo, Emery, and English 2006). This feeling of being on the periphery of the mainstream is fuelled by social, economic, and political factors that originate from outside the region but that have profound, long-lasting effects within it. Alienation is reflected in periodic exhortations that Northwestern Ontario should leave Ontario and join Manitoba, where, it is claimed, the region would be better integrated into provincial economic development. A sense of alienation is common across the provincial norths. This reflects the systematic underdevelopment that has occurred across these regions that sees economic development occur for the benefit of non-Native southern populations (Coates and Morrison 1992).

Regional leaders in Northwestern Ontario have requested policy and planning support from the provincial and federal governments to little avail over the years. Another proposed regional solution developed recently in Northwestern Ontario comes from the Northwestern Ontario Municipal Association (NOMA) in a report released in January 2007. NOMA (2007) identifies seven proposed areas of focus founded upon the primary recommendation for an underlying regional development authority to identify, promote, and develop economic opportunities for the region.

Beneath such general signs, however, one finds more subtle indications that the regional population is not homogeneous in either its problems or its aspirations. Many people still depend for their livelihoods on resource-based industries, such as forestry and mining, that are buffeted by social, economic, and political forces over which people in the region have little or no control (box 12.2). An example is to be found in the forest industry. In 2001, the industry employed 12,935 people, but by 2006 this number had fallen by at least 1,745 (Moazzami 2006). The sector has been hit hard by several factors, including increased competition from low-cost producers internationally, an increase in the value of the Canadian dollar relative to the American dollar, higher costs of electricity, the softwood lumber controversy between Canada and the United States, and rising fibre costs (Moazzami 2006). None of these issues originated in Northwestern Ontario; nor can they be solved solely within the region, although many ideas about solutions have been raised there (see Moazzami 2006; NOMA 2007).

That senior governments do not appear to have done much to help resolve these pressing difficulties increases the conviction that the concerns of people in the Northwest do not matter in Toronto or Ottawa. It is a small consolation to people in the region that they share their problems in the forest industry with other Canadians (see Canadian Broadcasting Corporation 2007).

> BOX 12.2 REVIEW OF ONTARIO'S MINING ACT
>
> After an unfortunate conflict between the Kitchenuhmaykoosib Inninuwug (K-I) First Nation and Platinex Inc. that saw the jailing of six community members for contempt of court over their refusal to allow the mining company access to their traditional lands, the Ontario government undertook to review and to modernize the legislation.
>
> The Mining Act has guided mining activity in the province for more than one hundred years. The social conditions that pertained when it came into force have changed radically. First Nations peoples are more and more convinced of their rights to protect their traditional lands, especially in the face of the expansion of resource extraction activities in areas north of fifty degrees latitude, which had been untouched until recently.
>
> The Minister of Northern Development and Mines, Michael Gravelle, writes that "I can assure you that the Government of Ontario is committed to meeting its duty to consult with Aboriginal communities on activities related to the minerals sector and ensuring that activities occur in a manner that is consistent with the Crown's obligations concerning Aboriginal and treaty rights" (Personal communication, 10 July 2008).
>
> A discussion paper entitled *Modernizing Ontario's Mining Act: Finding a Balance* (Ontario, Ministry of Northern Development and Mines 2008), appeared in August 2008. Discussions are ongoing, with no definite end date specified. Given First Nations' empowerment, the drive for economic development in Northwestern Ontario and local peoples' desire to affect provincial policy decisions, bringing the Mining Act into the twenty-first century will be a difficult task. Attempting to do so in a space of just over one month seems overly optimistic.
>
> By the authors, with permission.

The regional job losses of the past five years mirror other periods of decline in primary and related sectors. Yet at the same time, there have been gains in new sectors, fuelling the belief that as the nature of employment in the Northwest changes, some centres, such as Thunder Bay, will be successful in diversification efforts. Other, smaller communities have been bolstered in the past by new mineral finds, as was the case in Manitouwadge, when the opening of the Hemlo goldfields seemed to be the answer to the impending closure of a copper-zinc mine. New initiatives have also helped, as in the case of Atikokan, which was first revived by a coal-fired electrical generation plant when its iron-ore mines closed, and now by a major research centre as the coal plant is being threatened with closure because of provincial clean-energy policy. Success stories such as these encourage optimism among many that the economy of Northwestern Ontario will survive, diversify, and grow, even though the apparently successful communities such as Manitouwadge have faced major declines in population as the boom fades out and the bust begins.

In the next part of this chapter, we examine three cases: the Whitefeather Forest Proposal, sustainability in local communities, and provincial land use planning initiatives. These cases illustrate the nature and extent of alienation in Northwestern Ontario. However, they also demonstrate important aspects of the regional context: the rise of First Nations' empowerment, non-native people's perspectives on the roles of local residents in planning, and uncoordinated initiatives in land use planning based on southern Ontario perspectives.

THE WHITEFEATHER FOREST INITIATIVE

The Whitefeather Forest Initiative is centred on the Pikangikum First Nation, located about a hundred kilometres north of Red Lake in Northwestern Ontario (figure 12.1). Pikangikum is part of Treaty 5, which occupies a small area of Ontario and extends into north-central Manitoba. The First Nation and its traditional territory lie north of the area of active timber harvesting in Ontario and, thus, were not included in the large-scale land use planning exercises of the past forty years – the Strategic Land Use Planning process during the mid-1970s and Lands for Life/Ontario's Living Legacy during the late 1990s. Pressures to push timber activities – road construction, harvesting, and regeneration beyond the current northern limit – have been building in Ontario. The Whitefeather Forest Initiative is, in part, a response to those pressures experienced by the Pikangikum First Nation.

Before discussing this initiative, it is necessary to provide some background not only about the Pikangikum First Nation but also

Figure 12.2 First Nations treaties. From the *National Atlas of Canada*, with permission (http://atlas.nrcan.gc.ca/site/english/index.html)

about First Nations in Ontario and the Northwest. As figure 12.2 illustrates, virtually all of Ontario is covered by treaties with First Nations people. Although both treaty and Aboriginal rights are guaranteed in the Constitution, Sterritt (2002) indicates that both remain noticeably absent from legislation and policy that govern how land and resources will be used in Canadian provinces.

More specifically, in Ontario, Fernandes (2007) traces the failure of Condition 77 – as well as the revised Condition 34 – stemming from the timber-management environmental assessment completed in 1994. These conditions directed government and forest companies to be more cognizant of treaty and aboriginal rights. More specifically, Ministry of Natural Resources (MNR) managers were charged with examining opportunities for jobs for aboriginal people in the forest industry, for supplying wood to aboriginal sawmills, or allocating timber licences to First Nations. Failures to respect such treaty and aboriginal rights have stoked the resentments that First Nations people feel toward provincial and federal governments.

Local circumstances have exacerbated these resentments. The Pikangikum First Nation does not have all-weather road access; it has poor sewage treatment and water quality; and it has a high teenage-suicide rate. The reserve, at 1,808 hectares, is small but it is home to an estimated 2,300 people, one-third of whom are under the age of nine years (Independent First Nations Alliance 2004).

Living conditions in many of the remote Ontario First Nations communities continue to be a source of despair for residents. The Pikangikum First Nation has been ranked in the bottom third of First Nations in Canada in terms of socio-economic well-being (Armstrong 2001). Newspaper reports claim that in Pikangikum up to half the homes are in disrepair and, although occupied, are unsuited for habitation. Many have no running water, indoor toilets, or safe drinking water (Meadows 2007b). Community infrastructure is not at a level to meet the needs of residents: the provision of electricity depends on a diesel generator, for example. Widespread and devastating poverty on many Northwestern Ontario reserves has drawn international attention and the creation of an alliance among the First Nations in Northwestern Ontario and groups such as Save the Children Canada whose goal is to improve life on reserves (Philp 2007). NOMA (2007) emphasizes the need for municipal leaders in the northwest to work with First Nations leaders on economic-development strategies.

The Whitefeather Forest Proposal is the Pikangikum First Nation's response to poverty, population increase, and poor economic prospects. The proposal consists of a land use plan based on the traditional knowledge of the Pikangikum elders for resource development and environmental protection. The plan identifies areas for forestry and mining, for trapping, and for parks and protected areas. A critical component in the proposal is partnerships.[2] The Pikangikum First Nation seeks partnerships with government and non-government agencies as well as with pro-development and pro-protection organizations.

Another key component of the proposal is based on Pikangikum's location in the boreal forest, which puts Pikangikum squarely in the area of the NGO-driven Canadian Boreal Initiative (CBI) (CBI 2005) and the related Ontario-based Northern Boreal Initiative (NBI). It is in the latter that the Whitefeather Forest Initiative has its roots:

> Terms of reference for the 13,000 km² Whitefeather Forest were developed by the Pikangikum First Nation in cooperation with the Ministry of Natural Resources under the NBI policy framework ... While the Pikangikum planning process reflects the particular needs and circumstances of that community, the results of the Whitefeather Forest plan will be an indication of prospects for future land use planning under the NBI. The Ontario government has indicated that it is committed to "community-based land use planning," embodying a "strong conservation-based, orderly development approach" for all of the NBI associated First Nations. (Canadian Boreal Initiative 2005, 32)

The significance of the boreal forest in ecological terms, the various threats it faces, and the resilience of aboriginal culture combine to provide Pikangikum with another important dimension in the Whitefeather proposal. The community has applied for designation as a World Heritage Site for the Whitefeather Forest, in recognition not only of its boreal location but also of its cultural heritage.

The Whitefeather Forest Initiative offers several features that are noteworthy for our discussion of Northwestern Ontario:

- Community-based land use planning,
- Planning that is conservation-based,
- Partnerships with government and non-government organizations,
- Self-determination and self-help, and
- Sustainability in Local Communities

As we have intimated above, sustainability is clearly an issue for communities in Northwestern Ontario. However, one gets the impression that sustainability is conceived primarily in economic terms. We discuss here a case where people in local communities along the north shore of Lake Superior – specifically, Red Rock, Nipigon, Pays Plat, Rossport, Schreiber and Terrace Bay – articulate an understanding of sustainability that includes its social and environmental dimensions, as well as its economic dimensions. Two recent issues in the area illustrate the divergence in this respect between people in this area and governments.

In the mid-1990s, concerns grew along Lake Superior's north shore, and especially in Rossport, that the archipelago of offshore islands would be developed – perhaps for intensive tourism, perhaps for logging. Local people felt that they had a right to have their say in what was in their view their own backyard. Out of this concern came the Rossport Islands Management Board, a non-governmental organization composed of local Rossport citizens that took upon itself the twofold task of convincing the OMNR office in Nipigon that the district's land use guidelines needed updating, especially in respect to the Rossport Islands, and that they, the members of the Management Board, should be in charge of the update. Perhaps to their surprise, they won both points. They set out to revise the land use guidelines in the usual manner through inventories of life and earth science information. They spent a good deal of effort detailing current human use patterns and talking to current users.

A plan was prepared (Rossport Islands Management Board 1994) and presented to the OMNR district office in Nipigon and shared with interested parties in the north shore communities. The plan came down on the side of protection, arguing that the archipelago ought not to be developed in any substantive way. Small-scale operations, especially in ecotourism, might be appropriate. Logging by the mechanical clear-cut methods commonly used in Northwestern Ontario forests was not supported. However, the Ministry of Natural Resources did not accept the plan's recommendations or its apparent priorities.

Three aspects of this example are striking. First, OMNR managers in the area seem prepared to allow local people to attempt to determine their area's future, with some guidance, mainly in the terms of how to construct land use guidelines. Second, local people possess both the vision and the means to execute a project of this sort. Last, and in the absence of legislation or policy requiring OMNR to examine the plan in a serious way, OMNR was able to simply shrug off the plan's recommendations, with neither criticism nor apology. One might cite this example as one where local people had a better understanding of the breadth of sustainability than a provincial government department. At the same time, however, one must appreciate the paucity of institutional arrangements or options upon which local people could depend to have their plan implemented.

A second illustration concerns the prospects for, and people's concerns about, tourism on the north shore. Two different initiatives in the late 1990s provide the contexts for this discussion. One, the Lands for Life planning process, we consider in detail below; the other was the advent of the idea for a western Lake Superior

National Marine Conservation Area (box 12.3).[3] Both raise questions about tourism along the north shore of Lake Superior and provide opportunities for local residents to consider what role tourism might play in the sub-regional economy.

BOX 12.3   THE WESTERN LAKE SUPERIOR NMCA

Parks Canada began work to establish a NMCA in Lake Superior in 1993. It became apparent that an area bounded by Thunder Cape on the west, the international boundary on the south, the Slate Islands on the east and the shore of Lake Superior on the north offered the best mix of intact natural and cultural heritage.

A proposed area for the NMCA was taken to communities and prospective user groups for their comments and criticisms.

It was clear that many residents of the north shore were suspicious of both Parks Canada and the federal government. There was a good deal of misinformation circulated about what an NMCA was and what Parks Canada's intentions were for it. For some residents, especially those in Rossport, the idea that government would take control over the area was anathema. Similarly, boaters who regularly used the area proposed for NMCA designation were opposed, fearing regulations of the sort attributed to the US National Parks Service at Isle Royale National Park.

Parks Canada was able to counter the misinformation for most communities and residents, and the NMCA was officially designated on 25 October 2007.

By the authors, with permission.

Residents of the north shore communities were ambivalent towards tourism (Payne, Twynam, and Johnston 2001). On the one hand, they recognized that tourism would provide economic diversification where it was sorely needed. On the other, people here worried that tourists would be very different in their values and activities from local people, leading perhaps to pressures to curtail traditional recreational activities such as hunting.

As in their discussions about the future of the Rossport Islands, some residents emphasized several aspects of their views towards tourism:

- It ought to be small-scale, nature-based tourism;
- It ought to be based in the existing communities, rather than situated out in the islands;
- It ought to be compatible with local cultures; and
- It ought to be controlled by local people.

The information garnered from community members was submitted to the Lands for Life process and to Parks Canada as part of the latter organization's practice of acquiring public input on protected-area proposals. But residents might well wonder if their involvement mattered because little of substance has occurred in response to their efforts.

There are several striking aspects about this case. While being concerned about economic development, residents are equally concerned about the social and environmental effects that development may have on their lives. Their concerns about wider issues of sustainability are not unlike those expressed by First Nations people in Pikangikum: issues of quality of life, of environmental quality, of economic welfare and, above all, of local control. However, as we indicated in relation to the Rossport Islands Management Board's plan, no institutional basis existed to support such local desires for partnerships and control. Nor was there a mechanism through which either Parks Canada or the Ontario government could apparently provide meaningful feedback to the communities.

PROVINCIAL INITIATIVES IN LARGE-SCALE LAND USE PLANNING

The province of Ontario has a long-standing involvement in planning, especially for economic development in Northwestern Ontario, which is detailed in Lambert and Pross (1967) and, rather more critically, in Nelles (2005). We propose to examine provincial planning initiatives since World War II as they relate to Northwestern Ontario, focusing on the respective roles of the province itself, as well as local people, communities and First Nations.

Economic Regions and Growth Poles

The early days of planning around growth poles in Ontario must be viewed in a wider Canadian context.[4] Regional development was a public policy issue in Canada after World War Two, especially in

Atlantic Canada. Savoie (1992) and Beaumier (1998) offer strong critiques of how the problem of regional disparities was conceptualized and, more importantly, how government responded in the creation of programs and institutions to deal with them. The provincial norths had similar problems, albeit on a smaller scale.

Reed (1995) discussed "local dependency" and the role of the Ontario Ministry of Natural Resources in a case study of Ignace, in Northwestern Ontario. She succeeded in highlighting both the need for provincial intervention and the frustration among local residents about their powerlessness. Provincial intervention spoke to fairness in resource allocation, especially in relation to timber. It also spoke to the creation of economic wealth through job creation and infrastructure development. But local people, while appreciating possible jobs and infrastructure, found themselves locked out of a discussion that involved mainly resource companies and senior administrators in the OMNR. When opportunities to contribute occurred, primarily through the land use planning exercises outlined below, local people in the Northwest were frustrated to find that their concerns were given the same weight as those from people living outside their region. The dichotomy of provincial intervention and local frustration is a feature of the large-scale land use planning initiatives in the Northwest.

Ontario's initiative in growth-pole planning came from the Department of Economics and Development (DED). DED used growth-pole theory to propose to the government that it designate appropriate poles, for example, in Northwestern Ontario at Thunder Bay, and channel provincial funding into them, thereby encouraging private-sector investment and an expected "trickle-down effect" to other parts of the area.

However, as Nelles (2005) has emphasized, such centralized planning made the Ontario government uneasy. Ontario was led at the time by a Progressive Conservative government,[5] and while certainly progressive on some issues such as tertiary education, it believed in supporting the private sector in economic development rather than leading it. Furthermore, the government quickly realized that a policy commitment to growth poles would tie its hands, making the business of politics much more difficult.

This first attempt at provincial intervention in Northwestern Ontario was therefore abandoned. What is striking about it is the centralized nature of the planning process. There was no question that the provincial government, through its arm, the Department of Economics and Development, would conceptualize the problem and find ways to fix it. The idea that local communities, First Nations, or other regional bodies might initiate such planning processes was not considered. Hodge (1994) expresses the same view in this way:

> It is difficult to characterize this vigorous period in Canadian regional planning. Or was it regional planning at all, some would ask? There were federal/provincial task forces, development corporations and general funding agreements. But there was seldom a published regional plan. "Worst-first" concepts vied with "growth poles" when implementation was debated. And, not least, it was regional planning from outside the region. The planning was not generated and tended by a region's inhabitants. All this should not be unexpected, since economic development problems are complex and require, as well, the co-operation of several levels of government and many different program agencies. In the end, it seems institutional inertia probably triumphed.

Strategic Land Use Planning

During the late 1970s, Ontario initiated a land use planning process, termed Strategic Land Use Planning (SLUP), through the Ontario Ministry of Natural Resources that attempted to provide the forest industry with clear direction about which areas should be allocated for forestry south of fifty degrees north latitude. There is some confusion even today about the process itself. Some commentators (e.g., Payne and Graham 1987) heard echoes of growth-pole planning. Others saw SLUP as little more than an intra-ministerial attempt to co-ordinate policies and operations. The intent hardly matters now beyond the observation that what SLUP was about was not clear to participants. SLUP became a process about forests, parks, tourism, fish, wildlife, and Crown-land recreation because of the expectations that interested parties brought to the process.

Northwestern Ontario was one of three SLUP regions, the others being Northeastern Ontario and southern Ontario. Each region was divided into districts (OMNR administrative districts). Each district was required to meet its share of regional and provincial targets for timber, commercial fishing, hunting opportunities, and so on. The top-down nature of this approach meant not only that district and regional targets "rolled up" to provincial ones but also that the focus of political discussions was placed on the minister for natural resources and, ultimately, the provincial government in Toronto. Consequently, while there were local and regional public meetings in Northwestern Ontario to discuss targets and allocations, the important meetings took place in Toronto, where province-level organizations such as the Ontario Forest Industries Association, the Ontario Federation of Anglers and Hunters, the Wildlands League, the Ontario Prospectors Association, and the Northern Ontario Tourism Outfitters Association, as well as provincial government agencies, debated the issues. Local people and communities in Northwestern Ontario were largely ignored.

First Nations' people and communities were invited to take part in public meetings and to comment on draft proposals in much the same way that other stakeholders were. Not surprisingly, First Nations individuals, communities, and treaty organizations objected to the "stakeholder" label, arguing, as they would in subsequent large-scale land use planning exercises, that their treaty and aboriginal rights required that they speak to the provincial government as equals. When Ontario ignored these rights, First Nations involvement in the SLUP discussions ceased.

The process concluded in 1983 with the publication of district-level land use plans that would guide resource management decisions for the foreseeable future. However, even this part of the process caused confusion. Many thought these district-level documents were plans that the OMNR was required to follow. The minister, Alan Pope, clarified their status, indicating that these were "guidelines" rather than plans and that as guidelines, they were but one view of what might happen in an OMNR administrative district. Once again, Ontario edged forward on regional planning, only to back away when it became clear that a plan entailed provincial commitments to its content and that that might create political problems.[6]

LANDS FOR LIFE, ONTARIO'S LIVING LEGACY

The most recent land use planning initiative undertaken in Ontario was the Lands for Life process, which was completed in 1999.[7] Like SLUP, its area of focus was Crown land south of fifty degrees north latitude. Unlike the SLUP process, however, the Ontario government actively sought to focus debate about issues on regions where those issues were important. It did so by creating three "round tables" – one each for Northwestern, Northeastern, and Northcentral Ontario – composed of regional residents drawn to represent various stakeholders. In Northwestern Ontario, there were representatives from tourism operators, environmental groups, timber companies, prospecting associations, and the like. This group was charged with holding meetings, soliciting points of view, establishing planning direction, and developing a strategy, with the technical assistance of OMNR.

It is clear that the Ontario government had learned several lessons that, if applied in this planning exercise, could

- Involve a wide variety of interests,
- Encourage local and regional participation and even control, and
- Use government agencies to support, rather than to lead, the effort.

The first of these lessons is especially important because a heavy hand provincially and the strong involvement of provincial, nongovernmental organizations would push planning decisions to the provincial level, with little local or regional involvement.[8] Also, there was to be no top-down rendering of targets to be met at regional and district levels. Decisions were regionally based and there was no mention of districts, even for implementation purposes.

The revised process worked fairly well, except for two problems. Once again, First Nations were expected to be mere stakeholders, and they responded as they did under the SLUP process: they refused to participate in the Lands for Life process.

The second problem, much less obvious, was political. In locating the decision making for Lands for Life in the north, people living in much more populous southern Ontario were much less a part of the decision. This issue was especially important in Northwestern Ontario because here there were still many opportunities to establish large-scale parks and protected areas. Parks and protected areas were an important issue for southern Ontario residents. As a result, at the eleventh hour the premier of Ontario, Michael Harris, knowing that much of his government's support was in southern Ontario, brokered a deal with parks advocates to increase the area set aside for parks and protected areas in the boreal west. One might wonder whether the presence of "southern Ontario representatives" on each of the round table committees would have interfered politically with such negotiations. Clearly, the OMNR was committed to regional representation; taking the round tables beyond regional representation would have presented a whole new suite of problems.

Subsequent policy developments that have come out of Lands for Life include Room to Grow, an agreement related to the provincial Timber Class Environmental Assessment that involved the timber industry, a conservation coalition called the Partnership for Public Lands, and the Ontario government. Ontario Nature describes Room to Grow in these terms:

> In July 2003, the Room to Grow framework was legally enshrined through the Timber Class Environmental Assessment. The Room to Grow Policy contains three specific objectives:
>
> - To identify the gaps in Ontario's parks and protected areas system;
> - To benchmark and secure the long-term supply of wood necessary for industrial processing; and
> - To share permanent increases in wood supplies between gaps in parks and protected areas. (Ontario Nature 2007)

The Room to Grow document is quite specific about what it is meant to accomplish. It "sets out how permanent increases in wood supplies will be shared between new parks and protected areas to complete ecological representation, and more wood for the forest industry to support jobs and growth" (OMNR 2002, ii).

It is evident that the decisions taken very late in the Lands for Life process continued to direct land use allocations. People in Northwestern Ontario are largely marginalized by an institutional structure that is based in Toronto. The pattern evident in SLUP was repeated: First Nations' estrangement from the entire process and local and regional residents' views placed behind provincial level considerations.

Smart Growth

Ontario's Smart Growth began as a "provincial policy statement" (PPS, box 12.4) issued under Section 3 of the *Planning Act* in

1996. The policy statement was revised slightly in 1997 and more thoroughly in 2005 (Ontario Ministry of Municipal Affairs and Housing 2005). As such, it overlaps in time with the Lands for Life/ Ontario's Living Legacy land use planning process.

BOX 12.4   THE PROVINCIAL POLICY STATEMENT

"The Provincial Policy Statement [PPS] provides policy direction on matters of provincial interest related to land use planning and development" (OMAH 2005).

While the principles embedded in the statement apply throughout the province, it is clear from the description offered by the OMAH (2005) that its focus is communities and associated health, safety, and heritage issues. While these might be read as "environmental" concerns, the statement contains no mention of forests, a highly significant element of the landscape of northwestern Ontario.

The Provincial Policy Statement applies to the developed areas of the province, including those in northwestern Ontario such as Thunder Bay, Kenora, and Nipigon. In the rest of the region south of 50 degrees latitude, land use planning has been a matter for SLUP and Lands for Life and, more specifically, forest management plans.

By the authors, with permission.

It should be noted that both the PPS and the *Planning Act* are under the control of the provincial Ministry of Municipal Affairs and Housing, a government agency that has had little jurisdiction in Northwestern Ontario beyond organized municipalities lying principally in the southern part of the region. The Planning Act itself does not normally pertain to the unorganized and Crown land in the province, lands that make up the majority of territory in Northwestern Ontario.

Ontario saw a flurry of activity under the Smart Growth label just after 2000, and a piece of provincial legislation, the Places to Grow Act, in 2005. Most of the Smart Growth activity centred on the Greater Toronto Area (GTA), but a Smart Growth panel in

Northwestern Ontario did present a set of proposals to the government in May 2003. The panel's mandate was broad:

> The Northwestern Ontario Smart Growth panel will provide advice on a long-term growth strategy. In communities across the zone, it will help decision-makers promote and foster growth – growth that creates economic opportunities, sustains communities, attracts new residents, gives young families the option of staying closer to home and protects and enhances the northern way of life and natural environment (Ontario Smart Growth 2003, 5).

The panel stressed that the Smart Growth approach depended on three principles:

- Building consensus;
- Promoting and managing growth; and
- Seeing the "big picture" (Ontario Smart Growth 2003, 5).

Panel membership in the Northwest was heavily oriented towards business and municipal politics. Reed (1995, 135) points out that while municipal politicians might be regarded as representative of their populations, they tend to align themselves closely with business when economic times are hard. In these conditions, it is difficult to imagine that "consensus" would be broad. Moreover, the "big picture" that the panel fixes its gaze on is surprisingly narrow. Consider the panel's chief recommendation:

> ACTION ITEM 1: CLUSTER OF EXCELLENCE FOR FORESTRY AND FOREST PRODUCTS
> In northern Ontario, we should pursue the widest variety of opportunities to maximize the sustainable value of our natural resources. Examples of this include harvesting wood fibre, extracting mineral resources, developing cottage lots and commercial recreational operations on Crown land, bottling water, farming peat, capturing renewable energy and gathering wild rice. In Northwestern Ontario, arguably the most important opportunities are connected to the forestry and forest products cluster (Ontario Smart Growth 2003, 13).

A change in the Ontario provincial government, from Progressive Conservative to Liberal, interrupted the Smart Growth agenda in Northwestern Ontario, perhaps fortunately. The new government remained committed to the idea of Smart Growth, but until recently its attention was held by the pressure of growth in the GTA. In May 2007, the provincial government announced that northern Ontario would be the next focus of government Smart Growth efforts: "A Growth Plan for Northern Ontario, to be developed under the Places to Grow Act, will build on the Northern Prosperity Plan, the good work of the Northern Development Councils, as well as the recommendations from northern mayors" (Ontario, Ministry of Energy and Infrastructure 2007).

At this writing, the shape of the northern Ontario Smart Growth plan is not clearly known.[9] There may be opportunities to focus the plan on many places within northern Ontario and to involve both aboriginal and non-aboriginal peoples. But the government's statement contains echoes of the usual top-down, provincially determined priority setting.

In May 2008, the Ontario Ministry of Energy and Infrastructure released *Towards a Growth Plan for Northern Ontario: A Discussion Paper*. It identifies seven issues for consideration:

- Strengthening and advancing the resource-based industries;
- Growing emerging sectors;
- Fostering research innovation and commercialization;
- Increasing education and training opportunities;
- Retaining and attracting people and jobs;
- Supporting business development and entrepreneurship; and
- Making strategic use of infrastructure. (3)

The paper goes on to suggest ways and means of engaging northern Ontario residents in conversations regarding the appropriateness of, and priorities among, these suggestions. While the discussion paper appeared in May 2008, there is no firm timeline for gathering public input into the planning process. Moreover, at this writing, there is no apparent intention to link the Smart Growth process with other recent ventures to revise the Mining Act (box 12.2) or land use planning in the far north (box 12.5).

> **BOX 12.5. PROTECTING ONTARIO'S NORTHERN BOREAL FOREST**
>
> On 14 July 2008, Ontario's premier, Dalton McGuinty, announced that his government would "protect at least 225,000 square kilometres of the Far North Boreal region under its Far North Planning initiative" (News Release, Office of the Premier).
>
> At this writing, a planning team has been assembled within the Ministry of Natural Resources.
>
> While some took the announcement to mean more parks and protected areas, others have pointed out that Mr McGuinty did not say in his statement how the protection would be undertaken.
>
> Another important question, especially in light of our discussions above of past attempts to plan for land use in the north, is how the planning will be accomplished. Will the Whitefeather model be adopted by planning teams? Will the consultative measures in Lands for Life be utilized in the far north?
>
> Will there finally be legislation to guide such planning and to ensure that plans are implemented (Ontario, Office of the Environmental Commissioner 2007).
>
> By the authors, with permission.

In sum, the foregoing threads of land use policy and planning share several features. Provincial direction and control has generally been paramount in each effort, with the exception of Lands for Life. In that process, provincial interests asserted themselves only as the process wound up, and then the forestry Room to Grow policy ensured that those interests remained paramount. Provincial interest in sharing the decision making power has been evident in the Rossport Islands Management Board, in the Lands for Life round tables, and in the OMNR's Northern Boreal Initiative. These examples have been exceptional, and they have been overshadowed by provincial interests. Reed's (1995) characterization of the failure of new ideas in policy and planning as "institutional inertia" may be an appropriate description of land use planning in Northwestern Ontario since World War II.

CONCLUSION

Our consideration of Northwestern Ontario and the Lakehead Georegion began with the recognition that the entire area is peripheral to the core of Ontario-Toronto and southern Ontario. The experience of living in a peripheral area often leaves people feeling left out and forgotten. That certainly is the case for many in Northwestern Ontario. We discussed this feeling of alienation and its consequences – political and otherwise – early in the chapter. However, it is clear that alienation is not felt in the same way by all residents of this part of the province. First Nations' people have contended with alienation for generations. Our example of the Whitefeather Forest proposal illustrates that one First Nation – using the institutional opportunities available – has attempted to overcome its alienation and to take economic, social, and political control of its future.

Despite the well-intended attempts by the Ontario Ministry of Natural Resources to offer institutional support – through, for example, the regionally based boreal round tables – land use planning efforts in Northwestern Ontario perpetuate the core-periphery dichotomy in which people and communities in Northwestern Ontario are trapped. The government in Toronto simply does not understand why its attempts are not well-received.

If this state of alienation is to change, the Ontario government will have to open land use planning processes to local people, both Native and non-Native, and to local/regional non-governmental organizations. It will have to provide grants and incentives to assist local citizen and NGO groups to participate. Local people will have to recognize that they have resources such as colleges and universities that may be willing to work with them to foster broader, deeper and more technical interventions into what up to now has been Ontario's business.

NOTES

1 An example would be the growth of diamond mining at Attawapiskat, which is in northwestern Ontario according to our definition but which is serviced through Sudbury and Timmins, both, of course, having a far greater history of mining than northwestern Ontario.

2 At least one of these partnerships – that with the Partnership for Public Lands (PPL) – has become strained because, while the PPL favours protection, it is concerned about proposed forestry and mining activities in the Whitefeather Forest.
3 The National Marine Conservation Areas program is operated by Parks Canada, the federal agency whose mandate is natural- and cultural-heritage protection and presentation.
4 The term "growth pole" comes from François Perroux (1950), an economist whose work has influenced much of the policy developed to reduce regional disparities in Canada and elsewhere.
5 The Progressive Conservative party formed governments continuously from 1943 to 1985 in Ontario.
6 An interesting sidelight related to SLUP concerns the area in which the Whitefeather Forest proposal, discussed earlier, is situated. The area was called the West Patricia planning area. No timber allocations had been made there. Timber companies, however, were interested in moving logging operations into the area. The OMNR was to prepare a plan following the same process as was used in SLUP, but the plan was not completed.
7 Lands for Life became known as Ontario's Living Legacy when completed in 1999; after the Liberal party took office in 2003, the term "Ontario's Living Legacy" lost all currency.
8 Note that these land use plans were to be *strategic* and that the strategy to be implemented was that of the Ontario government and, specifically, the Ministry of Natural Resources.
9 The people of Ontario re-elected the McGuinty Liberals in the autumn of 2007. The development of a growth plan for northern Ontario was shelved at the election call, but will likely begin again, since the Liberals are continuing as the governing party.

REFERENCES

Armstrong, R.P. 2001. *The Geographical Patterns of Socio-Economic Well-Being of First Nations Communities in Canada*. Agriculture and Rural Working Paper Series Working Paper no. 46, Ottawa: Statistics Canada.

Beaumier, G. 1998. *Regional Development in Canada*. http://dsp-psd.pwgsc.gc.ca/Collection-R/LoPBdP/CIR/8813-e.htm.

Bray, M., and E. Epp, eds. 1984. *A Vast and Magnificent Land*. Thunder Bay: Lakehead University.

Bone, R.M. 2003. *The Geography of the Canadian North: Issues and Challenges.* 2d ed. Toronto: Oxford University Press.
Canada. *Department of Indian and Northern Affairs.* 2001. Ontario Treaties. http://www.ainc-inac.gc.ca/pr/trts/ont/ont_e.html.
Canada. *Department of Natural Resources.* 2007. *The Atlas of Canada.* http://atlas.nrcan.gc.ca/site/english/index.html.
Canadian Broadcasting Corporation. 2007. *Forestry.* 29 January. http://www.cbc.ca/news/background/forestry
– 2008. "Prospectors, Landowners Raise Concerns about Ontario Mining Act Meetings." 18 August. http://www.cbc.ca/canada/ottawa/story/2008/08/18/ot-mining-act-080818.html.
Canadian Boreal Initiative. 2005. *The Boreal in the Balance: Securing the Future of Canada's Boreal Region.* Ottawa: Canadian Boreal Initiative.
Chronicle Journal. 2007. "Numbers Already Affected in the West." *Chronicle Journal* 3 February 2007, E1, E2.
Coates, K., and W. Morrison. 1992. *The Forgotten North: A History of Canada's Provincial Norths.* Toronto: James Lorimer & Company.
Dickason, O.P. 1992. *Canada's First Nations: A History of Founding Peoples from Earliest Times.* Toronto: Oxford University Press.
Dilley, R.S. 1994. "Farming on the Margin: Agriculture in Northern Ontario." In *Geographic Perspectives on the Provincial Norths, Northern and Regional Studies Series,* no. 15. Lakehead University Centre for Northern Studies, ed., M.E. Johnston. Lakehead University and Copp Clark Longman, 180–98.
DiMatteo, L., J.C.H. Emery, and R. English. 2006. "Is It Better to Live in a Basement, an Attic or to Get Your Own Place? Analyzing the Costs and Benefits of Institutional Change for Northwestern Ontario." *Canadian Public Policy* 32(2): 173–96.
Fernandes, D. 2007. "The Legacy of Condition 77: Past Practices and Future Directions for Aboriginal Involvement in Forestry in Ontario." Unpublished paper, Faculty of Forestry, University of Toronto. www.nafaforestry.org/forest_home/documents/Fernandes2007-Legacy.pdf.
Hodge, G. 1994. "Regional Planning: The Cinderella Discipline." *Plan Canada,* Special Edition, 35–49.
Independent First Nations Alliance. 2004. *Independent First Nations Alliance.* http://www.ifna.ca/index.html.
Johnston, M.E. 1995. "Communities and the Resource Economy of Northwestern Ontario." In *Themes and Issues of Canadian Geography I,* eds. Christoph Stadel and Herman Suida. University of Salzburg, 107–15.

Kemp, D.D. 1994. "Global Warming and the Provincial Norths." In *Geographic Perspectives on the Provincial Norths, Northern and Regional Studies Series*, no. 15. Lakehead University Centre for Northern Studies, ed. M.E. Johnston. Lakehead University and Copp Clark Longman, 303–36.

Lambert, R.S., and Paul Pross. 1967. *Renewing Nature's Wealth: A Centennial History of Public Management of Lands, Forests and Wildlife in Ontario, 1763–1967*. Toronto: Department of Lands and Forests.

Martelli, J. 1993. "Regional Protest and Alienation: The Case of Northern Ontario." Unpublished MA thesis, University of Western Ontario.

Meadows, B. 2007a. "Global Warming Hurting Moose." *Chronicle-Journal*, 3 February 2007, E1.

– 2007b. "Pikangikum Plan Targets Infrastructure." *Chronicle-Journal*, 11 April 2007, A3.

Moazzami, B. 2006. *An Economic Impact Analysis of the Northwestern Ontario Forest Sector*. Northwestern Ontario Forest Council.

Nelles, H.V. 2005. *The Politics of Development: Forests, Mines & Hydro-Electric Power in Ontario, 1849–1941*. Montreal and Kingston: McGill-Queen's University Press.

Northwestern Ontario Municipal Association. 2007. *Enhancing the Economy of Northwestern Ontario*. http://www.noma.on.ca.

Ontario. Ministry of Energy and Infrastructure. 2007. "McGuinty Government Announces Growth Plan for The North." News release, 17 May. http://www.pir.gov.on.ca/english/news/2007/q2/n20070517.htm.

– 2008. *Towards a Growth Plan for Northern Ontario: A Discussion Paper*. Toronto: Queen's Printer.

– Ministry of Municipal Affairs and Housing. 2005. *Provincial Policy Statement*. Toronto: Queen's Printer.

– Office of the Environmental Commissioner. 2007. *Annual Report, 2006-2007*. Toronto: Queen's Printer.

– Office of the Premier. 2008. *Protecting a Northern Boreal Region One-and-A-Half Times the Size of the Maritimes – Backgrounder*. News release. http://www.premier.gov.on.ca/news/Product.asp?ProductID=2358-20k.

– Ministry of Natural Resources. 2002. *Room to Grow: Final Report of the Ontario Forest Accord Advisory Board on Implementation of the Accord*. Toronto: Queen's Printer.

Ontario Nature. 2007. *Ontario's Boreal Forest: Threats to Ontario's Boreal Forest*. http://www.ontarionature.org/enviroandcons/boreal/boreal_threats.html.

Ontario Smart Growth. 2003. *Shape the Future: Northwestern Ontario Smart Growth Panel: Final Report.* Toronto: Ontario Smart Growth. http://smartgrowth.gov.on.ca.

Payne, R.J., and R. Graham. 1987. "An Assessment of Northern Land Use Planning in Canada." *Journal of Canadian Studies* 22, 35–49.

Payne, R.J., G.D. Twynam, and M.E. Johnston. 2001. "Tourism, Sustainability and the Social Milieux in Lake Superior's North Shore." In S. McCool & N. Moisey, eds., *Tourism, Recreation and Sustainability: Linking Culture and Environment.* CABI, 315–42.

Perroux, F. 1950. "Economic Space: Theory and Applications." *Quarterly Journal of Economics* 64(1): 89–104.

Philp, M. 2007. "A Slap in the Face of Every Canadian." *Globe and Mail*, 3 February 2007, F1, F4–5.

Reed, M.G. 1995. "Co-operative Management of Environmental Resources: A Case Study from Northern Ontario, Canada." *Economic Geography* 71(2): 132–49.

Robinson, I.M. and G. Hodge. 1998. "Canadian Regional Planning at 50: Growing Pains." *Plan Canada* 38(3): 10–15.

Rossport Islands Management Board. 1994. *Strategic Plan and Land Use Guidelines, 1994.* Rossport, ON: Rossport Islands Management Board.

Rouse, W.R. 1994. "Climate, Climactic Change and Water in the Arctic Watersheds of Manitoba and Ontario." In *Geographic Perspectives on the Provincial Norths, Northern and Regional Studies Series*, no. 15. Lakehead University Centre for Northern Studies, ed. M.E. Johnston. Lakehead University and Copp Clark Longman, 337–57.

Savoie, D.J. 1992. *Regional Economic Development: Canada's Search for Solutions.* Toronto: University of Toronto Press.

Southcott, C. 2007. "The North in Numbers: A Demographic Analysis of Social and Economic Change in Northern Ontario." *Northern and Regional Studies Series* no. 15. Thunder Bay: Lakehead University Centre for Northern Studies.

Statistics Canada. 2007. "Population and Dwelling Counts, for Canada, Provinces and Territories, and Census Divisions, 2006 and 2001 Censuses." 100 percent data (table). *Population and Dwelling Counts Highlight Tables.* 2006 Census.

– 2002. *2001 Census Aboriginal Population Profiles.* Released 17 June 2003. Last modified 2005-11-30.

– 2001. "Population and Dwelling Counts, for Canada, Provinces and Territories, and Census Divisions, 2001 and 1996 Censuses." 100

percent data (table). *Population and Dwelling Counts Highlight Tables.* 2001 Census.

Sterritt, N.J. 2002. *Aboriginal Rights Recognition in Public Policy: A Canadian Perspective.* http://www.sterritt.com/papers/01-032002-njs.pdf

Weller, G.R. 1977. "Hinterland Politics: The Case of Northwestern Ontario." *Canadian Journal of Political Science* 10(4): 727-54.

Wightman, N.M., and W.R. Wightman. 1992. "Road and Highway Development in Northwestern Ontario, 1850 to 1990." *Canadian Geographer* 36(4): 366-80.

Zaslow, M. 1971. *The Opening of the Canadian North, 1870-1914.* Toronto: McClelland and Stewart.

– 1988. *The Northward Expansion of Canada, 1914-1967.* Toronto: McClelland and Stewart.

13

The Intersection of Landscape, Legislation, and Local Perceptions in Constructing the Niagara Escarpment as a Distinct Georegion

SUSAN PRESTON

SUMMARY

The Niagara Escarpment is unique among Ontario's georegions as the focus of Canada's first large-scale conservation plan. Its distinctive landscapes are the result of an ancient geological formation most recognized as a prominent ridge and known for its many rare species of flora and fauna. Aboriginal people were present eleven thousand years ago with diverse cultures when Euro-American settlers arrived in the late eighteenth century. Although distinct settlement and land use patterns are evident on the Escarpment, they are variations on larger regional patterns and continue to be influenced by the pressures of growth around the so-called Golden Horseshoe. These pressures drew attention to the need for conservation measures as early as the 1950s, and a campaign for action by the provincial government resulted in special protective legislation in 1973. The process of policy development through the following years brought this landscape into the public consciousness in an unprecedented way; thousands of residents participated in the planning process. Designation as a UNESCO World Biosphere Reserve in 1990 further asserted its singular identity. This chapter argues that as a result of these activities the Niagara Escarpment has become widely recognized in the public imagination as a distinct georegion.

Figure 13.1 Niagara Escarpment Planning Area and Plan Area, with permission

INTRODUCTION

The Niagara Escarpment is distinct for many reasons, including being among the first extra-regional special planning areas in Canada with a primary concern for conservation (figure 13.1). In Ontario it has been joined only by two others in the past few years: the Oak Ridges Moraine, which intersects with it, and the Greenbelt, which encompasses both of them. The special status of the Niagara Escarpment is the result of both public pressure on government and government action since 1973 in the form of legislation and policy designed to conserve areas of documented ecological significance, particularly in light of the rapid urban expansion of Toronto. While Toronto's growth has been a guiding force in provincial policy for the Escarpment, it does not appear to be the guiding force in public perception and values. They have been documented to be informed primarily by personal experience and a sense of local identity. The process of policy development has been very influential, however, in a widely held perception of the Niagara Escarpment as a single georegion. Like some Ontario georegions, this one crosses the boundaries of multiple spaces of governance. Unlike the others, however, its identity as a georegion is a comparatively new construct that is informed by – if not an outright product of – the growth of Toronto and its policy field.

This chapter is structured to characterize the Niagara Escarpment landscape in conceptual layers. I begin by characterizing the setting with a brief introduction to the biophysical environment, followed by an overview of human settlement history. The second part of the chapter considers the Escarpment in terms of how it is valued in policy arrangements, civil-society initiatives, and entrepreneurial activities. In the third part, the idea of an emergent georegion is discussed from the perspectives of policy, the scholarly literature, and residents, based on interviews in Escarpment communities. I conclude with suggestions for further research and planning.

The chapter is intended to demonstrate the gradual emergence of a public sense of the Niagara Escarpment as a single, linear landscape on a vast scale, one that contributes to many smaller regions, bioregions, and watersheds, but that has its own increasingly celebrated identity from end to end. It is a process that has led to the Escarpment becoming increasingly valued at local, regional, provincial, national, and international scales. Table 13.1 outlines some key defining aspects of this georegion: approximate population, geographic area, and major economic activities.

THE SETTING

The Niagara Escarpment has been described as the most prominent geographical feature of southern Ontario, and its unique biophysical

Table 13.1
Introducing the Niagara Escarpment Plan Area

Approximate Population
Est. 120,000 within Plan Area
Est. 500,000 within 1km of prominent elevations

Approximate Area
Est. 184,000 hectares within Plan Area

Major Economic Activities
Agriculture
Tourism and Recreation (eco- and agri-)
Service (urban)
Education (McMaster University and Brock University)
Mining (limestone)
Industry (steel)

Source: By the author, with permission.

characteristics have often resulted in a distinctive history and pattern of human settlement. Valuable sources that focus on the Escarpment's physical environment include the series of Leading Edge conference proceedings between 1994 and 2006, research produced under the auspices of the ONE Monitoring Program that was established as a result of the Escarpment attaining the World Biosphere Reserve status, the work of the Cliff Ecology Research Group at Guelph, and background reports prepared in support of the Niagara Escarpment planning process that are held in the library of the Niagara Escarpment Commission. (See Google for World Biosphere Reserve and Cliff Ecology Research Group; see also Niagara Escarpment Commission 2004; Niagara Escarpment Commission 2011; and Niagara Escarpment Commission and Ontario Ministry of Natural Resources 1976).

Tracing the history of human settlement and land use activity along the Niagara Escarpment calls for a wide-ranging review of the literature from archaeological studies to local histories, but there is still no single source that draws this history together. Very few works are explicitly focused on the human history of the Escarpment itself (but see Gillard and Tooke 1975). Instead, a patchwork approach that integrates many diverse fragments is necessary (box 13.1).

> BOX 13.1 SCOPE OF SOURCES NECESSARY TO DOCUMENT
> THE HUMAN HISTORY OF THE NIAGARA ESCARPMENT
>
> - local and regional histories
> - scholarly articles and texts
> - historical maps and atlases
> - thematic studies (e.g., transportation, resource exploitation, rural development, landscape heritage)
> - government documents, including policy
> - professional reports
> - inspirational works (poetry, photography, painting, etc.)
> - print media (newspapers, magazines)
> - promotional brochures (municipal, corporate, entrepreneurial, etc.)
> - cultural-landscape assessment
> - field surveys
> - Internet
> - interviews with the expert and lay population
> - interviews with agency and organization representatives
>
> By the author, with permission.

THE BIOPHYSICAL LANDSCAPE

In its entirety, the Niagara Escarpment is fundamentally a geological feature, dating 450 million years back, to the Silurian Period. During glaciation it was contoured into the Michigan Basin. Its form can be traced from near Rochester, New York, around the tip of Lake Ontario, and northwest to Tobermory. From there it continues under Lake Huron, emerging in Manitoulin Island, underwater again into Michigan, finally curving south through Wisconsin and into Illinois (figure 13.2). A collaborative research initiative was undertaken by Gordon Nelson and others in 1999 focusing on this Great Arc as a special landscape or sequence of related landscapes (for a collection of initial studies see Nelson and Porter 2002).

Sustained research on the biophysical characteristics of the Ontario portion of the Niagara Escarpment over the past few decades has revealed its characteristic features. These are inland and coastal cliffs

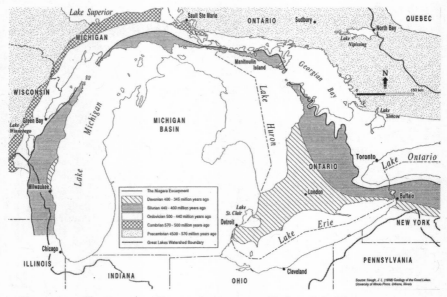

Figure 13.2 The Great Arc. The prominent topographic face of the Niagara Escarpment is indicated with a solid black line. From Nelson, Lawrence, and Beck (2000), with permission

in the north and south and a moraine-covered midsection with deep valleys and high rolling hills. In the south the Escarpment bisects with the Carolinian zone; the Dundas Valley re-entrant is the current northern limit of several plant and animal species. Other re-entrants, or erosion valleys, are prominent in the north, with two on the coast and two inland. In this northern section erosion along the coast of Georgian Bay has generated many spectacular caves and the famous flowerpot formations. The number of waterfalls along the Escarpment seems unresolved; one recent survey in the mega-city of Hamilton (former region of Hamilton Wentworth) tallied ninety-two in that area alone (Head and Nixon 2006). A methodologically rigorous, yet incomplete, survey in 2005 by the Hamilton Conservation Authority (2005) counted sixty-five.

The Escarpment has been shown to be important in providing unique and unusual habitat for flora and fauna, owing to the soil composition, microclimates, and topography. These features also contribute to its importance as a migratory flyway for birds of prey, at least as far north as Burlington. Over 40 percent of Ontario's rare flora are present (box 13.2), as well as twenty-five species of at-risk birds and several vulnerable species of mammals (box 13.3). As a result of the comprehensive landscape change all around it, the Escarpment has become a remnant haven for plant and animal life that was once more widespread. Consequently, considerable efforts have been made by residents as well as governments at all levels to establish protected areas for nature conservation. These initiatives are considered further in the last half of this chapter. Our overview now turns to human settlement history.

BOX 13.2   RARE FLORA OF THE NIAGARA ESCARPMENT

Southern Escarpment: cucumber-tree, paw-paw, green dragon (Arisaema dracontium), tuckahoe (*Peltandra virginica*), American columbo (*Frasera virginiana*)

Northern Escarpment: Rand's goldenrod (*Solidago glutinosa ssp. randii*) and roundleaf ragwort (*Senecio obovatus*)

The threatened American ginseng (*Panax quinquefolius*) occurs in rich sugar maple forests along much of the Escarpment.

Oldest trees in eastern North America (1,000 years): *Thuja occidentalis* (Eastern qhite cedar)

Significant species endemic to the Great Lakes occur on the Bruce Peninsula portion of the Escarpment, including lakeside daisy, dwarf lake iris, Hill's thistle, Provancher's Philadelphia fleabane, and Ohio goldenrod.

Ferns: 50 species recorded, including wall-rue (*Asplenium rutamuraria*), an Appalachian species rare in Canada. Most of the world population of the North American subspecies of Hart's-tongue fern (*Phyllitis scolopendrium* var. *americana*) occurs along the Escarpment.

Orchids: 37 species recorded in the northern parts of the Escarpment, including calypso orchid (*Calypso bulbosa*), Ram's-head lady-slipper (*Cypripedium arietinum*) and Alaska rein orchid (*Piperia unalascensis*).

Source: Niagara Escarpment Commission (2004), *About the Niagara Escarpment*. http://www.escarpment.org/About/facts.htm, with permission.

BOX 13.3   RARE FAUNA OF THE NIAGARA ESCARPMENT

Of the breeding species [of birds], 25 are considered nationally or provincially endangered, threatened, or vulnerable, including the bald eagle, red-shouldered hawk, black tern, Louisiana waterthrush, and hooded warbler.

Fifty-five mammal species and 34 species of reptiles and amphibians have been recorded. Rare species include the endangered north dusky salamander, the threatened Eastern Massasauga rattlesnake, the vulnerable southern flying-squirrel, and the rare eastern pipistrelle.

Source: Niagara Escarpment Commission (2004), *About the Niagara Escarpment*, http://www.escarpment.org/About/facts.htm, with permission

---

HUMAN SETTLEMENT

For purposes of discussion the history of human settlement along the Niagara Escarpment is organized here into three sections dealing with indigenous peoples, Loyalists and later Euro-Canadians, and industry and urbanization. While this discussion is intended to be loosely chronological, it must be understood that these artificial demarcations are not mutually exclusive, since, for example, the First Nations have vital communities in this landscape today.

Indigenous Peoples

Studies in southern Ontario, and specifically along the Escarpment, indicate the presence of humans in several locations as far back as the Paleo-Indian phase, 11,000 to 9,000 years ago. In some cases First Nations settlements were present when the Europeans arrived. Colonization history, even in this area, is complex and rife with injustices; today only a small fraction of land is reserved for the First Nations, for example the Chippewas of Nawash Unceded First Nation on Cape Croker in the Saugeen (or Bruce Peninsula) and the Wikwemikong Unceded Indian Reserve on Manitoulin Island. While these peoples have thrived in numbers, the Petun in the inland region between Georgian Bay and Lake Ontario disappeared shortly after arrival of the first Caucasian settlers in the mid-nineteenth century.

The Attiwandarons (now called Neutral Indians) had settlements in the southwest between the Great Lakes but are believed to have been displaced by the Iroquois of New York during the early 1650s. The area was subsequently a hunting ground for the Iroquois. In 1784 the British Crown "granted" them six miles (9 kilometres) on either side of the Grand River from Lake Erie to the headwaters because of their support of the King during the American Revolution. This Haldimand Land Grant, as it is known, did not extend east as far as the Escarpment. However, the Iroquois did maintain a major trail along the base of the scarp between Queenston and Ancaster, at which point it turned west to cross the Grand at what became known by colonists as Brant's Ford, the eventual site of the city of Brantford. The western extension of this fur-trade era trail is known as the Detroit Path and led at least to Detroit; from Queenston the trail passed through New York to Albany (Burghardt 1969).

Perhaps one of the best understood archaeological sites along the Escarpment is at Crawford Lake Conservation Area in Halton, where a fifteenth-century Iroquoian village has been reconstructed as an educational resource.

Loyalists and Later Euro-Canadians

United Empire Loyalists, who were predominantly of European descent, began settling the narrow stretch of land between Lake Ontario and the Niagara Escarpment in the last three decades of the eighteenth century. The central region between Lake Ontario and Georgian Bay was settled between 1800 and 1850; the northernmost section only after 1860. To the new non-Native settlers, the Escarpment was a blessing because its waterfalls provided power for mills, which Gillard and Tooke (1975) note was one of three primary reasons for new town sites in the settlement of Upper Canada.

Major settlements evolved along the lake shores at each end of the Escarpment, initially below the scarp face. Lake transportation was a primary focus for both Hamilton on Lake Ontario and Owen Sound on Georgian Bay. Historically the economy of the central and southern Escarpment was based in agriculture, as it was for most of southern Ontario. Distinct soils and microclimates resulting from the presence of the Escarpment, however, were found to be ideal for specialty crops, especially tender fruit in Niagara.

Survey plots for most of Ontario were highly regularized rectangular grids, and the farm fields typically followed this pattern. But along the more rugged parts of the Escarpment its regularity was interrupted, leaving a distinct swath of irregular field plots. Through this central zone between Lake Ontario and Georgian Bay the settlements were few and small, and this remains the pattern today (figure 13.3). Land that had been cleared by the first settlers was sometimes found to be too difficult for farming, and was often allowed to regenerate back into forest. In turn it became highly attractive as a lush retreat from the industrial cities. Passive and active recreation were distinct features of Escarpment landscapes from very early on, and have continued to grow in popularity.

Industry and Urbanization

Although the surface limestone was problematic for farming in many parts of the Escarpment, it was a ready source of materials for nineteenth-century construction, much of it for new rail beds and for the growing city of Toronto. But it also posed challenges for transportation and development owing both to elevation changes and to the costs of blasting into solid limestone in, or immediately below, the ground surface. In the early nineteenth century at least two major canals were proposed; by 1829 the first of four eventual iterations of the Welland Canal cut straight through the Escarpment to create a shipping route between Lake Erie and Lake Ontario. Another proposed canal was never constructed; it was to run along the toe of the Escarpment from Hamilton Harbour to Fonthill, where the vertical scarp gives way to more gradual slopes. At that point the canal could more easily traverse the Escarpment and turn south toward Lake Erie. When the railways came they also had to take account of the

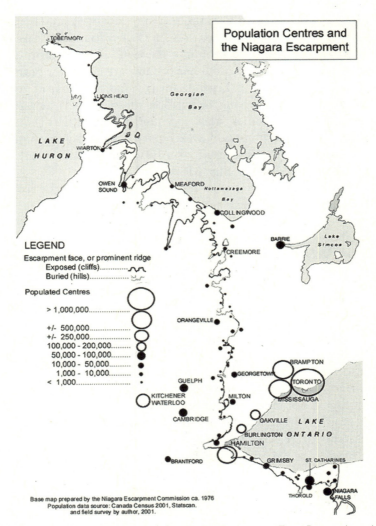

Figure 13.3 Population Centres and the Niagara Escarpment. From Preston (2003), adapted from NEC (1976), with permission

difficult topography, as did the roads and urban street systems. In the introduction to this volume, Nelson and Troughton comment on how the route taken by the railroads could determine the fate of new towns – indeed, changes in transportation technology from shipping, to rail, to highway, have brought both growth and decline along the Escarpment as well as elsewhere.

Industry was the driver in Hamilton until recently; similarly, shipping was the economic foundation of Owen Sound. In this post-industrial era, Hamilton's economy is slowly shifting more to the service sector, and Owen Sound has become attractive for both rural and ex-urban retirement. Ecotourism is a major thrust of its economic development, relying heavily on the waterfalls and rugged features of the Escarpment that surround the small city. Likewise, the port town of Collingwood, which once had a thriving shipbuilding industry, now focuses heavily on servicing tourists at the adjacent Escarpment resort of Blue Mountain.

Today the quarries that were the valued source of stone for early development are seen by the public in an entirely different light – as a blight on the landscape and a source of bitter dispute. Waterfalls now serve as scenic attractions and are part of the resource economy through recreation and tourism rather than energy. This is where the relationship to Toronto comes more clearly into view. Toronto has had and continues to have a significant direct impact on the Niagara Escarpment and its communities in several ways. It looks to the Escarpment for construction materials for endless development, and it also looks for picturesque respite from the city – for recreation and for ex-urban residential desires. The regular commutershed for Toronto now consumes over half the Escarpment, from at least St Catharines to Creemore. In the Escarpment's central rural areas the population is becoming increasingly non-rural; city folk own much of the land as weekend and retirement property, which has resulted in a notable decrease in family farms – and farming at all. Some of these properties have been left to naturalize or have been planted out under forestry programs sponsored by the provincial Ministry of Natural Resources. These are trends that began decades ago and have now reached a level where the effects are visible and widely recognized. They raise the issue of policy and related institutional arrangements, since the beginnings of policy attention specifically to the Escarpment arose out of concerns that urbanization was about to destroy this ribbon of green.

VALUING THE ESCARPMENT

A precursor to recognizing the Niagara Escarpment as a distinct geo-region is the attribution of value to it as distinct within its broader environmental context. This articulation of value can be seen in the emergence of special policy arrangements, in civic initiatives, and in place promotion. Each of these is discussed in turn.

Policy Arrangements

Today the Escarpment landscape passes through eight counties or regions, twenty-four municipalities or townships, and about one hundred hamlets, villages, towns, or cities. The boundaries of the Greater Toronto Area (GTA) extend over about one-sixth of its length. As a result, there are several layers of policy interest in this landscape and numerous perspectives and concerns regarding appropriate land use. The following sequence of events illustrates how the Escarpment was first discussed in the political arena, the values that were expressed, the clear relationship to concerns about Toronto's growth – and that of the whole Golden Horseshoe – and how the public at large was drawn into the discourse. This is summarized in table 13.2. This section proceeds from initial calls for conservation, through the establishment of the Niagara Escarpment Commission (NEC) and its planning policies, to more recent related initiatives, and it relies primarily on government documents and resources in the NEC archives.

Early Stirrings

In 1958 a group of elected local government representatives prepared a formal petition to have the Escarpment proclaimed a provincial park, describing it as "a symphony of rugged grandeur, broad vistas, historic lore, primitive beauty" (Tri-County Committee of Wentworth, Lincoln, and Welland 1958).

In 1959, the idea of creating the Bruce Trail was formulated by a small group of naturalists, an initiative that has been singularly influential in raising awareness of the continuous linear nature of the Escarpment. The Bruce Trail Association (BTA) was incorporated in 1963, and the eight-hundred-kilometre hiking trail officially opened in 1967. Founding members aggressively pursued provincial action to protect the Escarpment from 1960 onward.

In 1961 the planning director for the Town of Burlington and a co-founder of the Bruce Trail wrote that the Escarpment was "as moving as the romantic ruins of mediaeval fortresses; an unchanging and forceful element in a rapidly changing cultural landscape.

Table 13.2
Chronology of Events Drawing Public Attention to the Niagara Escarpment as a Distinct Landscape in Ontario

Year	Event
1958	Petition for Escarpment protection as provincial park: Tri-county Committee of Wentworth, Lincoln, and Welland
1959–67	Inception and creation of Bruce Trail by local naturalists
1961	Conference paper calling for protection of Escarpment by Burlington town planner, at Conservation Council of Ontario
1962	Dufferin Aggregates blasts through the Escarpment face at quarry along Highway 401 at Milton
1966	Report of Hamilton & Halton Conservation Authorities on quarrying, calling for protection of Escarpment
1967	Report of the Select Committee of Conservation Authorities with public consultation
1968	*Niagara Escarpment Study, Conservation and Recreation Report* (Gertler Report)
1970	Niagara Escarpment Protection Act
1971	Design for Development in Ontario: The Initiation of a Regional Planning Program
1972	*To Save the Escarpment: Report of the Niagara Escarpment Task Force*
1973a	Niagara Escarpment Planning and Development Act (with policy statement)
1973b	Parkway Belt Planning and Development Act (with policy statement)
1977	"Preliminary Proposals" (first draft, Niagara Escarpment Plan)
1978	Coalition on the Niagara Escarpment (CONE) civil society watchdog umbrella organization formed
1979	"Proposed Plan for the Niagara Escarpment" (second draft)
1983a	Report of the hearing officers based on unprecedented (for Canada) public consultation
1983b	Response to Hearing Officers
1983c	Final Proposed Plan
1985	Niagara Escarpment Plan adopted
1990a	Initiation of five-year review of Niagara Escarpment Plan
1990b	Niagara Escarpment Plan Area designated a UNESCO World Biosphere Reserve
1994	Revised Niagara Escarpment Plan released based on public consultation
1997	Escarpment Biosphere Conservancy land trust incorporated as charitable organization
1999	Initiation of second five-year review of Niagara Escarpment Plan
2000	Niagara Escarpment Planning and Development Act amended with public consultation
2001	Oak Ridges Moraine Protection Act
2002	Oak Ridges Moraine policy plan released
2005a	Revised Niagara Escarpment Plan released based on public consultation
2005b	Greenbelt Act passed; Greenbelt Plan adopted for Greater Golden Horseshoe incorporating Niagara Escarpment Plan Area and Oak Ridges Moraine Plan Area
2005c	Places to Grow Act 2005 passed for Ontario's Greater Golden Horseshoe
2006	Growth Plan for the Greater Golden Horseshoe released by province (from Places to Grow Act)

Source: By the author, with permission.

It threads together in a unity as distinguished as it is unappreciated, many diverse and disparate elements. It cuts across 'conurbation Canada' at precisely the right point to focus attention on the form of our major metropolitan areas; and winds its way to Tobermory as a veritable compendium of opportunities for the re-creation of our environment" (Pearson 1961).

Public attention was dramatically heightened in 1962 when Dufferin Aggregates blasted a huge gap in the Escarpment face at their Milton quarry, a gap that was clearly visible from the TransCanada Highway, 401. Only a few years later in a report on quarrying activities, the Conservation Authorities of Hamilton and Halton Regions wrote that the Niagara Escarpment "is one of the roots of *provincial identity*, yet paradoxically it needs vigilant protection." That report also proclaimed that "the Escarpment is without doubt the single most important scenic and recreational land asset of southern Ontario" (Beales et al. 1966).

This was immediately followed by the Report of the Select Committee of Conservation Authorities (Ontario Ministry of Natural Resources 1967), which was produced in part as a result of public consultation designed to identify specific public concerns. It set the tone for Niagara Escarpment planning as a special kind of regional plan, one that focused on conservation in concert with compatible development.

Mounting civic pressure on the provincial government led it to commission a report on the scope of and means for Niagara Escarpment protection. The terms of reference called for a study area extending two miles (3 kilometres) wide on each side of the "main axis" or prominent scarp features. Completed in 1968, the so-called Gertler Report included maps of the Escarpment from Queenston to Tobermory subdivided into zones based on recreational attractiveness, for the development of a natural park system (Gertler et al. 1968).

At the same time, the Metro Toronto and Region Transportation Study was completed, in which the Escarpment was conceived as a key feature of Toronto's open-space system. In 1970 *Design for Development: The Toronto Centred Region* also called for conservation of the Niagara Escarpment for public recreation (Ontario Ministry of Treasury, Economics, and Intergovernmental Affairs 1970). From these reports it can be seen that much of the provincial government's early interest in conserving the Escarpment came out of regional planning for Toronto.

Legislation and Policy

Interim legislation was passed in 1970, although with little resulting action. More public pressure resulted in the government-directed production of the task force report *To Save the Escarpment* (Government of Ontario 1972). Public meetings engaged thirty-five hundred residents, reinforcing or awakening a sense of this landscape as a distinct and important entity. Six months later, new legislation was passed establishing the Niagara Escarpment Commission and requiring studies leading to a policy plan. The policy document accompanying the legislation was part of the *Design for Development* series. An associated policy document in the same series and released on the same day focused on the Parkway Belt, an undeveloped corridor around Toronto intended as a limit to development and containing the urban infrastructure. It also argued for the preservation of the Escarpment, which it crossed, describing it as "one of the world's most impressive natural wonders" (Ontario Ministry of Treasury, Economics, and Intergovernmental Affairs 1973).

Numerous studies of the Niagara Escarpment were undertaken as stipulated in the legislation, launching a process that has intensified and continues today. Of particular interest was the debate about how to define the natural boundaries of the feature in order to determine the extent of the special planning area. Some argued that it should be limited to the obvious visible features, others argued for these features with an added buffer zone to reduce impacts and conserve ecological functionality, and still others were concerned primarily about how protective policies would negatively affect opportunities for development, and thus their own economic interests or those of their communities. This intense focus on defining a georegional boundary that transected many municipalities and privately owned properties had a significant effect on perceptions about just what the Escarpment is.

The preliminary plan proposals were released in 1977 with a focus on ecological protection and enhancement, sustainable resource use, recreational capacity and function, and the maintenance and enhancement of scenic qualities. It was heavily criticized for the extent of the plan area and its detailed prescriptive content.

A completely redesigned proposed plan taking a much more generalist approach was produced for public consideration in 1979. Public hearings on this document attracted 9,300 residents. It was revised again, and a final proposed plan put forward in 1983. By the time of the adoption of the Final Plan in 1985 the area directly affected had been reduced by 63 percent owing to arguments by local governments and individual landowners (figure 13.1). In keeping with the terms set out in the legislation, public plan reviews are to occur every five years. The first review was initiated in 1990 and took several years to complete; the second began in 1999, and the revised plan was released in 2005 (Niagara Escarpment Commission 2005). Each step of the policy process thus brought the Escarpment as a single provincial entity to the public's attention. The legislation and plan supersede all other provincial, regional, and local plans affecting the designated lands.

In 1990 the Niagara Escarpment plan area was designated a UNESCO World Biosphere Reserve in recognition of the plan and its objectives and the presence of two national parks within its boundaries. This designation is non-regulatory, but it does bring international attention to the goals of sustainable development and conservation. It has also been the impetus for an extensive, continuing program of environmentally oriented research.

Recent Related Initiatives

More recently three major initiatives of the provincial government have returned the public's attention to the Niagara Escarpment. First was the development of legislation in 2001 and policy in 2002 for the Oak Ridges Moraine, another transboundary landscape whose western edge intersects the Escarpment's midsection and extends eastward through Northumberland County. Central concerns driving this effort were similar to those informing Escarpment planning: the protection of water resources and the impacts of rapid suburban development especially related to Toronto.

The second initiative drawing public attention back to the Escarpment was the proposed route for the Mid-Peninsula Transportation Corridor, a major highway between Toronto and Niagara planned by the Harris Conservative government in the late 1990s that would require a massive cut in the Escarpment face somewhere in the Burlington/Milton area. The McGuinty Liberal government has "reworked" the controversial project as the Niagara to GTA Corridor in conjunction with their 2005/6 legislation and Growth Plan for the Greater Golden Horseshoe. An environmental assessment study was initiated at the end of 2006, and a limited public consultation was held in spring 2009.

The third major relevant provincial initiative is the Greenbelt Plan, adopted in 2005. This plan area embraces over 728,000 hectares integrating the areas subject to the Niagara Escarpment and Oak Ridges Moraine Plans with additional rural lands, stretching around Toronto in southern Ontario. Its extent was determined through assessment of the natural heritage and agricultural systems and the needs of existing settlement areas, and its objectives closely mirror those of the Niagara Escarpment Plan.

With all the various layers of policy affecting the lands along the Niagara Escarpment, the provisions of the Niagara Escarpment Plan prevail, with the exception that the Greenbelt Plan adds a provision for increasing efforts to establish protected areas. Extensive public participation in policy development through public meetings, hearings, written submissions, and internet access, as well as media coverage and the activities of local and regional groups of concerned residents have all combined to influence the public understanding and image of what the Niagara Escarpment is.

Civic Initiatives

The planning process was paralleled by the emergence of NGOs and activist groups concerned to protect the Escarpment and ensure that the government upheld its commitments. Many of them became organized under the Coalition on the Niagara Escarpment (CONE), a watchdog NGO that has provided the public at large with information on commissioner decisions and plan infractions and that has repeatedly stepped into the legal and policy arenas to challenge development plans that are inconsistent with the Niagara Escarpment Plan. CONE also raises and reinforces public awareness through a civic sponsorship roadside signage program welcoming travelers to the Niagara Escarpment World Biosphere Reserve (figure 13.4).

Local groups likewise draw media attention to activities understood to be destructive of the Escarpment's valued features. The Burlington area Citizens Opposed to Paving the Escarpment (COPE) worked to inform the public and contribute to the official review of

Figure 13.4 Roadside awareness program operated by CONE: Welcome to the Niagara Escarpment World Biosphere Reserve. Photo by the author, with permission

the proposed Niagara-to-GTA Corridor highway development. The Grimsby-based Friends of the Escarpment (FOE) formed to bring legal and policy action against logging the face of the Escarpment in the Niagara Region.

Activist and non-governmental conservation groups also undertake all manner of stewardship initiatives. Efforts by the Escarpment Biosphere Conservancy (EBC) land trust have extended public awareness of the Escarpment from the Bruce Peninsula into Manitoulin Island with several well-publicized acquisitions totalling over hectares (EBC 2009). The Bruce Trail Association (BTA) has acquired over twenty-five hundred hectares of land in trust as well, held by the Ontario Heritage Trust (BTA 2009). These activities have gained such momentum that the 2005 version of the Niagara Escarpment Plan included new provisions allowing for protected areas held by "conservation organizations." Additional Escarpment lands are held in conservation trusts through the Federation of Ontario Naturalists and the Nature Conservancy of Canada. The Conservation Authorities and the province have also invested in public lands along the Escarpment such that it hosts an extensive network of open spaces; in 2007 the NEC reported that the Niagara Escarpment Parks and Open Space System (NEPOSS) had a total combined area of 35,684 hectares in 131 parks and protected areas. These are all linked

Figure 13.5 Escarpment Country campaign, brochure cover (left) and roadside sign and highway billboard photos (right and below). Poster with permission of Milton Chamber of Commerce, photos by the author, with permission

together by the Bruce Trail. BTA members voluntarily steward individual sections of the trail and often work to raise awareness at a local level.

Place Promotion

Municipalities have promoted themselves in light of the special international status of the World Biosphere Reserve. Plans for two major interpretive centres were in the works for several years in Owen Sound (Escarpment Centre Ontario) and Dundas (Giant's Rib Discovery Centre). Funding has been a major hurdle for both, and if either is realized it will likely be in a more modest form than originally envisioned. With all of this attention, many municipalities have begun to anchor their public image on the Escarpment as a unique and celebrated place. Milton has even trademarked it and implemented a highway and road sign program proclaiming "Escarpment Country" (figure 13.5).

The private sector has not missed out on the opportunity to promote itself in relation to the Niagara Escarpment, Bruce Trail, and World Biosphere Reserve. Real estate developers have capitalized on this particularly in the Grimsby area, with magazine and billboard ads extolling their distinctive views and "green" lifestyles. A 2002 survey of promotional literature and even company names of wineries in the Niagara Region showed extensive identification with the Biosphere Reserve status, with the Bruce Trail, and with the ecological and picturesque landscape experience afforded by this special environment (Preston 2003). This imagery is projected to the public at large through mass media and the Internet, and it is also absorbed by locals in the course of everyday activities.

SPATIAL CONCEPTUALIZATIONS: THE EMERGENT GEOREGION

In this last section, I consider three things: how the Escarpment has been conceptualized spatially in policy, by non-fiction authors and academics, and by its current residents. Although the influence of the policy planning process has been profound in raising public awareness, my research shows that spatial perceptions vary dramatically between policy and affected residents, and that a third group – academics – tend to have yet other ways of conceptualizing this landscape.

The Policy View

Escarpment planners began with several attempts to divide the landscape into a sequence of units based on physical character, recreational capability, and attractive features (e.g., Gertler et al. 1968) but eventually shifted to take on a landscape-planning approach that saw the Escarpment as a single feature based on its geology, with a range of zones that could be present at intervals throughout. These were partly a matter of environmental quality and partly a matter of the development objectives of urban politicians, recreational objectives, and resource-use objectives. As a result, the Niagara Escarpment Plan has the landscape zoned according to the following characteristics and uses: natural, protection, rural, urban, minor urban, resource, recreation, and public land. Each zone type is identified in numerous places along a single contiguous feature from Queenston to Tobermory.

The Academic View

The first studies to be published about the Niagara Escarpment as a whole focused on geology: there was a thesis in 1936 and another in 1955 that was published by the Geological Survey of Canada in 1957. Looking at a selection of the literature since the 1968 Gertler Report, it seems that academics and authors of popular non-fiction literature have taken a different analytic approach to thinking about the Escarpment. They typically describe it as a sequence of distinct segments set end to end based on topographic character and relative location (Kosydar 1996); slope unit systems (Moss and Milne 1998); relative position-south, centre, north (Gillard and Tooke 1975); profile types (Reid 1977); forest zones and major features (Stephens and Stephens 1993); and municipal boundaries (Insight Canada 1989).

The Public View

Many nineteenth-century documents indicate the authors' personal appreciation for the Niagara Escarpment as a dramatic feature of local and county landscapes; none that I have seen suggest a public notion of it as comprising a distinct georegion or an awareness of its full geological extent. The 1875 Historical Atlas for Wentworth County, for example, traces it from New York State, through Queenston, Ontario, around the bay to Burlington and north to Flamborough, but no further (Page and Smith 1875).

In a set of interviews that I completed in three Escarpment communities in 2002 I found that most participating residents held a completely different sense of space, one that was grounded in personal experience. They typically had a sense of their local area as a series of concentric spaces from the immediate to the extra-local and then regional, and further, a sense that the edges of these spaces were strongly influenced by the visible and known features of the Escarpment. This inner ring includes the community-defining features that are familiar from their presence in everyday activities. They seem to be most often thought of in terms of named places, where names for places or features are known. This occurs most where the places are bounded, protected areas – typically conservation areas or other well-known local features.

The second ring, at a larger scale, is the extra-local or regional; this is typically very loosely defined by the regional topographic character in which the local area sits. Finally, the third ring is the area included in the Niagara Escarpment Plan. I state it this way because it seems that this provincial-scale awareness is primarily the result of the plan and the policy process, the Bruce Trail, and to a lesser extent, the World Biosphere Reserve designation. In other words, these institutional and civic initiative overlays have had a significant influence on people's recognition of the Escarpment as a large-scale feature based in geology and often characterized as "continuously forested." I suspect that this recognition is highest among Escarpment landowners or residents of communities affected by the plan. This was hinted at in a conversation with a shopkeeper in a hamlet that was ecologically part of the Escarpment landscape who believed they were not an Escarpment community because they were outside the plan area. Many interview participants were also aware of the feature's extension into Manitoulin Island, and just a few knew about the much larger feature of the Michigan Basin.

There is no question that the Niagara Escarpment was highly valued by many residents and visitors at the local and regional scale since at least the late-eighteenth-century documentary evidence supports this. The public sense of a continuous Niagara Escarpment as a special landscape or georegion in its own right appears to have emerged in the early 1960s and to have become codified in the public imagination by the 1980s. One of the great challenges – indeed a source of fiery dispute – in the policy-development process, especially in Grey and Bruce Counties, was determining the boundaries of this georegion. While the outside boundaries of the Niagara Escarpment are firmly articulated by necessity in policy, and perhaps slightly less rigidly among scholars and other authors, my sense is that among residents the boundaries are only specific only if their property is in the immediate vicinity.

What is important for the present discussion is that there is an increasing sense of a continuous, internationally recognized – even if at times locally contentious – feature and that this is a construct of the past fifty years. The actual image of the Niagara Escarpment on maps seen and discussed by thousands of people during this time, I think, can be credited with at least contributing to the social construction of the Niagara Escarpment as a distinct georegion. As urban pressures and environmental awareness increased during this period, the natural character of the Escarpment made it highly desirable for residential and entrepreneurial uses, and equally a focus for the activities of non-governmental conservation organizations and groups.

RESEARCH AND PLANNING: SOME CONCLUSIONS

Given the commitment to environmental research for the Escarpment under the Man and Biosphere program, it could be highly beneficial to coordinate existing environmental research with a program of investigation into the dynamics of population change and their impacts on this landscape. For example, what environmental processes are undergoing changes as a result of increased ex-urban settlement, including the division of farmland into fifty acre – and smaller – plots with estate housing? How do these residents view their impact on the Escarpment? And in the realm of planning, how effectively are municipalities applying the maxim that development must be compatible with the protection of water resources and other significant ecological and cultural features of the Escarpment?

While some landowners disliked the policy protection of this landscape for the constraints they felt it imposed on them, it now appears that the result of the protection in many cases has been an air of exclusivity and increased property values. Now the affluent appreciate assurances that their plot of paradise will not be spoiled by box malls and subdivisions. Some take great pride in being part of the Niagara Escarpment "happening," but some also seek to circumvent the regulations for their own amenity.

Diligence is the most tenuous aspect of planning here: the Niagara Escarpment Commission has exceptionally limited capacity to monitor or prevent infractions, and governments at various levels are rarely concerned to take on the active role of steward. Civil society makes remarkable contributions under these circumstances, but a greater commitment from governments at all levels is needed to understand the changing pressures and interests at work and to adapt planning practices to these changes in order to meet the objectives laid out in the Niagara Escarpment Plan. This includes the importance of recognizing that the interests of Toronto and the GTA must not be the only factors guiding planning in this georegion.

Escarpment communities each have their own histories, identities, and values, and these must also be recognized in the process.

REFERENCES

Beales, F.W., T.A. Beckett, D.J. Murray, R. Sherwood, G.F. West, and N.E. Wilson. 1966. "Quarrying and the Niagara Escarpment: A Brief to the Select Committee on Conservation, of the Ontario Legislature by the Halton and Hamilton Region Conservation Authorities Escarpment Committee." Unpublished report. Niagara Escarpment Commission Library, Georgetown, Ontario.

Bruce Trail Association. 2009. http://brucetrail.org/.

Burghardt, Andrew. 1969. "The Origin and Development of the Road Network of the Niagara Peninsula, Ontario, 1770–1851." *Annals of the Association of American Geographers* 59:417–40.

Escarpment Biosphere Conservancy. 2009. http://www.escarpment.ca/.

Gertler, Leonard, et al. 1968. *The Niagara Escarpment Study Conservation and Recreation Report.* Toronto Regional Development Branch, Department of Treasury and Development.

Gillard, William, and Thomas Tooke. 1975. *The Niagara Escarpment from Tobermory to Niagara Falls.* Toronto: University of Toronto Press.

Government of Ontario. 1972. *To Save the Escarpment: Report of the Niagara Escarpment Task Force.* Toronto: Queens Printer. Niagara Escarpment Commission Library, Georgetown, Ontario.

Hamilton Conservation Authority. 2005. *Unpublished Research Report on Waterfalls in the City of Hamilton.* http://www.conservationhamilton.ca/Asset/iu_files/WaterfallsreportApril05.pdf.

Head, Stephen, and Robert Nixon. 2006. *Waterfalls in the City of Hamilton.* Hamilton Naturalists Club. http://www.hamiltonnature.org/local species/waterfalls_intro.htm.

Insight Canada Research. 1989. *Niagara Escarpment Public Awareness Survey.* Unpublished report prepared for Ontario Heritage Foundation. Niagara Escarpment Commission Library. Georgetown, Ontario.

Kelly, Peter E., and Douglas W. Larson. 2002. "A Review of the Niagara Escarpment Ancient Tree Atlas Project, 1998–2001." In Shannon McFayden and Richard Murzin, eds., *Leading Edge '01 Conference Proceedings.* Georgetown, ON: Niagara Escarpment Commission.

Kosydar, Richard. 1996. *Natural Landscapes of the Niagara Escarpment.* Dundas, ON: Tierceron Press.

Matthes, U., J.A. Gerrath, and D.W. Larson. 2003. "Experimental Restoration of Disturbed Cliff–Edge Forests in Bruce Peninsula National Park, Ontario, Canada." *Restoration Ecology* 11(2): 174–84.

Moss, Michael R., and Robert J. Milne. 1998. "Biophysical Processes and Bioregional Planning: The Niagara Escarpment of Southern Ontario, Canada." *Landscape and Urban Planning* 40: 251–68.

Nelson, Gordon, Patrick Lawrence, and Catherine Beck. 2000. "Building the Great Arc in the Great Lakes Region." In *Leading Edge '99: Making Connections*. Conference Proceedings, eds., Stephen Carty et al., Georgetown, ON: Niagara Escarpment Commission.

Nelson, J. Gordon, and James Porter, eds. 2002. "Building the Great Arc: An International Heritage Corridor in the Great Lakes Region." *Occasional Paper* 29. Waterloo, ON: Heritage Resources Centre, University of Waterloo.

Niagara Escarpment Commission. 2004. *About the Niagara Escarpment*. http://www.escarpment.org/About/facts.htm.

– 2005. *Niagara Escarpment Plan*. Toronto: Queen's Printer for Ontario.

– 2011. *ONE Monitoring Program*. http://www.escarpment.org/education/monitoring/index.php.

Niagara Escarpment Commission and Ontario Ministry of Natural Resources. 1976. *Landscape Evaluation Study, Niagara Escarpment Planning Area*. Unpublished report. Niagara Escarpment Commission Library, Georgetown, Ontario.

Ontario. Ministry of Natural Resources. 1967. *Report of the Select Committee of Conservation Authorities*. Unpublished report. Niagara Escarpment Commission Library, Georgetown, Ontario.

– Ministry of Treasury, Economics, and Intergovernmental Affairs. 1970. *Design for Development: The Toronto Centred Region*. Toronto: Queen's Printer for Ontario.

– 1973. *Development Planning in Ontario: Government Policy for the Parkway Belt: West*. Toronto: Queen's Printer for Ontario.

Page & Smith. 1875. *Illustrated Historical Atlas of the County of Wentworth, Ont*. Toronto: Page & Smith.

Pearson, Norman. 1961. *Planning for Recreational Development of the Niagara Escarpment*. Paper presented at Conference on the Recreational Potential of the Niagara Escarpment, organized and sponsored by the Conservation Council of Ontario. Catalogued in Porter Library, University of Waterloo.

Preston, Susan M. 2003. "Landscape Values and Planning: The Case of Ontario's Niagara Escarpment." PhD diss., University of Waterloo.

Reid, Ian. 1977. *Land in Demand: the Niagara Escarpment*. Agincourt, ON: The Book Society of Canada.

Stephens, Lorina, and Gary Stephens. 1993. *Touring the Giant's Rib: A Guidebook to the Niagara Escarpment*. Erin, ON: Boston Mills Press.

Tri-County Committee of Wentworth, Lincoln, and Welland. 1958. *Petition for the protection of the Niagara Escarpment in Ontario*. Niagara Escarpment Commission Library. Georgetown, Ontario.

FURTHER READING

Bruce Trail Association. www.brucetrail.org

Citizens Opposed to Paving the Escarpment. www.cope-nomph.org

Coalition on the Niagara Escarpment. www.niagaraescarpment.org

Conservation Ontario. http://www.conservation-ontario.on.ca

Escarpment Biosphere Conservancy. www.escarpment.ca

Niagara Escarpment Commission. www.escarpment.org

Ontario Greenbelt Foundation. www.ourgreenbelt.ca

Ontario Heritage Trust, Natural Heritage page. http://www.heritagefdn.on.ca/userfiles/HTML/nts_1_2744_1.html

Ontario Ministry of Natural Resources, Niagara Escarpment Program http://www.mnr.gov.on.ca/en/Business/LUEPS/2ColumnSubPage/STEL02_165805.html

Oak Ridges Moraine Foundation. www.moraineforlife.org

UNESCO Man and the Biosphere Program. www.unesco.org/mab/mabProg.shtml

14

The Changing Natural Heritage of Ontario in Broad Perspective

STEPHEN D. MURPHY

SUMMARY

In this chapter I discuss the challenges Ontario faces with the fragmentation of habitats from urban and rural development. Rural development has had a steady impact in Ontario, at least through the World War Two and postwar eras, when mechanization and chemical management became more common, particularly in agriculture and forestry. The response was similar in the United States and Western Europe, where there were demands for much greater conservation of the natural heritage by such means as parks and protected areas. These demands led to debates over the roles of parks and protected areas, including the Temagami controversies, the Living Legacy, and more recently the new Parks Act of 2007.

All of this was paralleled by growing urbanization into the period from the 1960s to the 2000s. Even though urbanites are often assumed to lose connection with the natural heritage, the impacts of industrialization, transportation, and suburbanization have become so obvious in polluted waterways, channelized riparian zones, losses of urban forest, and the homogenization of natural features that there have been sporadic backlashes against urban development. Some of these backlashes stem from disasters related to weather events; for example, Hurricane Hazel spawned the expansion of conservation areas, notably in flood-prone river valleys. But some of them have been related to a growing recognition of cumulative impacts, which has led to the ensuing protection of the Oak Ridges Moraine and the new greenbelt, the Rouge Valley, and Duffins Agricultural Preserve.

Within urban areas, some ideas are gaining acceptance. Brownfields sometimes are being turned into parklands, rivers are being dechannelized, and burns are prescribed for savanna and prairie habitat. However, there is the fundamental challenge of restoring what has been lost for one hundred years in some cities. This challenge will be met through more sustainable use of resources but also through ecological restoration (the ability to repair most of the structure and function of ecosystems), remediation (the cleanup of contaminated lands and waters), and reconciliation (achieving the best possible ecological goals in areas where urbanization or development makes other options unlikely in the short term).

THE LEGACIES OF AGRICULTURE AND FORESTRY AND ONTARIO'S RESPONSE

Ontario's Resource Extraction History

In most respects, the impacts of humans on the natural heritage in Ontario are similar to the first phases of natural-resource exploitation and urbanization in North America. Ontario went through its agriculture and forestry phases in the nineteenth century. The major impact in the south was the clearing of forests and the drainage of wetlands for cropping and animal husbandry. Mining, fishing, and the use of rivers for hydroelectric power were by no means insignificant but not as widespread as farming and forestry. Ontario has experienced impacts predating European settlement, but it was not until the nineteenth century that they occurred on cross-provincial scales and at intensive levels. The mechanization of both forestry and agriculture meant that land was cleared very effectively for resource extraction without much thought being given to the environmental effects or conservation.

Ontario was very much part of the school of wise use advocated by Gifford Pinchot, although it is interesting that Pinchot's bête noir, conservation pioneer John Muir, and fellow naturalist John Burroughs visited Ontario frequently and advocated the creation of protected areas. By 1900, less than 15 percent of southern Ontario's forests, riparian zones, and wetlands were intact on a landscape scale, and less than 50 percent of the boreal forests were intact – much of the rest had been clear-cut in a manner that left the ecosystem unable to support the natural succession of boreal forests. Beyond simple numbers, there are cumulative and emergent impacts of resource use in the sense that fragmentation and human activity can affect ecosystem processes and species several times that of the actual area being exploited. Hence, it is not enough to say that a certain area of land and water is being used for economic activity and assess that impact. One also has to examine the multiplier effect of this activity on ecological processes. No one has yet come up with a way of doing this, beyond case-by-case studies, because effects are very much scale-dependent. One simple way of thinking about this is to realize that a mouse will react differently than a caribou or a bear will to someone farming a small plot or deforesting large tracts of land.

The studies that have been done tend to show that forestry on scales of one million hectares, as is common in northern Ontario, can eliminate habitat for large mammals on a scale of five million hectares because of the cumulative impacts of fragmentation. The impact of agriculture varies because some species adapt well to cropping, others are diminished, and the overall ecosystem processes are disrupted by plowing, fertilizers, and pesticides. For this reason, people have to create conservation farm plans and similar initiatives to reduce the impact of agriculture. And the forest industry has pushed to certify and improve ecological performance. Whether these initiatives are succeeding is a matter of ongoing controversy.

Protected Areas as a Response to the Legacies and Lessons of Resource Extraction

To give Ontario its due, the impacts of resource extraction were recognized relatively early on in comparison to much of the rest of North America. Algonquin Park was officially created in 1893, and Rondeau followed in 1894. The first Ontario Parks Act was passed in 1913, although reserves were generally created where land was "not suited" for agriculture or for urban settlements. Indeed, park creation was slow until 1954 when the Division of Parks was created within the Department of Lands and Forests, now called the Ministry of Natural Resources. Thus, between 1893 and 1954 eight provincial parks were created: Algonquin, Ipperwash, Lake Superior, Long Point, Presqu'ile, Quetico, Rondeau, and Sibley, which was later changed to Sleeping Giant. Between 1954 and 1960, sixty-four more provincial parks were established. The need was recognized to put as much, or more, emphasis on conservation as on recreation, and new categories of parks were created in 1967 and are being revised today to reflect International Union for Conservation of Nature (IUCN) classifications.

In 1996, perhaps too much emphasis was put on entrepreneurship, fees, and cost recovery on a park-by-park basis, and the parks became focused on revenue generation above all else. Since 1996, however, considerable effort has been put on conservation at landscape scales in partnerships with the Nature Conservancy (1996) and Ontario's Living Legacy (1999). While this program has protected nearly 2.5 million hectares of land, there was great controversy over whether it meant that unprotected areas were too vulnerable to unregulated resource extraction. The current total of 280 parks (7.1 million hectares) represents an ambitious expansion of the parks system. The new Ontario Parks Act (2007) and the Endangered Species Act (2007) emphasize ecosystem integrity and species conservation. Coupled with initiatives such as the Niagara Escarpment Act (1973), the Oak Ridges Moraine Act (2001), and the Greenbelt Act (2005), Ontario has a reasonably solid record of conservation at the landscape scale, although again, there are legitimate criticisms of government attempts to truncate the effectiveness of these acts.

In parallel with the creation of provincial parks, Ontario also embarked on watershed-scale management and protection under the aegis of conservation authorities. Agriculture and other forms of deforestation in Ontario contributed to smaller versions of the Dust Bowls prevalent in the central parts of North America in the late 1920s and early 1930s. Deforestation, soil erosion, soil loss, and flash flooding during the droughty 1920s and 1930s led Ontario to recognize the value of watershed protection. The Federation of Ontario Naturalists and farm groups led the lobbying effort that culminated in the Ontario Conservation Authorities Act in 1946. This was a joint municipal and provincial government approach that was largely organized along the ecological boundaries of watersheds – very unusual and very prescient for the time. Much emphasis was placed on good agricultural and forestry practices for soil and water conservation and flood control, but as urbanization advanced, there was increasing emphasis on the conservation of natural areas that otherwise would be exploited. This reaction to urbanization was couched in terms of soil and water management, but it did force urban areas and larger counties to confront their roles in the management and protection of natural areas.

When Hurricane Hazel resulted in the deaths of eighty-one people in the Toronto region and outlying areas in 1954, conservation authorities recognized that human settlements and floodplains were a deadly combination, and more emphasis was placed on keeping vulnerable urban development out of floodplains. Riparian zones and wetlands were both given greater protection, and municipalities had to recognize that "parks" were more than just recreational and cultural areas, even within urban boundaries. Interest grew in the creation and management of protected areas within urban envelopes. This was most pronounced in what is now the Regional Municipality of Waterloo, where environmentally sensitive policy areas (ESPAS) were created for the protection of valued ecosystems. Other municipalities followed suit in Ontario. However, as accelerated urbanization isolated these ESPAS, the Region of Waterloo recognized that landscape-scale protection was needed, and it now creates Environmentally Sensitive Landscapes. Blair-Bechtel-Cruickston and the Waterloo Moraine are the first examples of these; they were created after some resistance from private landholders but little from politicians. Despite such efforts in Waterloo and other regions such as Toronto and Ottawa, there are still many problems related mainly to urbanization.

CONTEMPORARY PROBLEMS RELATED TO URBANIZATION

The effects of urbanization began as soon as incorporated settlements grew beyond the size of farming centres. But they have accelerated in contemporary times, particularly in Ontario, a province where the human population has substantially migrated to urban areas. As I have discussed elsewhere – and indeed much of the middle section of this chapter is a reflection on some earlier ideas – a fundamental issue is to understand the many environmental interactions caused by these changes. Humans are a big part of this process. The chemical environment is altered by the use and venting of compounds from industry and transportation, and all organisms. The many interactions respond to changes wrought by human development. Often the impacts are subtle, since they may accumulate for many years before a sudden change occurs. A forest may slowly die and then change relatively quickly into a meadow. A stream once teeming with a diversity of organisms may become permanently imbued with excess nutrients and then saturated with a few species of algae.

These are noticeable impacts, but their origin – and prevention – lies in the smaller-scale interactions.

When discussing how something works, the unspoken corollary is the question of what you do with this knowledge. This issue is controversial because it often challenges existing cultural norms, for example, about how how an urban area should be managed. Analyzing how urban ecosystems work becomes controversial because "evidence" can be challenged, and there is usually disagreement over whether enough data have been gathered, whether the methodologies and analyses were appropriate, how comparable the data are between locations or between sampling times, how the risk to urban environments should be measured, and what standards of risk are acceptable. Science offers a valuable framework to test ideas, but it can require long-term studies of decades or more – too long a time when critical decisions need to be made.

A reading of the extensive literature leads to the conclusion that sound urban governance requires measuring local ecosystem structures and processes and projecting how they respond to urban development. If a city contemplates building single, detached housing ever farther from the core, what ecological structures will be harmed? Will forests disappear more or less completely, and how will pollution from all the additional cars harm terrestrial and aquatic systems? Will this diminish the quality of all life, including human life? Such questions are now being directed regularly at Canadian urban planning and governance authorities.

So far there are no easy answers, because cities create such extreme impacts. If urban sprawl triumphs, then much of the diversity and functionality of the displaced ecosystems vanishes. Conversely, if there is reduced sprawl and more intensification of cities around urban cores, the current transportation systems and building designs help create a dome of pollutants trapped in the core, harming human and other organisms alike. At this point it is important to merely acknowledge the complexity of the problem – a first step that can lead either to defeatism or to the political will to address the challenges.

Pollution

Pollution is the result of an unwanted excess of chemical compounds and particulates. Contributors include industry and transportation, which add dioxides (nitrous, sulfur, carbon), ground-level ozone (O_3), and heavy metals (lead, zinc) to the environment. They help to form the smog that is visible in urban areas and compounds that harm organisms through cumulative toxicity or mutagenesis and alter ecosystem functions.

It is relatively easy to measure ambient concentrations of pollutants but harder to measure their impact. Chemicals interact with one another, and individuals and species respond differently to varying concentrations. Moreover, cumulative impacts may take a long time to create a measurable impact, and it is often too late to repair the damage by the time the effect becomes observable. Road salts, fertilizers, and pesticides are particular examples of subtle pollutants that cumulate over a considerable time before their presence registers. Most pollutants create problems ranging from individual-scale changes, such as changes in frogs that become anatomically hermaphroditic, to changes in nutrient cycling that alter the species and the processes that characterize an urban ecosystem. There are often unnoticed changes to the fungi and invertebrates in soils and water that ultimately alter the plants and animals visible to humans.

Impervious Surfaces Causing Flooding, Erosion, and the Spread of Pollutants

On a sub-watershed scale, a major problem is how replacing recharge areas possessing vegetation and porous soils with impervious roads, sidewalks, and buildings disturbs surface water and groundwater and associated organisms. Even lawns can add to this problem, since they cause the soil structure to become so inverted that water capacity can be compromised and biological activity reduced. Impervious structures tend to increase the impact of storm events, making runoff an efficient carrier of pollutants. The addition of stormwater management ponds to urban developments was intended to reduce this impact, and it is clear that such ponds do mitigate storm events, but the problem is not so simple. Stormwater management ponds may only delay the release of pollutants into the storm drain system. If the ponds remain connected to riparian zones, for example, they can leach into groundwater. These issues, alongside stream canalization and other misguided treatments of riparian zones, mean that sub-watersheds may not drain properly and may thus fail to support ecological functions and mitigate natural hazards, and they may become sources of contamination. Run-off also carries health risks. Groundwater sources may become polluted and pose considerable risk for human consumption. Such pollution can result from inorganic or organic industrial contaminants, but it can also originate from fertilizers, which promote the growth of pathogens like *Escherichia coli* – a phenomenon more commonly recognized in agricultural regions.

Exotic Species

Some native species are endangered, become rare, and die off. Exotic species also come to dominate some ecosystems. Many threatened species are not likely to attract much attention or sympathy, since they are protozoans, fungi, or micro-invertebrates and possibly were never even identified or classified before they vanished. However, other more visible species are recognized as being endangered or otherwise at risk in urban areas. Of the five hundred species at risk listed in 2005 by Committee on the Status of Endangered Wildlife in Canada (COSEWIC), up to 5 per cent are in urban areas, not considering non-vertebrate and non-plant species. The number of species at risk varies among urban areas. Some cities are built at the terrestrial-aquatic ecotone, i.e., at the transition zones where these species meet and where many terrestrial, aquatic, and transitional species are found and can be at risk. Other cities, like St Catharines–Niagara Falls, are in southern climates that are home to a greater diversity of species.

Pollutants favour some exotic species and harm some native ones. The cumulative impact of changes to the landscape and nutrient and water cycles does likewise. Property owners deliberately or accidentally introduce exotic species. Again, these are late symptoms of problems that really needed to be solved long ago. By the time researchers can successfully measure changes in species' populations as well as changes in the larger communities and ecosystems and link these changes to impacts such as urban development or pollution, it may be too late, or at least very expensive and difficult, to solve the problems.

Some native species need no assistance because urbanization favours them. Some may be regarded as nuisances or pests, e.g., deer, coyotes, raccoons, skunks, muskrats, field mice, gulls, Canada geese, foxes, and perhaps bears and wildcats. Their presence is associated with problems of garbage foraging, property damage, fecal matter, and rabies. Indeed, even conservation efforts may create habitats that favour new pests or disease vectors like ticks that carry Lyme disease. However, interrelationships are complicated, and causes may be hard to pinpoint. For example, West Nile virus is encouraged less by conservation efforts, such as storm ponds, than by negligence regarding waste materials like tires, drains, and pop cans left lying around as places where mosquito vectors can breed.

Fragmenting Ecosystems

Possibly the largest challenge in any ecosystem is that at the larger, or macro, scale, development has proceeded largely unplanned and so has fragmented the ecosystem into small remnant pieces that may be only a few dozen to some hundreds of square metres in area. Species that dwell within fragments cannot survive if they try to move from one fragment to another, which impairs the necessary process of mixing genes from different local populations to create genetic variation and allow for future responses to changing environments. Often, one species of even a relatively mobile guild dominates under these conditions, and even on a watershed basis it is relatively easy to see the impacts of development, especially in urban areas (figure 14.1).

In this example, the impact is on the richness and abundance of breeding bird species in urbanized landscapes in the developed areas of the Oak Ridges Moraine in Richmond Hill, Ontario. The American redstart (*Setophaga ruticilla*) is a bird species that dominates this set of landscape fragments; hence richness and abundance = 1 (a low value). The remaining data on habitat area, habitat in landscape, total urban area, and fragmentation provide context: i.e., it is a heavily fragmented and urbanized area, and this is why one species, adaptable to this sort of bird's-eye view of an ecosystem, is so able to dominate such a simplified urban ecosystem.

Fragmentation also creates problems because much of the habitat lies against urban structures, leaving abrupt transitions between habitat and urban features. In such cases the urban perimeter around natural areas can be affected by drastic changes in temperature, humidity, light, nutrients, and water exchanges. In addition, physical features like wind can punch "holes" in remaining natural areas. Such perimeter influences are called "edge effects," meaning

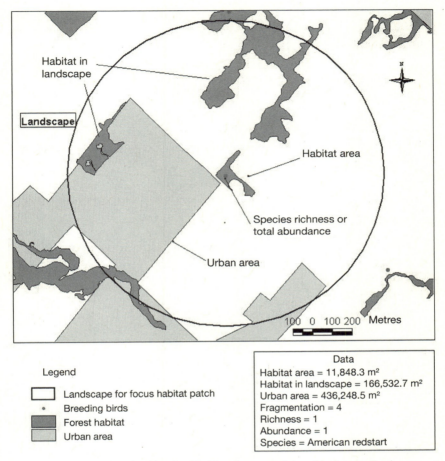

Figure 14.1 An example of localized habitat fragmentation and its impact. By the author, with permission

that there is exposure and transition between a habitat and its external landscape. This area is often hostile to species associated with the original habitat and attractive to exotic or different native species that will invade. These invading species can enlarge the perimeter, and in consequence an even larger area becomes hostile to the original habitat's species. However, because many variables influence the outcome of fragmentation, it is hard to predict how small a fragment has to be before a habitat becomes so altered that none of the original habitat remains.

Without intervention, many of the natural habitat fragments in cities may be doomed to vanish or become undesirable from an ecological or even an aesthetic perspective, as happens, for example, when a woodland becomes a weedy meadow. Some fragmentation has been caused by interventions that turned out to be devastating, even if they were done according to what was once assumed to be correct practice, for example, filling in wetlands for development from fear of diseases. In many Canadian cities waterfront and watercourse areas have been altered for a variety of purposes. Dams, berms, and channels have been built along watercourses for flood protection, sewage, stormwater management, or power generation. Low-lying areas have been raised to reduce flooding; wetlands have been filled for development or against perceived threats to human health. Ravines have been filled for development, traversed, or used as rail, power, and road corridors. The proliferation of buildings, yards, parking lots, and transportation and utility corridors has meant that little of the natural environment remains.

Shrinking natural areas in cities did inspire early attempts to conserve what was left. Such initiatives took the form of creating parks or planning housing in a fashion that encouraged open space. But consistent with a pervasive Victorian influence in North American cities, the focus in early parks was on open, mowed grassy areas. More recently, this approach has changed somewhat as more truly "natural" areas were saved and restored. To some extent, they have provided an outlet for people to enjoy what is left of "natural" fragments. But even these fragments remain at risk, as people trample them, bring exotic species on their clothes, encroach on natural areas with lawns, and allow their pets to roam and kill small organisms.

THE SHORT-TERM FUTURE FOR PROTECTED AREAS IN URBANIZING ONTARIO: RESTORING WHAT WAS LOST

The awareness of problems is of course a necessary first step in attempts to correct them. However, on its own, listing problems does little good in the absence of a strategy laying out systematic solutions. Here are some of the things that can be done. An immediate challenge is to mitigate problems and repair as much damage as possible. This means restoring native species and ecosystem processes. One main way to do so is through ecological restoration. In the strictest sense, ecological restoration would attempt to reestablish all the key ecological structures and functions of times past and so return us to a self-organizing and self-sustaining system. Normally, the reference state, or the ideal, is interpreted to be a pre-industrial landscape. But the essential problem is that cities so change the landscape and its ecology that restoration is probably impossible on this scale.

Most urban efforts focus on what is called ecological rehabilitation. The thrust here is to mitigate the worst impacts of pollution or other problems on a proximate level in order that some ecological functions and structures might be restored. For example, long-abandoned gravel pits could be re-shaped and species reintroduced so that the pits become small, deep ponds that would support a viable ecosystem. On former contaminated industrial or "brownfield" sites, heavy metals or organic pollutants could be removed through chemical or biological remediation, and acidification could be mitigated to promote the reintroduction of native species. Also on brownfields, old concrete foundations could be pulverized and turned into urban alvars – rocky outcrops with shallow soils or artificial rocks for reptile habitat. If properly designed, stormwater management ponds could remove pollutants and perhaps maintain some ecological functions. In more serious cases, ecological reconciliation may be practised. Reconciliation involves humans learning to share habitats with other species. Examples can be as benign or even inspiring as encouraging peregrine falcons to nest on tall buildings or as "boldly radical" as accepting field mice in our backyards or native plants in what used to be manicured lawns.

Regardless of the approach, ecologists work at micro-scales to mitigate urban impacts and redress imbalance and so return ecological health. Micro-scale experiments and knowledge are needed to ensure that we know how to restore ecological functions and structures. Soil is restored by removing excess nutrients and by planting cover crops that are harvested. Fungi and invertebrates may then be reintroduced (inoculated) into the soil. Plants can be seeded or transplanted from commercial or public nurseries. Streams may be restored by removing channels, taking old mill ponds or reservoirs "off-line," bypassing them with new stream channels, and re-introducing historical meanders. Wetlands can be restored by removing tile drains, since most cities are built on old farmlands that used such drains to make land arable. Integrated pest management can be used to mitigate exotics and give native species the chance to re-establish structure and natural ecosystem function.

All these approaches are often started on a local scale. In fact, an excellent start is right in people's own backyards. Students and professionals do experiments or demonstration projects on the scale of a few hundred metres of habitat. A neighbourhood adopts a forest. Boulevards and yards are planted with low-growing native species once bylaws are passed to allow this. The grounds of schools, commercial blocks, or industries are planted with native species. Trails may be closed, and isolated habitats may be reconnected, although there is a risk of facilitating disease and pest transmission. Innovations like green roofs may increase energy efficiency and reduce ecological impacts by impounding water or air pollutants.

Even "backyard" restoration can be relevant to the larger scale, since smaller nodes of restoration and conservation may have a cumulative impact. And the nodes can be connected; for example, a series of backyard naturalizations can benefit native species such as birds. A local focus often fosters ownership and a sense of care for these projects, and it may inspire others as they convince people that small steps can result in large beneficial effects. Local projects are also the most feasible ones in terms of politics, funding, and maintenance. However, if there is too much focus on micro-scales, the projects can also become disconnected ecologically. Indeed, the outcome may be that the easiest projects to fund or initiate get done, but at the expense of the larger, critical ecological issues that are ignored. Ecologically, this "backyard" style restoration could mean that a whole series of fragmented habitats are created, but ultimately it cannot be sustained without continuous human maintenance, contrary to the goal of creating an ecosystem that is self-organizing and sustaining over the long term. There is also a social-learning issue arising from over-reliance on micro-scale projects: because of a limited awareness of each small-scale effort, we may never learn from successes or mistakes and may consequently reinvent the wheel and repeat our errors.

The main point is that short-term, small-scale interventions are important but not sufficient. Ongoing efforts to date call for policies that encourage a priority-driven and integrated approach to local restoration and conservation efforts so that the needs within the city are connected. Likewise, since ecosystems have been artificially bounded by municipal and other political borders, new designations will be required to interconnect cities. Ecological restoration should also be a regional/county, provincial, federal, and international responsibility. There is a promise of better cooperation, but the danger also looms that such cooperation will fall victim to an overriding spirit of municipal, entrepreneurial-style competition.

THE LONG-TERM FUTURE: HOW PROTECTED AREAS WILL BE INCORPORATED INTO GREEN URBANIZATION IN ONTARIO

At the heart of the restoration ecology concept are the challenges inherent in the reconciliation of natural and built environments. Fundamental to these challenges is the "fragmentation" problem, which is pervasive in urban areas. For example, cities are primarily comprised of relatively low-density older suburbs, outer newer suburbs of often still lower density, and an outer rural zone that constitutes the edge of census metropolitan areas. The rural portions tend to disappear rapidly as suburbs expand. Prevailing forms of development have led to advanced fragmentation in terms both of natural areas and human habitation. Human habitat fragmentation is a consequence of urban sprawl and forces residents to travel far and often by single-occupancy vehicle to school, work, shopping, and recreation. In response to sprawling development, new models are presently being considered, models that would involve mixed zoned areas, much higher density, and better-planned transportation corridors and journey patterns from residences to other destinations. There remains, however, much disagreement over how well all this might work from an ecosystems perspective. A critical question is, what kinds of innovations might be favourable to urban ecology?

In the context of the longer-term future, interventions might improve the energy efficiency of individual buildings, thereby reducing overall energy requirements and the pollution originating from coal-fired and other power plants, which ultimately finds its way to cities. Dovetailing with the goal of energy efficiency are innovations that may serve multiple functions. For example, green roofs and living/green walls may attenuate the absorption and re-radiation of sun-generated heat (urban heat-island effects), resulting in lower energy use for cooling over summer months. They would also provide additional insulation to buildings and thereby reduce energy loss throughout the year. A further environmental advantage would be reduced water loss to storm sewers. Green buildings also promise to intercept and bioremediate or impound airborne and precipitation-borne pollutants. Additionally, they provide "ecosystems in the air" that serve as homes and sources for subpopulations of organisms, from fungi to birds and mammals. This is an impressive list of promises, and it remains to be seen whether green roofs are able to deliver all of these. Still, these promises do warrant hope that micro-scale approaches at the building-roof level can be a worthwhile first step in the environmental restoration of cities, especially in densely developed sectors.

Other innovations might consist of efforts to green an entire building or neighbourhood, rather than just the roof. The simple act of planting native trees and shrubs near buildings may reduce energy use in buildings and thus a neighbourhood's urban heat-island effect. New renewable energy sources may become viable and operate on a smaller scale; for example, each building could have an independent power source or a neighbourhood could share such a source. "Living machines" could be constructed; they involve self-contained (usually) aquatic ecosystems that are used to treat wastewater – rather than just treating it by chemical means in a waste treatment plant or by impounding storm runoff in a management pond. Functionally such innovations use both principles and the species/processes of aquatic ecosystems in building waste treatment facilities. Those who design and operate living machines probably will improve their efficiency so that they may become an increasingly viable option for treating wastewater and perhaps all household organic waste. A neighbourhood-scale version of a living machine might operate on the same principles as a neighbourhood stormwater management pond that captures runoff. Ultimately, the improvement of water quality at a neighbourhood level rests on more sophisticated predictive modelling of how water balances are likely to be altered by development changes.

At both a property/lot and community scale, work on greyfields and brownfields offers some possible direct and indirect benefits. Greyfields are normally abandoned or underused commercial buildings like early shopping centres or big-box stores that have closed. Brownfields are normally old industrial or heavy commercial sites that are contaminated, for example with gasoline from service stations or heavy metals from tanneries. Brownfields require more action, in that owners may be legally required to remove toxins before new development can proceed. The physical removal of toxins is generally rapid but expensive, since it involves chemical methods or some forms of microbial remediation. It is often used in cases of heavy metal contamination, because the elements in the compounds must be immobilized and removed, not detoxified. Slower and cheaper methods may also be used. An example is phytoremediation, wherein plants detoxify organic compounds so that a temporary ecological community may provide habitat for remediation with relatively low ecotoxicological risk, in other words, a low risk of inadvertently making the problem worse by making toxins mobile and able to spread to other parts of an ecosystem. By no means is there agreement on which approach is better, and the choice of remediation options rests on the preferences of policy-makers after consideration of factors related to site hydrology, geomorphology, and the chemicals involved.

Sites such as brownfields could be more fully restored ecologically if they were located in a landscape that would allow connection with otherwise isolated natural fragments. Alternatively, if the sites were economically attractive, housing or commercial development could cover remediation costs. In such cases, some ecological restoration might occur through building design, such as a green roof or landscaping, but the main environmental impact might be that the development of infill sites saved peripheral greenfields from development. Greyfield redevelopment may also prevent greenfield development.

Ontario has modified municipal and planning acts in order to introduce legislation specific to brownfield sites and other kinds of "infill" redevelopment projects. Cities themselves have also adopted infill-friendly bylaws and incentives. Overall, there appears to be an increasing interest in infill development, perhaps stemming from the ideal of a more sustainable city. Again, the most valuable impact may not be felt at the site of micro-scale restorations but may result from the ensuing preservation of greenfields further away. If most of the restored sites are near the city core or inner suburbs, there might be an ancillary benefit of intensification: less need to develop roads and subdivisions and thus avoidance of some of the environmental damages these developments cause.

Environmental and, more importantly, integrated urban planning from the scale of the neighbourhood to that of an entire urban watershed, must go beyond the innovations suggested here, though the objectives of more density and reduced dependence on long-distance travel by car are likely to be part of most urban environmental strategies. There may be more emphasis on co-housing, wherein individually owned units are clustered around shared amenities, including playrooms, full kitchens, and clusters of offices separated from the living areas. Co-housing may reduce the ecological footprint of home ownership by providing more incentive for reduced commuting to work, a willingness to engage collectively in initiatives like native species landscaping or energy-efficient buildings, and a sense of community cooperation and education that helps people reduce their resource use and waste production.

Cooperative housing as well as neo-traditional or New Urbanism styles of development may be beneficial from an environmental perspective, but these approaches are not without their critics. Some of the shortcomings they raise might be overcome by planning in conjunction with other efforts such as greenways. It has been argued that the successful implementation of a range of alternative, community-based, ecologically friendly arrangements, whether greenways or green roofs, is more likely because those living in co-housing or cooperative or New Urbanism housing units may be intellectually predisposed to further implementation of ideas that are not part of the status quo, though this is not always the case. Alternative housing developments may also spur the adoption of other innovations like altered street patterns to increase foot and bicycle transportation and make automobile trips less wasteful. Other critics call for a more ambitious, or even radical, approach, which involves actively discouraging dependence on automobiles and encouraging more reliance on public transportation.

The preceding paragraphs have deliberately focused on urban ecological planning. They have also stressed the role of locally based urban environmental measures, for cities are becoming one place where citizens may make a direct difference in combating large-scale issues like climate change. However, the message is clear that the ecological future will depend on political, social, and economic efforts, much of which are related to personal and institutional preferences and ethics. By no means will the adoption of such initiatives be simple, since they entail sociological challenges, political battles, and economic shifts that I have not addressed here. The crucial point is that urban ecology is not a silo wherein ecologists can do science ignorant of the non-ecological factors that affect the city system.

Ecologists must continue advancing scientific knowledge, but we also must link this to the changes in the urban milieu.

THE CHALLENGES FACING PROTECTED AREAS IN URBANIZING GEOREGIONS OF ONTARIO

No matter where one travels in Ontario, the direct or indirect effects of urbanization are measurable and significant. In the northern areas (boreal-dominated ecozones), urbanization is less sprawling, but impacts do arise from pollution transmission from US cities west of Ontario and from overall climate change at regional scales. In the south, these effects are compounded by the more direct impact of urban sprawl, especially in the Golden Horseshoe. They will multiply since Ontario's *Places to Grow* policy will, in the immediate term, have the unintended consequence of increasing urban sprawl as developers leapfrog the Greenbelt and move quickly into areas like Kitchener-Waterloo, Guelph, London, Brantford, Hamilton, Kingston, Peterborough, Barrie, Orillia-Lake Simcoe, and Collingwood. In the long-term the Ontario policy of requiring more intense development in urban cores will mitigate these effects, but there will be an initial struggle as the Ontario Municipal Board grapples with interpreting provincial policy statements.

Thus we can expect that almost all of southern Ontario below the Bruce Peninsula and to the east along a meandering line from Parry Sound through Huntsville to Ottawa and all points south will experience intense urbanization in cities or cottage country. However, Algonquin Park and the swamps of the northern parts of Hastings, Frontenac, Renfrew, Lanark, Lennox-Addington, and Leeds-Grenville will slow urbanization impacts because the Parks Act and acts related to wetland protection will not permit changing clay-lined swamps on top of the fingers of the Canadian Shield into warm "swimming pools" more amenable to real estate development. Further, efforts like the Algonquin to Adirondack (A2A) corridor and biosphere reserves may slow urbanization and conserve ecological functions in a large landscape area. The result could be a sea of small to large urban areas pock-marked among relatively undisturbed areas, at least for awhile. But these too will be affected by the cumulative impact of urban heat islands and regional and global temperature changes and other global climate changes. The effects may also include changes to surface and groundwater flow –

the actual result will depend on the degree of drawdown in urbanizing areas and channelization or, ideally, de-channelization through ecological restoration.

The areas that will face the greatest impacts, beyond those like Toronto that have already been affected, will likely be Waterloo, Wellington, Halton, Durham, Northumberland, Peterborough, Simcoe, and York. Some areas are protected, but some or all areas lie outside the Greenbelt, and all are along major transportation nodes near Toronto. The outcome could actually be positive if immigration to these locations is in fact concentrated in urban cores and high-density nodes of high-quality housing designs. This would encourage mass transportation use, although the municipalities and the province will have to commit themselves to leading development by first investing in mass transit like high-speed heavy and light rail. There will be commuting across large distances, but this will be less of a problem than it currently is if there is a world-class express mass-transit system. This scenario is idealistic in that it postulates that the two main impacts of sprawl and car use would be mitigated. It is pragmatic in that we have examples that work – look to the past and ongoing US effort in Portland and Seattle and efforts in locations like Dublin, Ireland, under similar pressures.

A real challenge will be to maintain agriculture in all of southern Ontario, since farmers will be tempted to sell out under the combined forces of the isolation of their land in a sea of urbanization and pollution; continued uncertainty about the weather, which will be exacerbated by climate change; and market fluctuations created by demand and the price of land, along with the consolidation of agriculture. The demand for land for recreation – and its urbanization impacts – will increase. A likely response will be to follow Waterloo Region's lead and create Environmentally Sensitive Landscapes that restrict development and create large protected areas that resist and reverse fragmentation impacts. There may be some attempts to emulate the Duffins Agricultural Preserve, but the experience with reserving land for economic – rather than ecological – reasons and the explicit maintenance of farming rights within Waterloo Region's ESL means that the political effectiveness of agricultural preserves will likely pale in comparison with the ESL approach.

ESL's save two functions with one policy-ecological function and agriculture – while maintaining land value. This value will be constrained in that runaway speculation-driven prices will not materialize, since restrictions on subdivision development work to favour those who wish to buy rural property as a place to live – and there will continue to be many – the prices will remain high, and those selling will get a fair price but not the windfalls that existed in the past and led to sprawl.

Far-eastern Ontario may see less of this, but one suspects that there will be some leapfrogging of development driven by Ottawa and even Montreal. How this would affect centres like Cornwall, Prescott, and Vankleek Hill is unclear, but some of these areas are still suitable for farming. Alternatives related to housing and recreation may not be attractive for quite some time because of the relative isolation of this region. Still, with places like Cornwall on the main Ontario-Quebec corridor and any advent of high-speed rail, pressure could rise sharply.

In the north, the direct impacts from urbanization will be minimal until enough people attempt to move into less-populated areas to escape high-density living. Areas like North Bay, Sault Ste Marie, and Sudbury are rehabilitating degraded lands and reinventing their economies as educational or high-tech hubs, so there will be some movement in those areas. However, their experience with the impacts of resource extraction and current models of southern urban development may mean that the northern areas learn the lessons faster and better than those in the south. Climate change may make living in the north more attractive to more people, but this impact is decades away. People already notice more variation in the weather; they will not notice a permanent shift in climate for quite a while longer as those shifts become stronger.

Resource extraction will continue in the form of mining and hydroelectric power, but forestry will likely decline or shift north. Biofuel demands may shift lumber plantations into poplar, willow, and switchgrass plantations, which are more useful than corn growing for biofuels. These new plantations can be ecologically functional if they are planted on local (five-to-ten-hectare) scales and rotated on a five-to-ten-year cycle so that plant and animal metapopulations will form and be maintained. This is unlike forest plantations, where the need to maintain weed-free conditions for decades and large-scale clear-cutting, tends to create ecologically sterile conditions prone to soil erosion and compaction. A reduced emphasis on lumber and pulp and paper may also mean less pollution of surface and groundwater. The farthest reaches of the northern Ontario will see challenges as roads and more hydro dams are proposed, the treeline slowly shifts north, and conflicts erupt between protected areas, First Nations claims, and private development. Areas around Quetico Provincial Park and Kenora provide examples where some of these issues are becoming a reality; expect it to be a problem all the way through Moosenee, Winisk, and Fort Severn.

In the end, Ontario's georegions will face the common problems of climate change, invasive-species immigration, and attendant land use conflicts, and the challenge of creating a governance structure that can maintain local accountability while still tackling landscape scale and inter-jurisdictional ecological issues. Climate change creates a particular challenge in the sense that it makes decisions on conservation and ecological restoration difficult. For example, is it better to plan for the past by preserving historical ecosystems or adapt to the future by mitigating impacts and anticipating changes like the expansion of the Carolinian Zone north and eastward? The boreal forest may also shrink with replacement by plains and then tundra. Ontario is fortunate in that it already has mechanisms like conservation authorities than can act beyond municipal boundaries, and it has shown a willingness to create cross-jurisdictional entities like the Niagara Escarpment, the Oak Ridges Moraine, and the Greenbelt to provide legislative teeth outside the usual Parks Act coverage to ecologically protected areas. While these efforts are not perfect and are still subject to politically motivated changes for the worse, Ontario's georegions have some strong mechanisms that can ensure they are protected while still growing in a managed sense.

This is optimistic to be sure and history teaches us that such optimism is not often borne out. But history is not a set of rigid patterns – although some people can be quite intransigent. The good news is that in the last half-century a culture of environmental protection and legislation has been created worldwide, a culture that is unprecedented. The ideological battles will continue for quite some time, and there will be setbacks often related to the unrealistic anticipation that an agreement (like the Kyoto Protocol) will be a panacea, when in fact an agreement is often a strategic move toward a longer-term achievement. Even the Greenbelt Act may work like this, despite the fact that it does actually achieve early benefits. The wider lens here is that all this involves debate about individual and group rights and have-not cities or even georegions. But since all georegions face or will face similar problems, albeit to various degrees, it is time to ensure that there is a coordinated landscape-scale effort for protection and managed growth. Ontario is painfully attempting to achieve this under the correct assumption that our ability to manage growth and the environment is already limited but will become more so with globalization and global climate change. The risk is not as stark as a collapse of georegional ecosystems or Ontario's civilization as a whole if we choose badly or do not act. It is more a matter of not being able to achieve a positive vision and of constantly being reactive and limited in our ability to innovate. Ontarians in all georegions would do well to remember that we depend on our georegions and the ecosystems within and across these boundaries. Let us continue to take steps to ensure we manage them appropriately and that our activities do not make effective management impossible.

FURTHER READING

Hough, M. 2004. *Cities and Natural Process: A Basis for Sustainability*. 2d ed. New York: Routledge.

Murphy, S.D. 2004. "A More Sustainable Urban Environment: Beyond Idealism and Cynicism." *Ontario Planning Journal* 19(1): 28–9.

– 2006a. "Applied Ecology and Restoration Ecology in the Twenty-first Century." In *Encyclopedia of Life Support Systems*, ed. A. Bodin. Geneva: United Nations Educational, Scientific, & Cultural Organization (UNESCO).

– 2006b. "Urban Ecology at Small Scales." In *Canadian Cities in Transition*, eds. P. Filion and T. Bundting, 178–91. Toronto: Oxford University Press.

– 2009. "Evaluating Ecological Restoration Projects." In *Ecological Restoration*, ed. S. Greipsson, 204–18. Dordrecht: Kluwer Academic Press.

Murphy, S.D., and L.R.G. Martin. 2001. "Urban Ecology in Ontario, Canada: Moving Beyond the Limits of City and Ideology." *Environments* 29:67–83.

Pickett, S.T.A., M.L. Cadenasso, and J.M. Grove. 2004. "Resilient Cities: Meaning, Models, and Metaphor for Integrating the Ecological, Socio-economic, and Planning Realms." *Landscape and Urban Planning* 69:369–84.

Pickett, S.T.A., M.L. Cadenasso, J.M. Grove, C.H. Nilon, R.V. Pouyat, W.C. Zipperer, and R. Costanza. 2001. "Urban Ecological Systems: Linking Terrestrial, Ecological, Physical, and Socioeconomic Components of Metropolitan Areas." *Annual Review of Ecology and Systematics* 32:127–57.

15

Changing Cultural Landscapes of Ontario

ROBERT SHIPLEY

SUMMARY

In order to grapple with the idea of change in Ontario's cultural landscape and how such changes unfold, we first have to be clear about what we mean by "cultural landscape" as well as a number of other key terms such as environment, healthy communities, ecology, and sustainability.

Most of the preceding chapters have explored the georegions of Ontario seeking a better understanding of the appropriate ideological basis for planning. If we define planning as the management of change, then the first chapters were essential in giving us a sense of the historical development of the province. The chapter immediately before this one (chapter 14) also served a vital purpose in describing the ongoing evolution of the province's natural heritage. What is left is arguably the most important aspect, attempting to understand how we deal with change in places where the vast majority of us actually live – cities and the areas immediately around them.

It is interesting and a little disconcerting that in the early twenty-first century many people continue to think that "environment" refers to places where there is still some naturalness, either unchanged by humans or at least resembling what things were like before so many of us got here. Clear water, breathable air, trees, wildflower meadows, and timid animals, those for most people are the elements that make up the Environment with a capital *E*. This notion is reinforced by the very way governments operate. We are told that "Environment Canada's mandate is to preserve and enhance the quality of the natural environment."[1] For its part the Ontario Ministry of the Environment is "responsible for protecting clean and safe air, land and water to ensure healthy communities, ecological protection and sustainable development."[2] We are reassured that the Ontario Ministry, "has been protecting Ontario's environment for over 30 years."[3] Some of us would contend that in very important ways the Ministry of the Environment has been doing nothing of the sort.

The *Oxford Canadian Dic*tionary defines environment as "the physical surroundings, conditions, circumstances ... in which a person lives [and] works." In other words the environment is where we are, not where we aren't. Most of us love to go into what we call the natural environment and, as Canadians, we believe implicitly in protecting those areas. In fairness our so-called environment ministries have done wonderful work in managing national and provincial parks, monitoring water quality, and cleaning up pollution. When it comes to the places where most Canadians live, on the other hand, they have been asleep at the switch.

For most of us the actual environment is the city or town that we refer to here as "cultural landscape." Once again we can turn to the *Oxford Canadian Dictionary,* which tells us that culture consists of "the arts and other manifestations of human intellectual achievement regarded collectively." The grandest collections of all the things humans do and have done are the settlements where we live. The dictionary defines landscape as "scenery," or what we see around us. The cultural landscape is comprised of all the spaces and structures we see around us from streets to stadiums to town squares and out into the farm fields and woodlots that make up the environment that we as humans have modified considerably.

There are three other significant concepts that the Ministry of the Environment claims to be protecting: healthy communities, ecology, and sustainability. According to the Ontario Healthy Communities Coalition, an incorporated registered charity, its mission is "to work with the diverse communities of Ontario to strengthen their social, environmental and economic well-being."[4] Now a world-wide movement, the healthy community idea was born two decades ago in Ontario and works to create the pre-conditions for health, rather than thinking of health only as the treatment of diseases. A key component of the approach is to ensure that people have vibrant and comfortable physical spaces in which to live and work. This means a strong sense of place, where they can recognize and feel at home in the environment where they actually spend their time.

Turning in our trusty dictionary to the word, ecology, we learn that it means "the relations of organisms to one another and to their physical surroundings." While we generally think of this as referring to insects and snakes in ponds, it also refers to human organisms and their physical surroundings, which are streets, sidewalks, and buildings. The health of urban ecology, in other words the cultural landscape, is every bit as important as the rural wetland and the wilderness.

Finally, we should think about the idea of sustainability. Sustainability is perhaps a better term than sustainable development, since the development part seems to lead many people to think that with a few adjustments to past practices humans can go on expanding their cultural landscapes forever. We should constantly be reminding ourselves that sustainability is intended to mean meeting our own needs without making it impossible for future generations to meets theirs. There has been too little attention paid to the fact that acting to ensure sustainability will inevitably mean curbing some of our desires and appetites, especially with regard to the use of the land, buildings, and resources that make up our built environment.

The one word that I have studiously avoided so far is "heritage," since it applies to the human-modified environment. I have avoided it because in our current political and planning system the term "built heritage" has unfortunately come to refer to a tiny subset of all the buildings we have inherited, and it almost totally ignores such other inherited factors as urban form, street and farm field patterns, traditional uses, and the memories of people that give these spaces meaning. Nevertheless, "built heritage" permeates our legislation and regulations, and we must live with it while trying to expand our understanding of what it really entails.

While the notion of protecting the natural heritage is now beyond serious question in Ontario, the idea of protecting or even properly managing the built heritage, that is, the urban spaces we inherited, seems odd to some, esoteric to others, peripheral to many, and almost beyond comprehension to many political and business leaders. We can take some comfort that past champions of sound management of the natural environment and responsible stewardship of natural resources once faced similar incredulity from those who were accustomed to think that rivers could be endlessly polluted, forests cleared, and wetlands drained. We can hope that with a better understanding of what it truly means to protect the part of Ontario's environment where people actually live, similar strides can be made toward a genuinely sustainable future.

THE SHAPE OF THE FOLLOWING DISCUSSION

In exploring the concept of change and how it works in Ontario's cultural landscape, we will consider a number of factors. First, are there many landscapes or just one? Second, we will look at the things that threaten the cultural landscape and the opportunities that present themselves. Third, the international context, both conceptual and legal, in which any discussion of the Ontario cultural landscape takes place, will be considered. Fourth, we will review the legislation that governs the management of land use in Ontario, which will lead us into considering a series of six dilemmas that face management of the cultural landscape. These include provincial vs local responsibility, property rights vs community values, heritage vs development, conservation vs tourism, skills requirements vs skills availability, and energy efficiency vs energy saving. Since any decisions that mediate among the preceding sets of polarities are ultimately made in the political arena, considerations of governance will be explored. The need for new paradigms and strategic approaches will be outlined, and finally the chapter will be concluded.

IS THERE ONE CULTURAL LANDSCAPE OR MANY?

Having addressed definitions earlier, we should deal next with the matter of number in relation to the term "cultural landscape." Taken together all the spaces in which humans live, work, build, manage, and modify the environment constitute the cultural landscape. For purposes of description, planning, and management, we also define discrete segments of that total landscape. In this way we might talk about rural as distinct from urban landscapes. We might also delineate a specific landscape that we decide is significant for historical or aesthetic reasons and protect it by limiting certain activities or building intrusions.

One of the ways we use the idea of multiple landscapes is in defining regional differences, as outlined earlier in this book. We come to understand that the different parts of Ontario have not only varied geology and topography but also different cultural landscape characteristics. Roads follow certain patterns, standard lot sizes vary, and building styles often reflect different ethnic traditions. For example, the two-hundred-year-old line between Scottish settlements in Wellington County and the German communities of Waterloo can be traced by looking at the masonry styles of farmhouses along the concession roads. The long river lots on the Detroit River tell of three centuries of the French land-tenure system. The series of perfectly spaced settlements in parts of the province can be best understood when we realize that every rail station had to be no farther than a day's wagon trip from the farthest farm and the station was the raison d'être for the towns' existence. The wealth of individual home owners can be measured by how far the bricks they used were transported.

While both the overall, singular use of the term and the idea of specific landscapes are useful, the remainder of this chapter will concentrate on the general notion of buildings and land use patterns. The cultural landscape will be taken to mean the settled countryside and the villages, towns, and cities, and the structures that constitute them.

THE THREAT TO AND OPPORTUNITIES FOR OUR CULTURAL LANDSCAPE

Forces that deteriorate, destroy, and render the built environment meaningless are categorized under the following four headings: natural decay, cataclysm, changing needs, and modernism. Each of these threats can either be countered or in fact present their own opportunities.

Once we build a structure for human activity, regardless of the materials used, the process of decay begins. Water and ice wear away at brick and stone. Dampness eventually rots wood. Protective paint peels, glass breaks, shingles are blown away by strong winds, insects burrow, and birds nest where we wish they didn't. The time varies, but all building materials have a natural decay cycle. Over thousands of years, however, humans have learned how to build cleverly and maintain and judiciously replace certain elements. As a result in twenty-first-century Ontario we have inherited a wonderful stock of private houses, grand institutional buildings, places of worship, bridges, and other structures that are still serving us well and that properly cared for, will serve us long into the future.

One of every seven buildings in Ontario was built before World War Two. One in fourteen was built before World War One. Since these structures have lasted this long, they often represent the best we have. When we look at our European cousins we see them still occupying buildings that are hundreds of years old. Natural decay, therefore, is a process that can be dealt with through regular maintenance and periodic rebuilding. Furthermore, building maintenance and restoration is a huge industry accounting for roughly half of all construction and millions of dollars of annual expenditure. It has the advantage over new construction of being more labour-intensive and generally of using more locally produced products, which increases its multiplier effect on the economy.

Buildings have a harder time standing up to cataclysmic assault. Floods, storms, fire, and most of all war can abruptly end the useful life of buildings. We have attempted to limit the destruction caused by these forces with our restrictions on building in flood plains, enforcing building codes that provide for structural integrity, fire-retardant walls and materials, sprinkler systems, and numerous other measures. In Ontario we are blessed with not having lost any buildings to war for almost two hundred years but in other parts of the world losses due to conflict have given rise to some of the international initiatives that we will explore subsequently.

Changing functions threaten existing buildings in at least three distinct ways: change in process or how the buildings or spaces are used, alterations in transportation patterns, and increased demand for higher density. Process change can be seen in the case of older industrial buildings where the transmission of power using belts and shafts once necessitated the building of multi-storey factories. Modern production facilities, however, are usually single-storey buildings with horizontal assembly lines. While this situation led to the abandonment of many sound old mills and factories, it also made these same buildings available for conversion to new uses. Once the concept of the loft condominium became popular scores of redundant industrial buildings across the province suddenly became prime real estate.

Alterations in transportation patterns threaten older structures and spaces in several ways. Activities such as road widening, which often destroy our tree-lined thoroughfares, are an example. But perhaps none is more important than parking requirements. Nineteenth- and early-twentieth-century buildings were designed and arranged during a time when people either walked or rode public transit to work and to carry on their daily business. The late twentieth century saw the proliferation of car use and the consequent need for parking. Thousands of perfectly serviceable buildings in Ontario have been demolished to create surface parking for cars. This has been a double blow because as we decreased the number of buildings in urban core areas to make way for cars to sit idle, we tended to diminish the viability of the business functions in the remaining structures. Planners, political leaders, and a growing number of citizens now realize that the age of car-dominated urban sprawl and the inefficient use of land can no longer be supported either environmentally or economically. This means that the value of the remaining older urban-core buildings is being rediscovered. What many people do not readily see is that not only the urban buildings but the old urban land use patterns are part of our inheritance. How we use the gap sites in our cities that were created by demolition to provide parking will be crucial to the future of what is now called re-urbanization (Bunting 2000; CMHC 1997b).

Increased demand for higher density threatens many existing urban cultural landscapes when properties are zoned to allow taller buildings than currently exist. An existing building which is in good condition and provides income to its owner becomes an obstruction when planning rules allow the property on which it sits to be used as the base for a much larger structure with considerably increased rentable or saleable floor space. While some people consider this potential for increased profit to be a natural or as-of-right advantage, it is in fact created by planning regulation and not by divine ordination. The same planning regulations allow not only for the maintenance of existing buildings, which is the case if the zoning is made appropriate only for those existing structures, but they also in legitimate circumstances allow for the compensation of perceived loss of property value.

The whole enterprise of increasing urban density must be carefully thought out. Intensification, which is the increasingly efficient use of land, has, in fact, been under way in the urban cultural landscape since the beginning of our settlements. With the exception of the ill-fated urban renewal schemes of the 1950s and 1960s, intensification did not involve the wholesale removal of existing buildings and their replacement by a different form of urban fabric. It was a far more organic process that saw the building of new structures between, behind, and beside the existing ones. A look at many Ontario streets with an eye for different architectural styles and an understanding that each is typical of a different time period shows that intensification is an old tradition. One can see a block of nine houses where three were built in the 1870s. Between them are three more built in the 1890s and between those, three more constructed in the 1930s. If we take this pattern and apply it to many large lot suburbs of the 1960s, we can see that if allowed, current owners would have the opportunity of dividing their properties and doubling or tripling the number of buildings without the need to build new infrastructure (water, sewage, streets) and without the need for any demolition.

Modernism is the fourth, final, and most intractable threat to the maintenance and continued use of the extant cultural landscape. The desire to experiment with new architectural forms and to try different ways of organizing space is in itself neither destructive nor opposed to conservation. Unfortunately, in the twentieth century the outgrowth and distortion of what began as essentially an architectural movement caused untold damage to the cultural landscape. Modernism was at its core an anti-urban sentiment. Many cities were in poor condition at the end of the industrial revolution of the nineteenth century, and the modernist approach was to either abandon them altogether or throw down the existing fabric and start again building new towns, tower blocks in park settings, bungalows on half-acre lots and variations on and combinations of the above. Repairing the existing structures as our ancestors had long been doing didn't seem to have occurred to *les enfants terrible* of the modernist school. Epitomizing this sentiment was a Canadian government official who said in the 1940s that it was a shame Canada had not been bombed like Europe so that we could build entirely new cities (Denhez 1997).

If this had simply remained the view of a subset of architects and planners, we may have enriched our cultural landscape with their experiments, the best of which would have survived and joined older buildings as part of the environment. However, the cat got out of the bag and became a monster. The idea was bad enough when it was expressed architecturally, as in the case of so-called urban renewal. Considerable tracts of the older parts of our cities were bulldozed and replaced by what were supposed to be better houses. That this was a misguided idea is borne out by the fact that today the neighbourhoods that were saved, such as Cabbage Town and Riverdale, are now considered the most liveable parts of Toronto, while Regent Park, the fruit of urban renewal, is being rebuilt.

The modernist ideal did not stay put in architectural practice. It spread to infect tax policy, public administration, banking, building inspection, real estate, marketing, and many other sectors. Tax policy, for example, developed in Canada with a blind bias toward the destruction of existing buildings. The cost of demolition can be written off against income immediately, but the cost of building maintenance is considered a capital expense and can be written off only at 4 percent annually. Parking lots on properties where buildings once stood have the lowest tax rate of any land use.

Public administration has similar biases against the care and maintenance of existing buildings. Grants are currently available for creating new housing units but not if those units are in existing buildings. Bankers for years were notorious for lending money freely to people to buy new houses but being stingy with loans for renovation. For many years the Ontario building code referred only to standards for new buildings, and the renovation of existing buildings had to conform to those measures no matter how inappropriate. Real estate marketing long steered people away from buying older buildings (Shipley 2000). Most of these things are now changing. Historic and just old buildings are now hot sellers. The building code has a section containing equivalent standards for older buildings. Nevertheless, for three generations in the twentieth century there seemed to be an attitude that buildings had a short twenty-five-year life and that everything in the cultural landscape would continually be replaced. The tax system remains intransigent on this issue.

THE INTERNATIONAL CONTEXT: BOTH CONCEPTUAL AND LEGAL

For much of the last millennium, since the European towns that clung around the ruins of their Roman predecessors began to revive and expand, constant building repair and reuse was the rule. Thousands of churches built from the eleventh to the fourteenth centuries are still in use, while many towns and cities have whole streets of buildings constructed from the sixteenth and seventeenth centuries. Building material was too precious and other resources too scarce for people not to be take great care in reuse and recycling. It was not until the Industrial Revolution, beginning at the end of the eighteenth and transforming not only Europe but North America in the nineteenth century that a conscious concern for buildings and structures as artefacts of the past began to crystallize.

The antiquarian movement probably began with those grand tours of Europe that became de rigeur for gentlemen after 1750. Returning to England, France, and Germany after experiencing the classical ruins of Rome, some of them began to look more seriously at the ancient stone monuments, castles, and churches in their own environments. At least two distinct schools of thought emerged. In France, Eugene Viollet le Duc, a prominent architect during the reign of Louis Napoleon, came to represent the concept of restoring historic sites to their former glory. This often meant demolishing parts of structures that had been added in different periods and imagining what the original builders may have intended if there was no actual evidence. Carcassone is the medieval walled town in southern France that best represents this approach with its pointed towers that inspired the Disney's Fantasy Land.

Perhaps not surprisingly the English took a quite different approach. Two Victorian figures, John Ruskin and William Morris, concerned in part over the vandalism that was destroying Stonehenge, formed a group called the Society for the Protection of Ancient Buildings (SPAB). Their philosophy was that ancient monuments should be maintained as far as possible to prevent further deterioration, but they opposed restoration, especially when it involved modern builders imposing their own ideas of what might have existed in the past. In the end they felt that if old structures could not be maintained, we should allow them to crumble as part of the natural cycle.

Both of these strains of historic conservation continue to exist in various forms and to influence heritage projects in Ontario today. When a site interpretive plan chooses to represent a certain date, as in the case of Dundurn Castle in Hamilton, where all the furniture and costumes of guides, as well as window restoration and colour schemes, come from 1857, the high point of the original owner's career, the Viollet le Duc approach is being followed. When a ruined wall is stabilized and left, as in the case of the Goldie Mill in Guelph, we see the legacy of Morris and Ruskin. Many other theorists and practitioners have followed but for the most part conservation decisions reflect some combination of the earliest polar views or some point on the continuum between them.

The wide-ranging changes that occurred to the cultural landscape during the Industrial Revolution of the nineteenth century happened at a scale and speed that was far greater than during the previous centuries. However, they did not really prepare people for the changes that rocked the twentieth century. The combination of vast building projects, on one hand, and the drastic devastation of modern warfare, on the other, threatened the cultural landscape as never before. Destruction of buildings on such a massive scale gave rise to something more profound than the modest conservation efforts of the Victorian era. First under the auspices of the League of Nations and subsequently of the United Nations, concerned thinkers from around the world began to gather and grapple with the challenge of maintaining what they came to see as the shared heritage of humankind. A series of charters and declarations slowly articulated not only the concerns about the world's cultural expression but also remedies for its conservation. The Athens Charter of 1931 was one of the first and the so-called Venice Charter of 1964 is one of the strongest and clearest expressions of the conservation ethic. Article 1 of the Charter states that the "concept of a historic monument embraces not only the single architectural work but also the urban or rural setting in which is found the evidence of a particular civilization, a significant development or a historic event. This applies not only to great works of art but also to more modest works of the past which have acquired cultural significance with the passing of time." Article 5 of the Venice Charter indicates that the "conservation of monuments is always facilitated by making use of them for some socially useful purpose."

In 1972 the United Nations Education, Social and Cultural Organization (UNESCO) brought together the many charters and declarations on historic conservation and created the World Heritage Convention (WHC). Member states that signed the convention undertook properly to protect and manage their particular part of the world's common cultural inheritance. While not as well known as UN treaties on human rights and environmental protection, the WHC is a genuine effort to ensure that each nation hands on its heritage to future generations. Canada became a signatory in 1975, and the various provincial acts are the local legal expressions of Canada's acknowledged international obligation to protect and conserve the cultural landscape.

LEGISLATION GOVERNING LAND USE MANAGEMENT IN ONTARIO

The Ontario Heritage Act

The Ontario Heritage Act (OHA) was originally passed in 1975 and has been updated and amended since then. Other pieces of legislation that impact decisions concerning the cultural landscape include the Planning Act, the Municipal Act, and the Building Standards Act. While the OHA is the poor cousin among the province's laws and is often given less weight than other statutes when there are conflicts and sometimes disregarded altogether in spite of being the local expression of an international legal obligation, we will examine it first.

The Heritage Act is primarily a piece of enabling legislation that sets out processes but few principles. It allows and encourages, but does not require, municipalities to take measures to conserve their heritage. The act follows a very common approach that grew from the nineteenth-century preservation movements described above. Sometimes known as the "three legs of the stool" idea, it advocates (1) the identification of heritage resources, (2) protecting those resources by imposing a process to review proposed changes and if necessary prevent alterations that compromise heritage value, and (c) providing financial assistance. Municipalities may appoint committees of citizens to advise councils on heritage matters.

The original Heritage Act made some implicit assumptions. First, the identification and conservation of the cultural landscape was best done at the local level, with little guidance or resources from above. What structures or spaces were important was a matter of local opinion. Second, any part of the cultural landscape worth saving was likely to be uneconomical and in need of external funding assistance. Third, a building or a piece of land to be conserved was essentially a compound outside of which anything was allowed. In urban areas this sometimes meant saving one building and demolishing and levelling everything else that once formed that structure's context.

The identification leg of the stool has come to be called "designation" in Ontario. Currently municipalities can create inventories that are fairly straightforward lists of properties that might be considered for the more formal designation process. Even being on the list is intended to provide some protection that will be discussed below in the section on the Planning Act. Structures and sites can be designated if they meet at least one of criteria of architectural, historical, or contextual significance. "Architectural value" means that the building is unique, special, or even a good local example of a particular style. "Historical value" means that the site has some connection with a famous person, event, or tradition. "Contextual value" refers to how the site or building teaches or informs us about the past. Individual buildings can be designated, as can whole streets, neighbourhoods, or districts. In almost all cases it is the outside visible parts of structures that are designated, and only occasionally are interior features such as wall paintings or staircases covered.

In order to designate a property or district under the OHA, a municipality must pass a bylaw that contains a standard set of information, including a legal description, a value statement, and a list of character-defining elements. Properties designated in the past do not always have bylaws containing all of this information. In the case of districts, municipalities are also required to complete a plan with design guidelines. The designation is supposed to be registered on title and goes with the property when it changes hands. To a large degree the purpose of designation is the recognition and celebration of our heritage.

The second leg of the preservation stool is protection. Under the OHA, when an alteration to a designated building or a structure within a heritage district is proposed, the owner must obtain a heritage permit. Alteration can mean anything from window repair to demolition. The plans are usually reviewed by the Municipal Heritage Committee, which can negotiate changes but eventually makes a recommendation to the municipal council, which has the final decision. Although it was not always the case, a council can refuse demolition. Rather than being a freeze on change, a kind of pickling of buildings, the designation and review process is intended to provide a forum for the management of that change. Resources such as Mark Fram's 1988 book, *Well Preserved,* can be used to assist with this process. In heritage districts the design guidelines are there to help steer property owners and committees toward acceptable change that preserves the values stated in the original plan while allowing reasonable upgrading and expansion.

The third leg of the conservation stool was to be financial support. Originally the idea was that if people were going to be required to repair historic brick work, replace old wooden shingles with proper reproductions, or keep their wooden window frames, then the cost difference between the best or proper materials and the cheap alternatives – concrete blocks, asphalt shingles, and vinyl – should be offset by the community because of its desire to conserve those heritage features. This is the case in other countries where tax breaks, grants, incentives, or some combination of these measures are provided in support of conserving the cultural landscape. It has been the conspicuous missing leg of the stool in Ontario that, as we will see below, has left conservation in a precarious position in this province.

The Ontario Planning Act

In a number of instances cultural-landscape conservation can be accomplished using provisions of the Ontario Planning Act. In fact the Planning Act is in some ways even more vital, since virtually all land-use decisions turn on planning considerations. In many cases Heritage Act provisions relate directly to certain planning procedures. For example, when a planning application is made for the rezoning of a designated property, it can trigger a request for a Heritage Impact Assessment (HIA). HIAs are evaluations of the effect a given development may have on recognized heritage resources. Conversely, when a Heritage Conservation District is approved, the zoning should be brought into line so that permitted land uses will support the goals of the district.

Like the OHA, the Planning Act is primarily a procedural document setting out how to go about planning. Official plans (OPs), which are intended to set the context and general goals for communities, are required to be produced in every municipality. If they are to have any influence on decision making for cultural landscape issues, there must be specific relative policies in OPs. For example if Heritage Impact Assessments are to be required for any development proposal that may impinge on a designated site or district, then it must be so stated in the municipality's OP.

Other parts of the Planning Act that can be used to influence decisions around heritage sites are the provincial policy statement, community improvement plans, zoning, holding provisions, density bonusing, interim control, site plan control, minor variances, plans of subdivision, consents, and, finally, development permits. Appendix 1 briefly outlines how each of these aspects can be used.

The Municipal Act: Tax-Incentive Financing

While it is the Planning Act that outlines the rules concerning Community Improvement Plans, it is under the Municipal Act that funding of various kinds can be put in place. Other financial incentive ideas will be discussed below under the heading of Heritage Development.

Property Standards Act

The Ontario Heritage Act gives municipalities the power to determine and to enforce minimum property standards in the case of designated structures, in order to prevent "demolition by neglect," which has been an all too prevalent strategy employed by unscrupulous property owners in the past. Having been refused a demolition permit, owners simply abandon the building or even purposely stress it and wait for it to be condemned. If proper and specific official plan policies are in place a municipality can enter the property, make repairs, and charge them against the property owner.

THE SIX DILEMMAS

We have considered the nature of the cultural landscape, the international and conceptual context, and the legal framework that attempts to guide informed decision making when heritage resources are concerned. However, around such a sensitive, value-laden, and volatile subject there are bound to be strong differences of opinion. Here I characterize these as a series of polarities or dilemmas, setting out the competing views and offering some possible resolutions to the debates.

Property Rights vs Community Values

One of the most common objections to heritage considerations being taken into consideration in land use decisions is that owners of private property in a capitalist society should be able to do whatever they want. No public policy, it is argued, should stand in the way of property owners maximizing the economic potential of a building or site, making alterations as they see fit, choosing the cheapest maintenance method, or modernizing. Such changes might include installing windows that are different from the originals, covering the building with vinyl or aluminum siding, or even demolition. This view of absolute property-owner rights contains at least two profound misunderstandings.

First of all, it fails to recognize that many other restrictions are placed on private citizens in general and on property owners in particular. We live in a society replete with curbs on private behaviour designed to protect the common good. There are speed limits on roads and highways, restrictions on where people can smoke, and rules about who can offer legal or medical services.

The list could go on, but the opponents of regulations aimed at protecting heritage values say that all these measures are based on science such as the evidence of the harmful effects of second-hand smoke. Heritage, on the other hand, is just a matter of opinion. Some people like old buildings, while others prefer the new, and they should be allowed to opt out of such rules if they have different tastes. But consider the regulations against building in flood plains. The rules are designed to prevent structures being destroyed by flood water and also to maintain soil stability in areas subject to inundation. If a person owns property in a designated flood plain they can't opt out when they decide they would like to build a house. They are either in the designated area or they are not, and the definition of a flood plain is a historical one as much as it is a scientific one. Based on historical records the flood plain is defined as the area that might be subject to worst flood likely to happen within a hundred-year period.

The designation of an individual property or a district as having heritage significance is determined according to a set of criteria that conform to clear international standards and are as objective as scores of other similar measures, including medical diagnoses. Heritage significance as determined under Ontario law is not a matter a taste or individual preference. It is a matter of fact.

Legal Precedence

Legal precedence is the second factor that is misunderstood by opponents of heritage and planning regulations relating to the conservation of the cultural landscape. It has been clearly established in Canadian case law that maintaining structures and sites that have a cultural value to the community at large can take precedence over private property rights when the correct procedures of identification and designation have been observed. The law is straightforward in recognizing that while buildings are property and the owner has the right to enjoy the benefits of that property, at the same time aspects of the site also in a sense belong to the community at large.

One of the best examples of this concept involves churches. Church sites and buildings belong to the faith group that built or purchased them. However, these sites have been exempt from taxes since their creation, which means that the rest of the community has paid for the water and sanitary services, snow removal, and street maintenance, in many cases for over a century. This becomes an issue when, as is increasingly the case, the congregation dwindles and the faith group wants to sell the site. At that point the rest of the community, through their council, has a legitimate claim to wanting to maintain the building, which may be a significant component of the cultural landscape. A Supreme Court of Ontario decision has confirmed this principle. The same concept applies to any other significant components of the cultural landscape that have been identified.

Provincial vs Local Responsibility

Because the responsibility for land use regulation was assigned to provincial governments under the British North America Act (1867), the onus to respond legislatively to the signing of the World Heritage Convention fell to provinces. In 1975 when Ontario drafted its Heritage Act the powers of the day decided that the responsibility and authority to designate significant heritage sites was most appropriately placed at the local level. It was reasoned that local people knew their own heritage best, that modes of planning and operation are local and that approaches to decision making differ from place to place. The legislation consequently took the form of enabling but not requiring municipalities to undertake the identification of heritage sites. It also mandated municipalities to appoint committees originally named Local Architectural Conservancy Advisory Committees (LACAC). Once designated, however, a building could still be demolished by the owner after a waiting period (Shipley 2000).

The heritage conservation outcomes over the intervening thirty years have been uneven to say the least. Most large municipalities did constitute committees and began designating sites, and many even developed specialized staff positions for heritage planning. Some even created lists of properties they wanted to designate in the future. On the downside, many municipalities either never appointed committees or let their membership lapse, and many either designated few sites or none at all. Even where some heritage planning was undertaken, it was generally given low status in the municipal hierarchy, and decisions were easily influenced by those making property-rights arguments. For example, it was common across the province not to designate properties if the owner objected. This meant that the list of recognized heritage sites in many municipalities did not actually reflect the most important buildings but only those that the owners allowed to be recognized. Even when designated, many historically significant structures were subsequently demolished. A study conducted in 2002 found that over four hundred designated and listed sites had been destroyed between 1985 and 2000 (Shipley and Reyburn 2003). An earlier study had determined that almost 20 percent of structures listed on the Canadian Inventory of Historic Buildings had disappeared between 1970 and 1998 (Carter 1999).

Provincial heritage legislation in Ontario was shown to be rather ineffectual in actually conserving the shared cultural legacy. In 2005 the legislation was finally updated after years of consultation. Some of the main features of the revised act were that the power to actually prevent demolition was given to municipalities and the mandate of what were now called Municipal Heritage Committees was broadened to include advising councils on all heritage matters and not just architectural conservation.

It was actually left to the province's Planning Act to make one of the greatest recent strides in conserving the cultural landscape when in 2005 Ontario issued a new Provincial Policy Statement (PPS) under the act. The PPS is designed to go beyond the procedural nature of the Planning Act itself and to give direction on what the province expects from the outcome of the planning system. The PPS now states that significant heritage resources and cultural heritage landscapes "will" be conserved. The PPS, while a strong statement of what the senior government desires, is not definitive, since it encourages a number of outcomes such as urban intensification and protection of aggregate resources that may still lead to conflicting claims, but at least the need for cultural conservation is recognized in the legal framework that guides decision making. It is felt by many legal minds that the PPS no longer allows municipalities to ignore or opt out of heritage planning.

Architecture in the Pickle Jar vs Culture in the Landscape

Both the origins and the traditions of heritage conservation in the nineteenth century and the nature of the Ontario heritage legislation in the late twentieth century seemed to presuppose a kind of elitist attitude toward architecture (Delafons 1997; Murtagh 1997). The types of structures that were valued and protected in the last century often reflected a social as well as an architectural snobbery. Courthouses, Christian churches, the grand houses of the wealthy, and military sites became synonymous with capital H heritage. On the other hand, factories, workers housing, and buildings connected with women's or minority history were often ignored. Ornate Victorian porches and gingerbread were celebrated, while dams, bridges, and lime kilns were treated as invisible. This approach to heritage has been called "finial retentive."

Under the Ontario Heritage Act there are three criteria for designation: architectural value, historical value, and contextual value. Even today many municipal committees are almost entirely focused on architectural importance and building details. At the same time the approach to protection leads to a kind of game-preserve mentality. Once a building is designated many well-meaning heritage advocates want to freeze it in time and resist any change. It is like putting a few animals in a park and feeding them and assuming that it is open hunting season on everything outside the park fence. Where there is an opportunity for committees to exercise some power they are apt to act fanatically, feeling that everywhere else architectural atrocities are being committed. The results are often ugly battles that cause other property owners to avoid designation.

All of that fails to respond to the advice of the Venice Charter and other foundational UNESCO documents that remind us that the term "heritage" applies "not only to great works of art but also to more modest works of the past which have acquired cultural significance with the passing of time," and that the conservation of sites "is always facilitated by making use of them for some socially useful purpose." As well, we are counselled that what is important is not just single structures but "the urban or rural setting in which is found the evidence of a particular civilization, a significant development or a historic event."

In other words, our heritage is all of our cultural landscape and not just selected fragments of the environment. We give special recognition to some sites, a tiny minority in fact, in the same way that a graduating class elects a valedictorian. One person speaks for and represents the rest, but everyone else gets a degree and has the opportunity to live on, serve the community, and reach their own potential. Designated buildings are representatives of our heritage, not the only ones allowed to live.

Heritage vs Development

For a number of reasons there is a widespread notion in Canada that development means the building of new structures either where none ever existed before or where previous buildings have been demolished. Prominent Ontario jurist Marc Denhez has pointed out on numerous occasions that this is no accident. He explains that in the years after the Depression and the Second World War the Canadian income tax rules were structured to ensure the planned obsolescence of the built environment (Denhez 1997). The architects of this system believed that it would be possible to stimulate constant growth only by perpetual destruction and rebuilding. They built in depreciation so that building costs could be recovered over a twenty-five year period, they disallowed the expensing of maintenance, and they gave preferential tax treatment to parking lots. The result is that what has come to be considered normal is in fact a strong tax bias against conservation and reuse of the cultural landscape.

As a result most efforts to save and recycle the built assets of communities have been seen as anti-development. We can either have heritage, which means stopping all change and costing public money, or we can have development, which creates jobs, generates wealth, and signals progress – or so goes the argument. What such a view completely fails to recognize is that over half the construction spending in Ontario is renovation. Renovation is more labour intensive than new building and uses more locally produced materials. This means that the economic benefit of reuse and renovation is greater for the local community. And if we realistically define development as the creation or enhancement of useable residential, commercial, industrial, or institutional space, then renovation and reuse are every bit as much legitimate development opportunities as new construction. In other words, the dichotomy that has been perceived between heritage and development is a false dichotomy.

Many writers have examined the approaches and building techniques that can extend the useful lives of buildings almost indefinitely (Kincade 2002; Powell 1999). At the same time, studies have shown that the renovation and/or adaptive reuse of older buildings can be highly profitable (Mason 2005; Rypkema and Wiehagen 2000). In Ontario there is a thriving business in heritage development, with costs that are sometimes, but not always, higher being more than offset by greater return on investment (Shipley et al. 2006). Even the problems of site contamination, which often threaten to impede efforts at adaptive reuse, can be overcome (De Sousa 2002). The tide of opinion has turned to the extent that many municipalities are now prepared to offer tax incentives, flexibility on parking and other requirements, and relief from development charges in order to encourage investment in revitalization projects that reuse older buildings and benefit urban revitalization efforts. When municipalities offer such incentives they are usually rewarded economically in the long run (Barber 2003).

If all of this were not enough to convince people to move past any notion that heritage is not opposed to development but is in fact a very sound and profitable kind of development, there are important additional arguments in favour of conserving the cultural landscape. A growing body of evidence links conservation with social benefits (Graham 2003). The most economically successful communities are more and more being recognized as the ones with the most liveable and pleasing environments. Typically they are the ones with the strongest sense of place, tradition, and character. They are the communities with the most social capital.

An abiding problem in many communities is that those who set themselves up as the guardians of the cultural landscape, the local heritage committees and historical societies, do not recognize their allies in the heritage development business.

Energy Efficiency vs Energy Saving

Perhaps the last of the dilemmas that appear to plague efforts to conserve and reuse the built elements of our cultural landscape is the notion that old buildings are energy inefficient. There has long been a focus on current energy bills for heating and cooling of buildings, which has led to important advances in materials, building techniques, and design that aim to improve the energy performance of new structures. Those advances are admirable and the move to LEED certification (Leadership in Energy and Environmental Design) is to be commended. However, according to Rypkema and Moe of the US National Trust for Historic Preservation, if we are interested not just in saving on current energy bills but in the much broader matter of overall energy sustainability then we have to consider the bigger picture (Ripkema 2008).

The greenest building, they tell us, is the one that is already built. This is understood clearly if we think of embodied energy. Every brick in an existing building was created by burning fossil fuel, every timber was transported to the building site using fuel, and all the nails were manufactured in coal-fired shops. As long as the building stands, that energy is captured and held or "embodied" in the structure. When we demolish the building we not only use energy to do so, but we loose the embodied energy forever. Then we burn more irreplaceable energy to replace the old building. While the new building is arguably more energy efficient, because it takes decades of savings on the monthly heating/cooling bill to make up for the energy used to construct it, we never recover the embodied energy lost by the demolition. The same large-scale equation is true when we keep the old building but replace the wooden windows with vinyl or metal. It takes decades for the heating bill savings to make up for the energy used in manufacturing those windows from non-renewable resources. Wooden windows, on the other hand, are reparable and are made from a renewable material, but when the metal and vinyl windows bend and crack, they will have to be thrown out. Replacing entire buildings or parts of buildings only appears to be cheaper because we live in world where true costs are not calculated properly.

Rypkema points out that, so-called green buildings are not synonymous with sustainability, while conserving the cultural landscape is in itself a sustainable activity. Development without heritage conservation is not sustainable. We have learned to recycle things as small as pop cans and as ubiquitous as paper, but we have not come to terms with how to recycle the largest artefacts that humans create, out cities, towns, and villages and the buildings that constitute them.

CONCLUSION

While there may be considerable differences across the various regions of Ontario in terms of geology, tradition, resources, and location, the changes that take place in the cultural landscape are largely driven by the same forces and subject to the same policies and political system. What we have to keep in mind are the following four factors. First, the same forces of nature and human activity are constantly eroding the physical expressions of our culture. We have to understand those forces and take them into consideration. They present as many opportunities as they do threats.

Second, the international community not only has vast experience in conserving its shared heritage, it also has set out common principles and guidance for all countries. Canada and Ontario are signatories to these global undertakings to hand on the cultural landscapes we have inherited and to convey them in better condition than we received them. To that end we have created laws and regulations that we have not observed to the degree we should. We have neglected our responsibility, under-funded the support mechanisms for heritage, and created unfair and distorted economic conditions that discourage proper stewardship. In multiple local decisions we have favoured private rights over community values in spite of legislative and judicial direction. This situation can be corrected only when the provincial government exercises more leadership in setting common standards.

Third, in many cases the very people who see themselves as champions of the cultural landscape either do not recognize and support their allies or obsess so much on detail that they miss the totality of the cultural heritage. Among the allies of cultural conservation are the builders and developers who risk their money and pour out their own creativity to breath new life into old structures. On the other side, the finial-retentive and protected-treasure approach will not in the end result in true conservation.

Fourth, we must come genuinely to learn about the role that the built heritage has to play in the search for human activity that is sustainable.

APPENDIX

Table 15.1
How the Planning Act Can Be Used to Protect Heritage Resources

Planning Act Section	Description of and Purpose of the Section	Selected Examples of How Sections Can Be Used, Including Strengths and Limitations
Section 3, Provincial Policy Statements	Policy statements are made under section 3 of the Planning Act to clarify the provincial interest. A statement of particular importance in the conservation of heritage is the following: "2.5.1. Significant built heritage resources and cultural heritage landscapes will be conserved." (Note: the statements says "will be conserved.")	For the purposes of heritage conservation, Provincial Policy Statement 2.5.1 may be the strongest legislative instrument available. It is clear and unequivocal in its direction, but also there is a growing body of case experience from the OMB and the courts that, while not unanimously supporting the "heritage" view is informative in the direction it gives. Municipalities not undertaking these measures might be considered not to be carrying out their legislated responsibilities.
Section 16, Official Plans	All municipalities in Ontario are required both to have OPs and to update them every five years. Official Plan policies and Official Plan Amendments implement the provincial planning interests. OPs are general policy documents setting out the direction and principles of development and are very appropriate documents for the expression of heritage goals and principles.	A typical Official Plan might contain statements such as "the Town will seek to preserve, protect and enhance natural physical features and biological communities, and cultural heritage resources [and] a key strategy of this plan is to protect land resources including landscape features, systems and areas that perform important natural functions or which provide economic and recreational opportunities. Included in this category are natural and cultural heritage resources, recreational lands and agricultural lands."
Section 28, Community Improvement Plans	CIPs refer to "the planning or re-planning, design or redesign, re-subdivision, clearance, development or redevelopment, reconstruction and rehabilitation," of an area, and the provision of "residential, commercial, industrial, public, recreational, institutional, religious, charitable or other uses, buildings, works, improvements or facilities."	Section 28 is particularly significant in that it allows a council to make special grants to projects that are part of Improvement Areas. There may be cases where a Heritage District Plan could be usefully structured as a CIP. Alternatively, there may be some interesting heritage objectives that can be achieved through the CIPs themselves. This is a good

Table 15.1 (continued)

Planning Act Section	Description of and Purpose of the Section	Selected Examples of How Sections Can Be Used, Including Strengths and Limitations
Section 28 (continued)		tool for dealing with areas such as brownfields and downtowns where special planning provisions can facilitate wider goals.
Section 34, Zoning	While official plans set out general direction and policy, it is zoning that carries the weight of law in land use matters. A number of aspects of buildings can be prescribed by the zoning, including setback from the property line, massing, parking requirements, height and so on. Zoning must be fair, uniform, and predictable, so it is often a heavy and blunt instrument when used to try to accomplish heritage conservation aims. Zoning can also regulate use, and that can be both a help and a hindrance to conservation.	One of the best uses of zoning is to assist in negotiation of mutually desired developments. When a property owner or developer applies for a zoning change or approval, it can be a time to discuss desired results: a planner armed with good, clear official plan policies can expedite and facilitate approvals for owners ready to voluntarily support those policies. The following is an excerpt from a zoning bylaw that resulted from such a negotiation: "The permitted uses shall be located only within the existing heritage structure [and] the external appearance and height, comprising the entire north, west and east facades and hip roof, and all associated building materials and architectural features of the two-and-one-half storey heritage structure shall be preserved and maintained."
Section 36, Holding Provision	Under this section a municipality can designate an area where it can specify the use to which lands, buildings, or structures may be put at some time in the future. In order to do this the municipality must have provisions in effect in its official plan.	The advantage of using this section has been to postpone decisions on certain approvals while other opportunities are explored. Since "H" is the symbol used to identify holding zones on maps, it has often been mistaken for "heritage" zoning. Indeed it has been used for that purpose.

Table 15.1 (continued)

Planning Act Section	Description of and Purpose of the Section	Selected Examples of How Sections Can Be Used, Including Strengths and Limitations
Section 37, Height and Density Transfer	Municipal Councils may enact bylaws, if they have the proper policies in place in their official plans, to permit a higher density and greater building height on a given property in return for some undertaking on the part of the property owner. This provision can be used in order to compensate a property owner for the arguable loss of potential value when a property is not developed in a way that it might otherwise have been.	An example: if a property is currently zoned low-density residential and a building on that property is given historic designation, there is no loss of value. However, if in the future adjacent properties are being redeveloped at a higher density and their current buildings being demolished and replaced by larger ones, then it might be argued that the designated building that is subject to a veto on demolition has not achieved its potential value increase. In such an instance the municipality, under s. 37, could award the owner a density-increase privilege that could be used elsewhere or sold.
Section 38, Interim Control	Local municipal councils can direct that a review or study be undertaken in respect of land use planning policies and may pass a bylaw to be in effect for a period of time of the study but not to exceed one year.	The interim control bylaw prohibits the use of land or buildings within the defined area except for such purposes as are set out in the bylaw. This is a fair and useful way to proceed when undertaking a heritage conservation district plan or other secondary plan that may ultimately have an impact on allowable building form.
Section 41, Site Plan Control	Under this section of the act a municipality can specify that an area be subject to Site Plan Control. This means that any development within the area usually requires plans or drawings to be submitted in support of a site plan application and may be approved subject to certain conditions.	Areas subject to control include such matters as "parking facilities and driveways, walkways, lighting facilities, walls, fences, hedges, trees, shrubs or groundcover, garbage facilities, easements, grading, provisions for the disposal of water from property, and the colour, texture, and type of materials, window detail, construction details, architectural detail and interior design." Site Plan Control can often provide opportunities to enhance the surroundings of valued heritage features.

Table 15.1 (continued)

Planning Act Section	Description of and Purpose of the Section	Selected Examples of How Sections Can Be Used, Including Strengths and Limitations
Section 45, Minor Variances	When property owners want to make a seemingly small alteration that may conflict with the letter of the zoning regulation, they can apply to the municipality's Committee of Adjustment to be allowed to undertake the work as a minor variance from that zoning.	Minor Variances for the appropriate use and development of land, buildings, or structures that constitute heritage property must be considered within the context of the provincial interest, official plan policies, and the zoning bylaw.
Sections 51 and 53, Plans of Subdivision and Consents	Plans of Subdivision have been one of the main mechanisms of development for the last generation. For the most part they have occurred on "greenfield" sites and in terms of heritage have dealt mainly with rural buildings and archaeological sites. But with a desired move toward re-development in existing urban areas, these will potentially collide even more often with the built heritage. Consents pertain particularly to rural properties where councils feel that a full plan of subdivision is not necessary.	With the proper OP policies in place it is perfectly legitimate for municipalities to require as a condition for Plans of Subdivision, that heritage structures be maintained. An example of the wording from such a condition is, "Notwithstanding the heritage policies contained in Policy XX of the OP, the stone structure referred to as the ... House and located in Lot YY, Concession Z, shall be retained, conserved and incorporated into the permitted development for the area designated as Medium Density Residential 2 (Street Townhouses)."
Section 70.2, Development Permit System	This is an alternative to zoning and one that is used in place of zoning in the UK and elsewhere. This could be a good alternative to zoning for heritage properties and in heritage conservation districts.	Zoning is a restrictive approach in which there is a clear list of things that are permitted and a restriction of anything that is not. The Development Permit system is a performance-oriented approach in which one is allowed to do anything so long as it meets a performance standard.

Source: By the author, with permission.

NOTES

1 http://www.ec.gc.ca/default.asp?lang=En&n=ECBC00D9-1.
2 http://www.ene.gov.on.ca/en/about/index.ph.
3 http://www.ec.gc.ca/default.asp?lang=En&n=ECBC00D9-1.
4 http://www.healthycommunities.on.ca/ohcc.htm.

REFERENCES

Barber, S. 2003. "Municipal Tax Incentives in Victoria British Columbia: A Case Study." *Plan Canada* 43(2): 20–2.

Bunting, T.E. 2000. "Housing Strategies for Downtown Revitalization in Mid-Size Cities: A City of Kitchener Feasibility Study." *Canadian Journal of Urban Research* 9(2): 45–175.

Carter M. 1999. *CIHB Revisited*. Ottawa: Heritage Resource Associates Inc.

CMHC. 1997a. *Conventional and Alternative Development Patterns – Phase 1: Infrastructure Costs Report*. Prepared by Essiambre-Phillips-Desjardins Associates.

– 1997b. *Conventional and Alternative Development Patterns – Phase 2: Municipal Revenues Report*. Prepared by Hemson Consulting.

Delafons J. 1997. *Politics and Preservation: A Policy History of the Built Heritage 1882–1996*. London: Chapman & Hall.

Denhez, M. 1997. *Legal and Financial Aspects of Architectural Conservation*. Toronto: Dundurn Press.

De Sousa, C. 2002. "Measuring the Public Costs and Benefits of Brownfield versus Greenfield Development in the Greater Toronto Area." *Environment and Planning* B 29:251–80.

Fram M. 1988. *Well Preserved*. Erin, ON: Boston Mills Press.

Graham B. 2003. "Heritage as Knowledge: Capital or Culture." *Urban Studies* 39:1003–17.

Kincaid, D. 2002. *Adapting Buildings for Changing Uses: Guidelines for Change of Use Refurbishment*. London: Son Press.

Mason, R. 2005. *Economics and Historic Preservation: A Guide and Review of the Literature*. A Discussion Paper Prepared for the Brookings Institution Metropolitan Policy Program, Washington, DC: Brookings Institution.

Murtagh, W.J. 1997. *Keeping Time: The History and Theory of Preservation in America*. New York: John Wiley & Sons.

Powell, K. 1999. *Architecture Reborn: Converting Old Buildings for New Uses*. New York: Rizzoli.
Reeve, A., B. Goodey, and R. Shipley. 2007. Townscape Assessment: The Development of a Practical Tool for Monitoring and Assessing Visual Quality in the Built Environment. *Urban Morphology* 11(1): 25–43.
Rypkema, D. 1994. *The Economics of Historic Preservation: A Community Leader Guide*. Washington, DC: National Trust for Historic Preservation.
– 2008. "Heritage Preservation and Energy Sustainability." Address to the Annual Conference of the Architectural Conservancy of Ontario, Collingwood, May 2008.
Rypkema, D., and K. Wiehagen. 2000. *Dollars and Sense of Historic Preservation: The Economic Benefits of Preserving Philadelphia's Past*. Washington, DC: National Trust for Historic Preservation.
Shipley, R. 2000. "The Impact of Heritage Designation on Property Values." *International Journal of Heritage Studies* 6(1): 83–100.
Shipley, R., and K. Reyburn. 2003. "Lost Heritage: A Study of Historic Building Demolitions in Ontario, Canada." *The International Journal of Heritage Studies* 9(2): 151–68.
Shipley, R., S. Utz, and M. Parsons. 2006. "Does Adaptive Reuse Pay: A Study of the Business of Heritage Development in Ontario, Canada." *The International Journal of Heritage Studies* 12(6): 505–20.

16

Retrospect and Prospect

GORDON NELSON

INTRODUCTION

The original intention of this concluding chapter of Beyond the Global City was to offer an overview of the findings, along with some general suggestions or recommendations for the future. This approach followed from the book's goal of providing a fuller understanding of the complexity of Ontario and its various georegions as a basis for improved planning. *Beyond the Global City* was seen as an antidote to the Toronto-centred flat-earth model of the province. This model ensued from the commitment of the provincial government and other leaders to neo-liberalism, globalization, and rapid economic growth, notably, during the last three decades. Our hope was that this book would reveal the challenges and opportunities offered by a richer view of the varied natural, social, and economic histories and current status of the various georegions of Ontario.

It was not, however, our original intention to present detailed analyses and recommendations for the future of Ontario and its georegions. Rather, we saw this book mainly as a source of greater understanding of the province among its citizens, educators, young people, and decision makers. This book was intended as a framework and a catalyst for organizational and individual thought about the past, present, and future of the province and its georegions. *Beyond the Global City* would provide substance for thinking about the georegions of Ontario in other than the Toronto-centred terms that now apply.

However, events have overtaken these original intentions. Neo-liberalism, globalization, and the Toronto-centred model ran onto the reefs of market failure and economic collapse in 2008 when this book was in its final stages. Housing, financial, commercial, and other activities fell into steep decline. Jobs disappeared, and proposals were made with the aim of returning to the rapid economic growth and consumption in play before the meltdown. It behooves us to comment on these developments in the light of the learnings in this book, and I do so towards the end of this chapter. In making these comments I refer to comparable discussion and recommendations in the 2008–9 Annual Report of the Environmental Commissioner of Ontario entitled Building Resilience (Environmental Commissioner of Ontario 2008–9), as well as comments by critics such as the economist Joseph Stiglitz (2007), the historian Tony Judt (2010) and the environmental administrator Ruckelshaus (2010).

OVERVIEW

Beyond the Global City: Understanding and Planning for the Diversity of Ontario is, so far as I know, unique in its up-to-date regional coverage of the province. Other informative books have been published on aspects of the province, such as *Patterns of the Past: Interpreting Ontario's History* (Hall et al. 1988); *Ontario 1610–1985: A Political and Economic History* (White 1985); *Looking for Old Ontario: Two Centuries of Landscape Change* (McIlwraith 1999); and *Making Ontario: Agricultural Colonization and Landscape Re-Creation before the Railway* (Wood 2000). Such books provide valuable insight into the ecological, social, economic, regional, and political history of Ontario but do not offer the broad and systematic historical and regional analysis of *Beyond the Global City*.

Informative volumes have also been published that concentrate on individual cities, areas, and regions in Ontario, such as *Kingston: Building on the Past* (Osborne and Swainson 1988); *Hamilton and Region* (Dear et al. 1987); *Special Places: The Changing Ecosystem of the Toronto Region* (Roots et al. 1999); and *Peterborough and the Kawarthas* (Adams and Taylor 1985). These and other more focused studies have been important in preparing *Beyond the Global City*, as have other publications listed in chapters 1 and 2 as well as elsewhere in this book. Without such rich and supportive scholarship our work would have been impossible.

This analysis and interpretation of Ontario's complexity began in 2004 when the province was divided for research purposes into eleven georegions with distinctive geologic, hydrologic, biologic, and cultural characteristics. Each of the georegions was assigned for study to a person or persons deeply knowledgeable of its history and distinctive natural and cultural qualities. The basic motivation for these studies was concern about Toronto's accelerating growth and its impacts on the environments, economies, and social systems of other provincial georegions. The built metropolis and its surrounding network of roads, parks, recreational lands, and other infrastructure were sprawling into nearby georegions and absorbing them into its umbrella in ways often seen as not to their or the province's advantage.

All of this was occurring with the direct and indirect support of the provincial government, which introduced green belt, smart growth, and other legislation and policy to steer metropolitan growth. A big part of the provincial government's motivation was the idea that it was building a metropolis of international stature, an economic engine whose continued growth would spread benefits to all parts of the province. This thinking was very much in line with underlying development concepts such as trickle-down theory, globalization, and neo-liberalism, with their commitments to high growth rates, a free-market system, less government regulation, low taxes, especially for industry, and a leading role for business and commerce in increasingly interrelated global cities such as Toronto.

In this context one of the main purposes of our georegional analysis was to demonstrate that Ontario should be seen as far more than a "flat earth" vision of Toronto surrounded by smaller urban centres, farmlands, and northern wild places that were ultimately to be incorporated into its metropolitan and global system. And, in fact, the chapters in this book show that Ontario is made up of a varied mosaic of georegions that in many cases are not receiving the provincial recognition and support that the authors of *Beyond the Global City* think they require for strong locally influenced futures.

Each of the georegions has individual challenges and opportunities that lie beyond the growing global city model and require more attention by the Ontario government. These challenges and opportunities are addressed in each of the eleven georegional chapters and will not be repeated here. What I will do, however, is to present some selected economic, social, technical, environmental, and governance highlights that apply to the various georegions before putting forward some theoretical discussion, general suggestions, recommendations, and further food for thought.

Retrospect and Prospect 391

SELECTED HIGHLIGHTS OF THE ELEVEN GEOREGIONS

While the following highlights are selected to identify major challenges and opportunities in each of Ontario's eleven georegions, they should not be seen only as applying to individual georegions. Many environmental, social, economic, and other challenges, while most noticeable in certain georegions, are applicable to varying degrees in all of Ontario. In this respect the highlights selectively identify major challenges and opportunities in individual georegions. However, they also collectively illustrate the overall impacts and challenges of the global, neo-liberal, high-growth and metropolitan-centred model of Ontario during the past four decades (Saul 2006). In reviewing the following observations the reader should be aware that they are those of the editor who had access to the evolving chapters in *Beyond the Global City* and is responsible for this interpretation of findings and significance.

Of the eleven georegions, one, the Great Lakes, is immersed in struggles to control water quality, water quantity, exotic species, and healthy accessible coasts, as well as growing demand for water pipelines, bigger freighters, deeper canal systems, and other challenges to an international waterway governed uneasily by two nations and numerous states, provinces, and local governments.

Land use pressures, particularly along the lower Great Lakes, are damaging and destroying wetlands, dunes, and other critically important ecosystems, with Ontario and Canada lagging well behind the Great Lakes states and the United States in coordinated coastal planning (Lawrence, this volume, chapter 3).

A growing concern for the Great Lakes is increasing demand for water, particularly in large and growing coastal cities such as Chicago, Detroit, Cleveland, and Windsor. These communities are drawing increasingly on rivers or ground water leading into the Great Lakes. Their diminishing effect is exacerbated by the construction and planning of water pipelines to supply inland communities such as London and the Waterloo area. These challenges are augmented by the deterioration of water quality in inland watersheds and by the likely lowering of lake levels owing to climate change.

The Great Lakes have been managed by governments in Canada and the United States on an ecosystem basis since the 1970s. This has led to considerable improvement in water quality but has not had much impact on the growing number of invasive species, such as the zebra mussel, or on the rising water demands noted previously. In 2008 the states, provinces, and two federal governments around the lakes agreed to control the water supply by prohibiting withdrawals to areas located beyond the states and provinces bordering the Lakes. The implementation of this new agreement will, however, be a major challenge, particularly if climate change and other factors significantly reduce supplies in relatively highly populated areas close to, but outside, the Great Lakes basin, especially in the United States.

Ontario and Canada will have to be vigilant to protect their interests in the lakes. Existing cooperative planning organizations such as the International Joint Commission (IJC) and the Environmental Commissioner of Ontario need to play stronger surveillance, monitoring, and reporting roles requiring more staff and funding. The question of how to regulate water withdrawals and other pressures is becoming more prominent given current inability to control greater water use among communities and ultimately for the lakes as a whole.

Two other georegions, the southwest, or Carolinian (Hilts, this volume, chapter 5), and the Huron (Caldwell, chapter 6), are struggling to sustain their agricultural and rural character; the Carolinian is beset to a greater extent by numerous relatively large sprawling cities and towns such as Kitchener-Waterloo-Cambridge, London, and Windsor. Along with the decline of the automobile and other industries in both these georegions, and especially in the Carolinian, an expanding web of roads, transmission lines, and other infrastructure, as well as urban sprawl, is fragmenting woodlands, wetlands, and biodiversity, while also threatening the most valuable and productive farmland in the province and among the most productive in Canada.

One significant innovation in the Carolinian Canada Georegion was the creation of the Carolinian Canada Coalition in the 1980s (Johnson 2007). The coalition initially performed a broker role by encouraging private conservation stewardship on farmlands around the receding woodlands and wetlands, with their unique biodiversity nationally and numerous species at risk. Provincial funding was provided to farmers agreeing to best-practice guidelines for protecting remaining natural areas. The provincial funding died out, however, and the coalition metamorphosed into an education and communications organization promoting awareness of the ecological and social services provided by natural areas and the importance of private stewardship and best practices. In recent years the coalition has identified "The Big Picture," a network of existing and potential natural areas, the basic framework required to maintain and enhance woodlots, wetlands, and their flood control, water quality, soil stabilization, and other ecological services to the Carolinian Georegion (Johnson 2007).

In 2008 the Carolinian Canada Coalition incorporated and gained government recognition as a charitable organization. It also produced both a strategic and a business plan built around the adoption of the three basic roles of networking and the facilitation of programs and projects, serving as a catalyst for action, and promoting research. The coalition's success owes much to the composition of its board, which is more or less evenly divided among government agencies, private organizations, and individuals cooperating in planning and implementing ecosystem recovery, farmland stewardship, and other programs. As a facilitator focusing on delivering programs and projects through collaboration, the coalition can serve as a model for similar work in other georegions.

Lessons about the value of cooperation in stewardship can also be derived from experience in the Huron Georegion, which is especially challenged to maintain its relatively low rural population and agricultural small-town character in the face of the expanding metropolis. In Huron, committees have been set up at municipal levels that include government and citizen members and that work together to plan and implement nature conservation and sustainable agriculture. Stewardship councils have been established on a province-wide basis by the Ontario Ministry of Natural Resources, and they seem to have been especially successful in Huron Georegion. Best-practice guidelines for coastal landowners and managers have been produced in Huron, with the private Lake Huron Coast Institute playing a catalytic role.

Huron Georegion has also been one of the areas in Ontario most challenged by the push to create large pig farms in the 1990s and early 2000s. Some of these farms housed thousands of pigs that year-round produced waste equivalent to that of some human communities. Huron was prominent in measures to limit the impact of such farms on water quality and on nearby settlements through the design of a set of guidelines and requirements for the establishment and operation of pig farms. These guidelines include controls on the means of disposing of liquid waste, especially in winter when it was often spread on frozen snowy fields, with consequent runoff and water pollution in the warming spring.

Many of our studies reveal that the challenges of Toronto's growth are not by any means new. Decades ago railways from Toronto headed out to capture the Hamilton area, as well as London. In the late 1930s and early 1940s the construction of the Queen Elizabeth Highway through the fruit belt lands along the western shores of Lake Ontario led to the creation of the urbanized Golden Horseshoe between Toronto and Buffalo. The loss of much valuable orchard and woodland is dimly remembered as time passes and as some remaining fragments regain value for wine production on a global scale, albeit at the cost of rising demand for water and an increasing impact on water quality.

Muskoka and Georgian Bay were accessed from the growing city of Toronto in the early nineteenth century. Yonge Street was originally built north through the Holland Marsh area and the surrounding bush to Penetanguishene on Georgian Bay. In the 1840s and 1850s, lumbermen, prospectors, and miners moved out of Toronto to exploit the bush. Many of the prospectors came from Ohio and other parts of the United States. They and others were impressed with the canoe routes, waterways, hunting, fishing, and scenery. Their favourable impressions spread, and others returned to tour and for recreation even as the forests were being cut down and the area opened up to largely abortive agriculture via rail lines and colonizing roads financed from Toronto and by the Ontario government.

By the late nineteenth and early twentieth centuries, Muskoka was linked to Toronto commercially as well as through campers, cottagers, and tourists who took trains and steamers to their summer domiciles on the shores of Muskoka's many lakes. Much was invested in lodges, roads, and recreational infrastructure such as Algonquin Provincial Park. Muskoka has, in recent decades, become a second home and playground for many international celebrities, so that it is substantially tied to Toronto and the globalization network. And while winter recreation has been increasing and the resident and retiree population slowly rising, Muskoka loses much of its population in the winter. It is still predominantly a one-industry and one-season place with all the economic and environmental challenges that entails for local folk and visitors from other parts of Ontario, challenges likely to be deepened by current and anticipated economic conditions. In this respect Luka and Lister (chapter 8) see currently unforeseen opportunities in linking recreation and tourism in Muskoka with a need for deeper understanding of nature and the environment among an increasingly urbanized population in Ontario.

Along the shores of Georgian Bay old fishing and wheat ports such as Wiarton on the lower Bruce Peninsula and nearby Owen Sound and Collingwood have passed through their nineteenth- and twentieth-century involvement with fish-packing plants, grain elevators, and other infrastructure and now are largely committed to recreational marinas, condominiums, and the infrastructure of recreation. More and more people living in these Georgian Bay communities are transplanted retirees, perhaps in excess of 20 percent in places such as Owen Sound. These folk live among a legacy of the past that constitutes a major ongoing challenge for the future. The resources of the forests, the soils, the coastal wetlands, and the near and offshore waters have been severely affected by the heavy hand of historical colonization and exploitation. The lake trout and other fisheries are gone; efforts to restore them are under way but inadequate as yet to the task.

Peterborough began as a separate settlement and grew to be a functional centre for the surrounding region, especially during the nineteenth-century lumber era and early industrialization up to the 1940s (Bain and Marsh, chapter 7). Peterborough Georegion has since declined industrially, in large part because of competition from the growing industrial cities of Toronto and the Golden Horseshoe. From an early date, however, it was a region of interest to recreationists and tourists, particularly from Toronto. More recently Peterborough has become the home of Trent University and a centre for biotechnology and tourism.

Early Europeans were mainly Irish, although subsequently Scots, English, and others moved into the growing town as well as the surrounding Kawartha Lakes area. Much effort was put into logging, along with sawmills, grain mills, and other industrial activity, stimulated to some degree by the construction of the Lake Ontario to Georgian Bay Trent/Rideau Canal system. This system, however, soon proved unable to compete with the new roads and railroads of the mid- to late nineteenth century. A highway network now links the Peterborough Georegion with the Toronto metropolis, and plans are being considered to restore strong rail links with the big city. Whether this will happen, whether it will yield significant benefits to the Peterborough Georegion, or whether such ties will be foregone, and it will pursue a more independent course is in the balance; the position taken by the provincial government will likely be decisive.

The Kingston/Upper St Lawrence and the Ottawa Georegions have been less affected by the growth of Toronto, being located some distance from it physically, historically, and culturally. Kingston and the Upper St Lawrence Georegion are hundreds of kilometres away to the east, mainly on the hard crystalline acidic rocks of the Canadian Shield (Osborne, chapter 9). Here the forests, streams, and fish of Lake Ontario were favourable to exploitation, especially early on. Some of the limestone and glaciated soils beyond Brockville were favourable to agriculture, but often only marginally.

The main advantage for Kingston and the Upper St Lawrence communities was their location along the great river between Montreal and Lake Ontario. Early and mid-nineteenth-century ports were numerous along the river banks as well as the northern Lake Ontario shore, including Cornwall, Johnstown, Brockville, Gananoque, Kingston, Cobourg, Port Hope, Frenchman's Bay, Toronto, Hamilton, and Niagara. In the nineteenth century the flow of traffic, along with agricultural and industrial development, favoured business in all these towns to some degree. In the 1820s, 1840s, and 1880s major improvements were made in the St Lawrence Canal system, circumventing the huge historic rapids and permitting larger and larger vessels to track upstream and across the lake without stopping at the small ports, which, then, slowly declined.

Two major improvements were made in the late 1950s and 1960s. One was the construction of the great St Lawrence Seaway, which made it possible for very large ocean-going craft to bypass not only Kingston but even Toronto and Hamilton in moving through the Welland Canal to the upper Great Lakes. A second major bypass was the building of the 401 highway. This route was constructed from Montreal to Toronto, Windsor, and Detroit on the high ground away from the St Lawrence, leaving the valley communities and Kingston stranded on the shore, linked by the historic and now relatively little-used old King's Highway #2. The impact of these two bypasses on Kingston and the St Lawrence valley communities has been heavy in terms of the decline in industrial development and employment. In addition, the construction of canals and dams has contributed to losses in fish habitat as well as water quality, wetland, and riverine ecosystems. Some of these impacts have fallen on First Nations lands on either side of the St Lawrence.

And this tale is not yet over, for proposals are now under consideration to increase the size of the canals through the St Lawrence to provide for the passage of increasingly large ocean-going craft from Europe and other parts of the world. These proposals may not be viable economically in terms of anticipated traffic and would have significant effects on river and Great Lakes ecosystems, as well as the economic and social well-being of the St Lawrence River and other communities. The proposed changes could also increase the deterioration of the Great Lakes' water quality, heightening the risk of the escape of exotic organisms such as the gobi, a Black Sea fish now interfering with the food chain, especially in Lake Erie. Careful economic, environmental, and social-impact assessments are needed here, with opportunity for ample participation by local people. In this respect, the establishment of the Frontenac Axis Biosphere Reserve near Gananoque and St Lawrence Islands National Park has been a catalyst for much more vigorous networking and collaboration among conservation, agriculture, recreation, and other sectors in planning for sustainability in the St Lawrence valley.

In leaving the Kingston and Upper St Lawrence Georegion we would be remiss in not understanding the potential of the City of Kingston to act as a major conservation and development hub relatively independent of Toronto. The city has a rich past as reflected in its historic buildings and architecture. In spite of industrial decline Kingston has gone forward economically and socially on the basis of its urban amenities, its slower lifestyle, and the strong supporting presence of institutions such as Queen's University, federal prisons, and the military. These institutions employ thousands of people, and their indirect benefits, for example through student and military cadet spending, are substantial. Historic architecture, the Royal Military College, the old downtown, and the renovated waterfront are strong attractions for tourists.

In chapter 9, on the Kingston Georegion, Brian Osborne offers some alternative strategies applicable to the Toronto-centred global growth model for the great city itself, as well as Kingston and other urban areas in Ontario. Leading examples, some of which are already in early play, are the smart-city, slow-food, and creative-city movements, which aim to generate nearby markets and economies.

Provincial decision makers need to consider more carefully such alternatives and offer support for them in relevant georegions of Ontario.

Ottawa also stands apart from Toronto because of distance, the presence and support of the federal government, and its bilingual character (Seasons, chapter 10). Ottawa has benefitted from its research community, including universities, agencies such as the National Research Council, and strong computer and electronic industries. The City of Ottawa is also bastioned by its large contingent of federal administrative, legal, military, and other departments, as well as tens of thousands of federal employees, far exceeding those of any other georegion in Ontario. Operationally, the Ottawa Georegion actively extends well into the province of Quebec. Many federal employees live in Quebec and commute to Ottawa and vice versa, because some government offices are located across the river in Gatineau. Many Ottawans seek recreation summer and winter in places like the Gatineau Hills.

The big concerns with the Ottawa Georegion are, first, its cross-river and cross-boundary relations with Gatineau and Quebec, and second, stagnant or declining communities in the surrounding countryside. These communities have older, less well-educated and less economically and socially flexible populations than the central city of Ottawa. The residents depend largely on agriculture, often marginal agriculture, and on cottaging and tourism. Their young people are not stimulated to innovate, stay, and help their communities to grow. Many leave for bigger opportunities elsewhere, in part because of lack of financial means to attend college or university. To a considerable extent this observation also applies to many financially challenged families and students in other georegions such as Toronto, Muskoka/Georgian Bay, and Northeastern and Northwestern Ontario.

The ultimate question for the Ottawa Georegion is the extent to which it will eventually harden into an essentially federal entity. Its historic growth, current character and general well-being are due to its selection as the capital of Canada. Ensuing administrative and financial linkages and benefits largely made the Ottawa Georegion a place apart from the rest of Ontario. As Canada grows along with the economic and social ties between Ontario and Quebec, it is possible to foresee the creation of a separate cross-border district similar to that of Washington, DC. Such an arrangement is foreshadowed by the National Capital Commission, which coordinates park and other infrastructural development in this georegion.

Northeastern and Northwestern Ontario are perhaps the most similar of Ontario's georegions. Much of this is due to shared natural characteristics such as a cold northern climate, the boreal forest, and the rugged, crystalline glaciated rocks of the Canadian Shield. The two georegions also have relatively large First Nations populations that still live and in many cases still wish to live close to the land in their traditional domains. The Northeast and the Northwest do, however, differ in their ethnic make-up and to some degree in their social and economic patterns. The Northeast has a strong francophone constituency and a Roman Catholic background, which is reflected in the strong separate school system (Lemelin and Koster, chapter 11). Mining has been, and is, a mainstay in the Northeast, while lumbering tends to be more significant in the Northwest. The populations in the Northeast and the Northwest are both concentrated toward the south in cities like Sudbury and Thunderbay. Claybelt farming and gold, silver, and other mining led to communities like Kapukasing in the Northeastern mid-north. But most of the central and northern parts of both the Northeast and the Northwest are isolated wildlands occupied, to a large extent, by First Nations (Johnston and Payne, chapter 12).

Ever since the late nineteenth and early twentieth centuries, Northern Ontario has been viewed as a resource hinterland by Toronto financiers and business interests and by the provincial government. Relatively little manufacturing of forestry or mineral products has been conducted there, and leasing systems have favoured large-scale cutting and mining by big resources companies with support from the Ontario and federal governments. These companies and forest and mining developments have been constantly challenged by boom and bust in demand, usually ensuing from global economic fluctuations. In boom periods the companies extend their operations into new ground and often diminish or mothball their operations in bust periods, with adverse effects on one-industry towns, local people, communities, and sustainability. Large-scale cutting and mining also bring major environmental disturbances through dam construction and damage to wildlife habitat, such as that of caribou and migrating Central and South American breeding birds.

The expansion of forestry and mining companies also disturbs and dislocates First Nations communities, and considerable tension and conflict ensues. Among the suggested solutions is allocation of more control to local governments and more cooperative ventures among government, industry, local communities, and First Nations. In many quarters there seems to be a growing desire to soften the grip of the Ontario Ministry of Natural Resources (OMNR), which is basically responsible for governance of much of the Northeast and Northwest and makes decisions that can be at odds with those of First Nations or of local municipal governments in the southern part of the two georegions. In the Northwest the power exerted by the OMNR and the Ontario government more generally has led to a feeling of alienation and potential bonding with Manitoba.

One recent major move has been the decision by the Ontario government to set aside much of the northern part of both georegions for conservation and First Nations but few details on this are available as yet. This provincial decision seems likely to be divisive and contentious politically. Many companies and more southerly northern communities may be on one side and conservationists and First Nations on the other, although intragroup differences could be greater than foreseen.

In summary, except for the increasingly rapid expansion of the Toronto Georegion, Ontario's georegions have remained comparatively stable at the core with some fluctuations in their character and boundary zones as a result of economic, social, technical, environmental, and political change. Among the more noticeable changes is the evolving emergence of the Northeastern and Northwestern georegions out of what was essentially one Northern Ontario in the early 1900s. And it may be that the recent Ontario government decision to set aside the far North in both georegions will result in an entirely new georegion in ensuing years. A case can be made for a Far Northern Georegion on environmental, socio-economic, and ethnic or cultural grounds. Precedents for such an arrangement can be found in Quebec following the James Bay and other power projects in the 1970s and "land claim" settlements with the Inuit in the Arctic in the 1990s. At that time the substantially self-governing territory of Nunavut was set aside for the Inuit people.

A recent addition to Ontario's georegions has been the Niagara Escarpment (Preston, chapter 13). It was established by act of the Ontario government in the 1970s. The main motivation was environmental and involved calls for protection of the limestone, cliffs, and scenery of the Escarpment against the growing demand for construction materials. In the 1980s and 1990s the Niagara Escarpment Georegion was institutionalized through the development of a provincial act, a series of management plans, and a network of government services to support its development. The restrictions that government regulators place on local inhabitants are considerable and still resented by many of them. Others outside the Niagara Escarpment area in Toronto and Hamilton do, however, support this regulatory system and the associated extension of the Toronto Georegion's "green infrastructure" because of the high level of protection the arrangement affords to the accessible, rugged cliffs and headlands, forests, and summer and winter recreational opportunities along its more than seven hundred kilometre length.

Another georegion may also emerge along the edge of the Canadian Shield between approximately the Kingston area and western Muskoka and the Georgian Bay shores. The case for this georegion is being made mainly on environmental and recreational grounds somewhat similar to those for the Niagara Escarpment. The long ribbon along the edge of the Shield has been named The Land Between. This area was logged, mined, and cut over in the nineteenth and early twentieth centuries. It has since succeeded to attractive forest cover around numerous lakes amid pink-rocked majestic scenery. One concern is that lumbering and cottagers from Toronto and nearby urban areas will continue to move into this georegion, lining its lakeshores, fragmenting its forests, and facilitating declines in biodiversity, water quality, and other environmental values. The success of a call for some kind of special protection for The Land Between remains to be determined, but initially, at least, it seems worthy of consideration. In any event the emergence of The Land Between must ultimately be seen as due to a substantial degree to pressure from devotees in Toronto, including transplanted retirees wishing to protect cottage and camping country.

This brings us to the great city itself, an expanding urban agglomeration of some four to five million people out of an Ontario population of approximately eight or nine million (Sportza, chapter 4). From a small early-nineteenth-century nucleus of a few hundred Loyalist settlers absorbing, dislodging, and scattering a thin group of French and aboriginal people, Toronto has grown directly by encroaching on nearby lands and communities and indirectly by extending its environmental, recreational, economic, social, and political system into the Golden Horseshoe, the Niagara Escarpment,

Georgian Bay and Muskoka, and Northwestern and Northeastern Ontario. Its influence is constantly being pushed into the Carolinian Canada and Huron Georegions.

Carolinian Canada will be heavily affected by the recent Greenbelt and smart-growth initiatives of the Ontario government. These are attempts to intensify metropolitan growth within a smaller area while extending environmental protection and recreational space around the big city. One already discernable result is the leapfrogging of development into cities at the outer fringes of the greenbelt such as Kitchener/Waterloo and Cambridge. And such leapfrogging has defeated several previous attempts to develop greenbelts and limit Toronto's growth.

Associated with these land use planning measures are proposals to extend and improve transport in and out of the Toronto Georegion. Metrolink is a provincially sponsored initiative that is involved in studies and proposals to improve traffic and interconnections within the Toronto Georegion as well as with outlying centres considered not yet firmly linked into its network. So we have proposals to link Pearson Airport to the city centre by various types of rapid transit, a choice proponents say can be made on the basis of quick environmental-impact assessments that very likely will not give affected communities and individuals the opportunity to make informed decisions.

Another example is the proposal to improve transport along the Golden Horseshoe between Toronto and Buffalo by constructing a new highway or rail line atop the Niagara Escarpment. This would supplement the Queen Elizabeth Way, which was built below the Escarpment along the shores of Lake Ontario in 1939. Like the Queen Elizabeth Way, the proposed Toronto-Niagara project would fragment farmland and natural spaces during its construction and operation and subsequently attract a ribbon of new residential and commercial developments along its length. These impacts would be especially significant for the Niagara Escarpment itself. It would then be hemmed in by two major highways as well as criss-crossed by transverse routes that have left no major stream valley undeveloped, the last being the controversial road recently built through Red Hill Valley on the south edge of Hamilton. A "top of the Escarpment" thruway or rail line would also further fragment the woodlands and farmlands of the western part of the Carolinian Canada Georegion.

Amid this litany of concerns, some recent moves by Toronto and the provincial government are viewed favourably by many local people. A leading example is the passage of the Oak Ridges Moraine Act in the early 2000s. This act has put into play a land use plan and supporting financial arrangements that will limit urban encroachment on the morainal lands and conserve their forests and natural qualities, particularly the groundwater supplying many streams leading into Toronto and related communities.

On the other hand, the Oak Ridges Moraine Act is an excellent example of the Ontario government's tendency to plan for province-wide issues on a relatively specific geographical basis. In this respect numerous other places in the province were known to have morainal urban encroachment challenges at the same time as Oak Ridges and Toronto. An outstanding example is the moraine system west of Waterloo, Kitchener, and Cambridge in the Grand River Valley and the Carolinian Canada Georegion. This system and others have been under study by agencies in the Ontario government. Many observers expected these studies to lead to province-wide legislation and/or policy applicable to moraines with significant forest, acquifer, biodiversity, recreational, and agricultural values throughout Ontario. However, recent news releases indicate that the studies have concluded that existing law and policy is sufficient to protect such areas, in contrast to the decisions taken with respect to the Oak Ridges Moraine Act and the establishment of a special planning regime and supportive funding.

In his 2008–9 annual report, the environmental commissioner of Ontario expressed concern about the province's tendency to plan or protect "by exception," as in the case of the Oak Ridges Moraine legislation. The commissioner sees the Oak Ridges Moraine Act, 2001; The Greenbelt Act, 2005; and the Lake Simcoe Protection Act, 2008, as reflecting a disturbing trend to protect notable natural heritage features – and their water supply or other ecological services – on an individual basis rather than to implement broader-based safeguards under the Provincial Policy Statement, which sets out requirements for land use planning and management for the entire province (Environmental Commissioner of Ontario 2008–9, 22)

Here it is appropriate to recall the long-standing interest of some leaders, residents, and friends of Toronto in the creation of this city region as a separate province or a region with provincial powers.

This status has been perceived as giving the metropolis the control over taxes, land use planning and management, finances, and other policies and practices needed to override many of the barriers perceived as inhibiting its growth as a great global city.

Andrew Sancton's book *The Limits of Boundaries: Why City Regions Cannot Be Self-governing* (2008) is quite relevant here and, in fact, arises from Toronto's interest in separate status in the 1990s and early 2000s. Sancton reviews relevant experience with the restructuring of municipal, state (provincial), federal, and city region boundaries such as Berlin, Hamburg, and Madrid and concludes that such efforts are rarely successful. Among the major reasons are difficulties in reaching agreement with all the affected parties.

Visions of a relatively independent Toronto region or province in the 1990s saw proponents call for boundaries including much of central and southern Ontario, extending from approximately London in the west through Hamilton and Niagara Falls to Barrie on the north and Port Hope in the east. The big question, as Sancton points out, is why people in these diverse areas would be interested in being incorporated in such a Toronto-centred province. What advantages would accrue from surrendering substantial powers to Toronto? Where would the boundaries be drawn and with what justification? Why would the current province of Ontario be interested in such an arrangement given its responsibilities for the rest of a less prosperous, residual Ontario? And beyond the province lies the federal government and its questions about the implications of such a change to national political representation and the national governance system. A recent book by Michael J. Trinklein (2010) entitled *Lost States: True Stories of Texlahoma, Transylvania and Other States* bears Sancton out in describing largely unsuccessful efforts to create new states and political entities in the United States.

The foregoing discussion lends strong support to our central argument that much better understanding and consideration of the other georegions of Ontario is needed as part of broader provincial planning and policies. A reading of John Sewell's excellent studies *The Shape of the City: Toronto Struggles with Modern Planning* (1993) and *The Shape of the Suburbs: Understanding Toronto's Sprawl* (2009) shows the regular formulation and deficiencies of the traditional approach to provincial planning. This largely centres on Toronto and the encouragement of its growth as a financial, economic, and metropolitan engine, with some concern for recreational opportunities and natural and environmental protection. In master-planning exercises carried out periodically since the 1940s, city and provincial planners have worked in uneasy association with developers to provide for the growth of the city. This has involved, for example, substantial historical and ongoing financial support for the sewage and water infrastructure required for large new subdivisions located outside the political boundaries of the Toronto urban area, as well as provincial planning for major roads and throughways such as the Queen Elizabeth and the 401, 403, and 404. These routes have run well beyond extant Toronto boundaries and worked to tie outlying areas and their citizens increasingly into the Toronto-centred network with its associated inner-city focus, congestion, and gridlock, as well as economic, social, environmental, and political costs to outlying regions.

With provincial support, city planners put forward green belts in 1909 and then more recently, in the 1940s, 1950s, and 1970s. It has always been difficult making these work to control sprawl, largely because the opportunity has always been left open for developers to leapfrog opportunistically along the Queen Elizabeth or beyond the 401 and perhaps in future, the Metrolink. Today too many people, planners, and decision makers do not understand or wish to understand this pattern of planning and its role in the ongoing sprawl of the metropolis. And here we see sprawl as including not only the outward sprawl of the city and its fringe but also scattered growth in lands beyond the city fringe.

A new approach is needed that would involve more understanding and appreciation of georegional needs and opportunities and much greater consideration of province-wide policies for common issues such as Ontario's moraines. However, careful assessment is needed here, with ample opportunity for civic input, since provincial policies have been found generally effective and acceptable in cases such as flood-plain and hazard zoning but much less so in cases such as provincial control over aggregate lands and quarrying. Yet hopes for such critical environmental and other assessments are dimming as municipal, provincial, and federal interest rises in reducing environmental assessments and public input into plans for development projects, which these governments see as urgent because of the economic slowdown and threats to jobs and development. A grand example of this is the 2008 Green Energy Act, passed quickly by the province to circumvent assessments and possible obstacles by lower levels of government and citizen groups, expedite the spread of wind towers and other technology, and promote more jobs in the province.

OTHER IMPLICATIONS

The authors of most of the chapters in *Beyond the Global City* have put forward some suggestions and recommendations for future planning of their georegions. Here we will consider a limited and workable set of general recommendations applicable to all georegions of Ontario. These recommendations are in line with the vision of Toronto as a major part of a sustainable set of interrelated provincial georegions.

We can turn first to chapters 14 and 15, by Murphy and Shipley. These authors address natural and cultural issues, respectively, on an Ontario-wide basis. Murphy identifies a number of issues that are to varying degrees common to most or all of Ontario's regions. He does this by providing a brief history of rural and, especially, of increasingly intense urban development in the province. Both rural and urban development are seen as fragmenting forests, wetlands, and natural habitats, polluting water, damaging or destroying riparian zones, reducing the biodiversity and resilience of natural systems, as well as lowering air quality. Murphy traces the historical public response to these issues through the creation of an increasingly diverse array of parks and protected areas. These include the establishment of the watershed conservation authorities in the 1940s and 1950s. Other more recent creations, notably in the Toronto Georegion, are the Oak Ridges Moraine conservation program, the Rouge Valley parkland, and the Greenbelt. At a general level, however, environmental challenges face many cities in Ontario. Murphy emphasizes two basic approaches: ecological restoration and cooperative planning. Citizens, planners, and decision makers throughout the province would deal with economic, technical, social, and environmental issues in an integrated rather than a specialized or narrow way.

Murphy discusses the meaning of restoration, notably in urban environments where strict restoration or replication of the original habitat or ecosystem is generally impossible. He makes a case for ecosystem remediation that, while imperfect in terms of what has gone before, does provide for useful natural habitats and improved ecological services, for example, improved air quality and the amelioration of "heat island" and other climate changes in urban areas.

He notes that the restoration or rehabilitation of gas stations and other industrial and commercial "brownfield sites" can provide space for residential and other urban intensification. Overall, he sees an array of prospects and consequences ensuing from greater acceptance and use of urban ecology in planning. Murphy also touches on the issue of climate change, which will affect all Ontario's georegions in varying ways and degrees but which cannot be explored in any detail in this book.

Murphy focuses on more urbanization throughout Ontario. Basically, he sees large-scale urbanization as proceeding in the Toronto Georegion but also in many other centres in the southern part of the province, including London, Kitchener-Waterloo-Cambridge, and other centres in Carolinian Canada, Peterborough, and Ottawa, as well as other centres in southeastern Ontario and North Bay, Sudbury, and other urban areas in the mid-north. As long as immigration and other forces remain relatively high, this extensive urbanization will occur, even with current intensification policies. His call for more environmental or green initiatives and planning as an integral part of urban policy therefore becomes even more salient, as does the need for more province-wide policies, for example, with an improved Ontario Provincial Planning Statement (PPS).

Shipley addresses the neglected field of cultural, or built, heritage on a provincial basis. He poses the challenging question of whether there is one cultural landscape or many. He does not intend to belittle the important task of identifying significant historical and architectural structures and landscapes; rather Shipley wishes to draw attention to the overall value of all of Ontario's built or cultural landscape.

This heritage represents historical, architectural, economic, and social conditions of earlier times and is consequently of considerable educational and symbolic value today. More broadly and intrinsically, however, Ontario's built heritage and cultural landscapes represent substantial past economic investments whose value is too easily lost because tax laws and institutional guidelines are favourable to demolition rather than conservation. This concern is exacerbated by the Ontario government's strong support of intensification in urban development in order to reduce sprawl and its unwanted impacts on the environment and surrounding farms, villages, and towns. In proposing measures to improve planning and decision making, Shipley calls attention to the Ontario Heritage Act, the Provincial Planning

Act and municipal official plans. He supports use of Heritage Impact Assessments to evaluate the effects of development proposals on existing valued landscapes, as well as policies for the conservation of built heritage in municipal official plans. Measures in the Ontario Planning Act and the associated Provincial Policy Statement, including community improvement plans and development permits, could also be used more forcefully by municipalities. Shipley also recommends greater municipal use of tax and other incentives to support conservation, restoration, and adaptive re-use as much, or more, than entirely new development.

Shipley ends his chapter by addressing a number of underlying dilemmas involved in protecting our built heritage and providing for more effective, fair, and balanced overall development in urban and rural Ontario. They include heritage vs development and energy efficiency vs energy saving. These two dilemmas are of considerable interest to planners and decision makers, as well as to the "average citizen's" general understanding of the need to support more effective built heritage conservation and restoration.

The heritage and development dilemma refers to the general belief that development means new building either in open spaces or those opened up by demolition. Many attempts to protect or even adaptively reuse existing structures are seen as anti-development or anti-progress. This, at the root, is a most misleading perception. According to Shipley, in Ontario today more than half of construction spending involves renovation or adaptive reuse. Shipley notes that the lives of older structures can be extended for lengthy periods and often reused at greater intensity to meet new opportunities. This leads to the dilemma of energy efficiency vs energy saving. The greenest building can be seen as the one already built, because it avoids the loss of all the sunk energy embodied in old structures as well as the use of much new energy in erecting new ones. In this respect conserving the built environment or cultural landscape is a sustainable activity. It requires learning to "recycle the largest artefacts that humans create, our cities, towns and villages and the buildings that constitute them."

Beyond the Ontario-wide recommendations made by Murphy and Shipley lie others that arise directly or indirectly from the various georegional chapters in this volume. One recommendation that arises in a number of chapters of *Beyond the Global City* is the desire for more effective bottom-up and collaborative community planning. The practice of and the need for community planning is enunciated in chapters on the Huron, Peterborough, Northwestern, and Northeastern Georegions and implied in the chapters on most others. But as several authors intimate, effective cross-sectoral community planning is a challenging task because it has to face the focused thought and language of specialties. People often are not accustomed to or trained for cross-disciplinary and cross-sectoral thinking, and it can be relatively expensive in time and money. The federal government has a community futures program that assists communities in these respects and has had some success in Ontario, for example, in Huron Georegion.

Ontario's provincial government could usefully consider ways of assisting with more effective collaborative community planning as part of a program in which greater attention is devoted to the needs of its georegions. One way would be to encourage the formation of integrative region-wide coalitions of concerned government and non-government organizations and groups such as the Carolinian Canada Coalition. Another way would be to encourage more cross-sectoral and cross-disciplinary forums or summits to cooperate in charting ways forward.

Certain groups need to be targeted with respect to the message in this book, particularly students in public and high schools, as well as colleges and universities. Ever since curriculum changes were introduced in the schools by Ontario governments in the mid- to late 1990s, students have been given inadequate opportunities to gain in-depth knowledge of the environmental and human history, character, challenges, and opportunities of different parts of Ontario and the province as a whole. This situation should be rectified as quickly as possible, since young people, including the offspring of immigrants, will soon be making decisions for the province. These decisions are likely to be less than satisfactory for many of Ontario's ecoregions if they are made without knowledge of the type presented in this book. Other groups that should be targeted are the media, the civil service, and political decision makers. Three major Ontario newspapers are prepared and published in Toronto, and they certainly focus on that georegion, but many citizens and decision makers in the metropolis are left with low levels of appreciation of Ontario's other georegions, their needs, and their importance to Ontario.

An improved understanding of the complexity of Ontario would likely cause decision makers to give much greater consideration to policies and programs of long-term benefit to all or many of its georegions. Such policies and programs could include

- Investing in government institutions that provide stability in communities and georegions challenged by the decline of business and private employment. Such institutions would include public and high schools that have frequently been closed because of "low enrolments," with the result that students and teachers and their daily commercial transactions are then lost to places elsewhere. The same applies to government office closures. Examples are closures resulting from municipal amalgamation and encouraged or forced by 1990s provincial governments in jurisdictions such as Chatham-Kent. Here two counties, at least twenty-four townships, and many associated government jobs and local expenditures were lost when all were amalgamated into one municipal regional government.
- Providing incentives to colleges, universities, and research institutions to diversify into outlying georegions and communities. Some colleges and universities have taken this route in recent years, sometimes with the assistance of municipal or other governments. One highly successful example is the establishment by Wilfrid Laurier University of a Brantford campus centred in the formerly vacant Carnegie library in the historic administrative square in Brantford. Campus growth has led to renewed investment and revitalization of the City of Brantford. More recently the University of Waterloo has established branches in Cambridge and in Kitchener in cooperation with McMaster University
- Strengthening environmental-assessment procedures and regulations, which have become more and more bureaucratically pro forma. Support and funding for them declined following the election of the Conservative governments of the 1990s. Without serious and civically responsive assessments more imperfect decisions can be made. A leading possibility at the time of writing is government interest in investing substantially in large-scale infrastructure such as roads, pipelines, and other hard services in order to compensate for loss of jobs in private industry. Some pundits have called for the quick start of "shovel ready" projects with the aim of rapidly responding to these losses. But environmentally, socially, and economically sensitive decisions are not likely to be made without strengthened government commitment to environmental

and heritage assessment programs and associated opportunities for local government and citizen participation. And these should be accompanied by strong regulations and management.
- Making green infrastructure a large part of future infrastructural development. This should be done in two ways. First by ensuring all significant hard infrastructure proposals have a green component, i.e., that they include reforestation, restoration, and other means of offering the ecological services that nature provides. Green infrastructure projects should also include proposals to restore damaged ecosystems by rebuilding woodlands, wetlands, and the natural corridors needed to link fragmented ecosystems as well as improve water quality, reduce floods, ameliorate climate change, and provide recreational, educational, and research opportunities near home areas. Many of these employment and learning opportunities could benefit younger people, displaced workers, and new Canadians.
- Establishing more reliable and supportive arrangements for families and individuals whose qualifications and local opportunities leave them at the edge of or outside the job market in numerous georegions (Krugman 2009). In light of the current economic difficulties concern is being expressed about the inadequacies of unemployment insurance and other social support programs, which often were downloaded to the financially strapped municipal level by the governments of the 1990s. An example is provided by social assistance programs for single mothers, older people, and the disabled. To reach these people the province needs to move beyond unemployment insurance, since it often does not apply effectively to the socially challenged or indeed, to private business, notably small business. The current Ontario government has promised to restore provincial support for municipal social assistance programs by 2012. However, in terms of long-term sustainability, more basic and durable support needs very serious consideration, especially for young people struggling to find their way educationally and economically, as well as for older workers and retirees whose prospects for employment and pensions have been dampened by the economic downturn.

Other suggestions could be put forward, but a major need is for a mechanism that will work in the long term to ensure the province takes a more balanced approach to its various georegions. As the major economic, social, financial, and population centre in Ontario, Toronto certainly deserves special attention, but not to the neglect of other georegions. Toronto's needs are unlikely to be neglected or downplayed by the provincial or the federal government, because it is a political-power base, a major centre for commerce, media, and local and provincial governments. The same cannot be said for many other georegions.

To ensure that messages from all georegions are regularly and effectively raised with the provincial government we suggest three alternative mechanisms for consideration by responsible authorities and citizens. The first is an Ontario Georegional Consultative Committee that would advise the premier regularly about the status and needs of the various georegions. This committee – as well as the other two alternatives to be discussed later – would provide the premier with advice that could be related to that given in the partisan political arena of Ontario's elected legislature. The advice could be requested by the premier or offered by the committee, and the work of the committee would be reported to the public. The form of this reporting would have to respect the integrity of the advisory process, as well as the "need to know" of the public. The committee could be comprised of knowledgeable people from each of the georegions, people who would not be overtly partisan in their views. Developing an appointment procedure to ensure relatively unbiased membership would be a challenge that might involve referral of final decisions to a judicial body. Appointments could be made for three to four years, with members serving no more than two terms. The Georegional Consultative Committee could report regularly on an annual basis and periodically at the request of the premier.

The two other alternatives are an Ontario Georegional Forum and an Ontario Georegional Commissioner. These two alternatives avoid some of the procedural and appointment challenges of the Consultative Committee. The Ontario Georegional Forum could be appointed with a small, competent staff, which would plan and hold annual meetings that officials and concerned persons from each of the georegions could attend. The topics of the annual fora would be selected so as to provide for discussion and reports on progress, challenges, and needs in some or all of the georegions of Ontario. The results of the forums and any associated studies by staff or consultants would be reported on regularly.

The third alternative of an Ontario Georegional Commissioner is comparable to the existing office of the Environmental Commissioner of Ontario. The Georegional Commissioner could operate in similar ways: conducting studies and enhancing public awareness by public speaking and other means. One of the responsibilities of the Georegional Commissioner would be the preparation and publication of an annual report on a topic relevant to the well-being of some or all of Ontario's georegions.

A RECONSIDERATION OF SOME RELEVANT THEORY

We are now coming close to the end of our journey. Yet in many ways this book has merely opened doors to further consideration of Ontario as a mosaic of georegions, rather than as predominantly a province focused on the prime economic engine of Metropolitan Toronto. As noted earlier, this perceptual framework is a reflection of globalization and neoliberal thinking, which envisions the world primarily as an expanding market of great global cities, centres of production, international exchange, and sources of wealth for surrounding regions (Harvey 2005).

Other influential ideas relate to the driving concepts of globalization and neoliberal thought. They include trickle-down theory, creative destruction and the regional state. In trickle-down thinking the vision is that wealth generated by the great cities will spread outward and downward to stimulate enterprises, urban centres, and rural areas within and outside the metropolitan umbrella. Growing doubt about the efficacy of this model has arisen in recent decades, and there has been considerable evidence against the consistent occurrence of such benefits in challenged cities such as Detroit, Chicago, and Los Angeles, and in other European, African, and Asian urban agglomerations. To read the chapters in *Beyond the Global City* is to see a number of instances in which trickle-down has not provided its postulated benefits, for example, Hamilton and Brantford in the Carolinian Georegion or Northeastern or Northwestern Ontario.

Creative destruction is another concept in the neoliberal and globalization arsenal. The economist Joseph Schumpeter is apparently father of this notion (1942/2009). He sees major technical and associated changes triggering new economic and social development and laying waste to existing systems and older technology. The argument is that the successive introduction of the train, the car, and the computer destroyed earlier communication, economic, and social patterns but led ultimately to higher levels of productivity and well-being.

Creative destruction envisions people rising from the ashes to ever higher levels of technical achievement and economic sophistication. Less well elucidated are the negative effects of commitment to this model, which include the dislocations and suffering attendant on the rapid destruction of the old economic and social systems. There is also the often unquestioned assumption that ever higher levels of technical and economic progress are possible and will bring benefits to most of society and certainly to its stronger and harder-working members. Such assumptions are at issue now in Ontario as its georegions enter a period of major economic stagnation and possibly significant decline. This impending economic and social change is to a considerable degree a result of the collapse of the inadequately regulated global and neoliberal financial system, which was built on the rapid international movement of large amounts of incompletely understood financial information and investment packages, as well as cheap oil, of which more will be said later. All this, in turn, was based on the rapid communication processes made possible by the introduction of the computer and related systems. The impact of this financial and economic letdown may be very heavy in global-city regions such as Toronto. Official unemployment in the summer of 2010 exceeded 9 percent, among the highest in Ontario. Dislocations will have major effects on investment, employment, and social conditions in this georegion and also in georegions closely tied to it such as Muskoka–Georgian Bay, Peterborough, and Northeastern and Northwestern Ontario.

Anything approaching a full analysis and interpretation of Schumpeter's ideas about creative destruction and its implications is beyond our immediate purposes. The pursuit of more creativity has been advocated by Martin and Florida (2009) as a way forward for Ontario, although without much consideration of its potentially destructive effects. Moreover, it can be argued that substantial levels of creativity have developed in many sectors and georegions of the province, to the point where creativity has produced an ongoing series of destructive, economic, social, and other effects whose occurrence and impact were not adequately foreseen by Schumpeter or his followers, as a result of the work he produced many decades ago. Our system seems to be having trouble coping with unemployment and other effects of complex creativity and the frequent residential and other relocations now more or less expected for people in the workforce.

Seen against the passage of time, many of Schumpeter's expectations of capitalism are now questionable; for example, he discounted the likelihood that disparities in income and earnings could increase with creativity and economic growth. Yet such disparities have been a major concern of recent decades; the disparities have risen rather than lowered in the past two decades, and earnings at the top have risen to levels many times the earnings of most citizens.

This raises another question about Schumpeter's vision. He foresaw that the growth of capitalism and income would largely cover the costs associated with creativity and destruction such as unemployment or health costs. He had little specifically to say about the latter in terms of the huge demands these costs are now making in Ontario and other parts of North America, much of this accruing from an aging society, something he again did not foresee in current terms.

The same applies to his failure to consider the environmental costs of creativity and destruction, such as the decline in air and water quality and climate change and their implications for human well-being. In making these remarks my purpose is not to criticize Schumpeter unfairly from a great time distance. Indeed, Schumpeter actually believed that capitalism as he understood it – that is, as a historic process of creation and destruction – could not continue indefinitely.

Another concept that recent financial and economic changes have shown to be shaky is that of Ontario as a regional state. This concept was put forward vigorously by the economists Thomas J. Courchene and Colin R. Telmer in their 1998 book *From Heartland to North American Regional State: The Social, Fiscal and Federal Evolution of Ontario*. Essentially their argument was that largely as a result of the actions of 1990s Harris governments, Ontario had moved from what was conceived of as the heartland of an interrelated Canada to an increasingly independent economic power in North America and globally.

In the heartland era before the innovations introduced by the Harris governments, Ontario was seen as an economic bellwether for the stability of Canada. Ontario's financial, economic, and social decisions closely matched those of the federal government in Ottawa. In this way Ontario acted as a heartland supporting east-west trade and other linkages, as well as federal nation-building through medical, research, educational, and social-support policies and programs across the country.

This more or less harmonic situation began to break down slowly in Ontario and other provinces before the Harris crowd appeared on the scene in the 1990s. One strong influence was the apparent decline in the power of national governments. The introduction of the free market and neoliberal ideas worked against the traditional regulating and oversight role of central governments and put more power in the hands of commercial and industrial enterprises and lower levels of government, including the provincial and municipal governments in Canada.

On the other hand, in part for ideological reasons and in part to reduce debts in the 1980s, the federal government cut numerous environmental and social service programs while downloading regulatory and other powers to the provinces and business. Later provincial governments, such as the Harris government, followed suit in passing often poorly funded responsibilities on to local governments. Taxes were also cut to lower costs for business and industry in attempts to make the province more competitive internationally. These moves were seen as especially important in lowering costs relative to those of the United States, creating a "level playing field" and enhancing prospects for north-south trade. These measures were greatly stimulated by the 1989 North American Free Trade Agreement (NAFTA) with the United States and Mexico.

As a result of these and other steps, in the late 1990s Courchene and Telmer concluded that Ontario was becoming increasingly powerful in relation to other provinces and the federal government. Its north-south trade was accelerated vis-à-vis the traditional east-west exchanges, and it was emerging as a regional state. In this respect Ontario was seen as emulating developments in Europe. Particular examples were the emergence in Spain and southwestern France of a cross-border trade and commercial alliance centred on the Barcelona region. A similar trans-border region grew up around Stuttgart in Germany, Lyon in France, and Milan in Italy (Courchene and Telmer 1998, 274). Much of this thinking came in the early 2000s, at a time when Ontario was still relatively prosperous, in large part because of the province's burgeoning auto industry, which was closely integrated with counterparts in the United States. It was, therefore, relatively easy to assume this growth would continue and that Ontario and the financial and economic engine of metropolitan Toronto would drive the province to become a regional state in the sense of its European precedents.

Today this vision is definitely shattered. The assumption that Ontario and Toronto would continue to grow rapidly has fallen by the wayside. The United States is in bad shape financially, economically, socially, and politically. The automobile and other mainstay industries are awash with products not only from Canada and the United States but from competitors overseas. Business, commerce, and employment are faltering, and the outlook is bleak for any Ontario regional state. In fact, the federal governments in both the United States and Canada are assuming major financial, regulatory, and other powers that were only recently seen by neoliberals as likely gone forever. Furthermore Ontario's fiscal and economic decline is such that it is now eligible for transfer payments from the federal government as part of the traditional east-west equalization programs that have held the country together since the 1960s. And overshadowing all this are questions about the survival of US globalization, or thrust to empire (Wright 2009). Add to this the economic and political rise or India, China, and other countries, their interplay with the West, and related questions of enterprise and freedom in the global sense (Attali 2006).

FURTHER THOUGHTS ABOUT THE FUTURE

While all this is occurring, the Ontario government is faced with a situation in which the costly demands of Toronto's Georegion are increasingly difficult to meet. Trickle-down did not work consistently in grander times and is likely to be less effective in the near- to mid-term as productivity falls and the need for provincial and federal support increases. In deep recessions in the past, financially challenged people have left big cities to find more secure niches in smaller towns and the countryside.

The other georegions in the province are going to need Ontario's strong support, especially with infrastructure costs and employment opportunities that currently are overwhelmingly directed to Toronto. For example, in 2008 the York-Toronto Centre subway extension was estimated to cost close to a billion dollars, yet the total 2008 allocation for roads, bridges, and other municipal infrastructure in communities outside Toronto was $400 million (*Lake Erie Beacon*, 25 April 2008, 1).

More examples could be given, but enough has been said to show that the Ontario government – in association with Ottawa – should play a stronger role in promoting understanding and support for all its georegions. The basic purpose of *Beyond the Global City* is to provide this greater understanding and, in the process, identify some implications, and provide suggestions and recommendations for each georegion and for the province as a whole. It is interesting that the decision to produce this book was made before the global financial and economic meltdown that now provides unanticipated justification for our work.

It is not possible to delve into this unanticipated situation in any greater detail here. Yet we do feel it appropriate to comment on one of the major proposals or visions put forward to help put Ontario back on a high-growth trajectory in future. The proposal for a more creative Ontario has been made by two business professors from University of Toronto, Roger Martin and Richard Florida (2009). The proposal arises from a request by Ontario's premier, Dalton McGinty, for advice on how the province should chart the way forward.

The premier made his request early in 2008, when it was clear that Ontario's great dependence on automobile production and other commodities was being challenged by markets that were waning in the face of an overproduction of cars. The premier and the authors of the report recognized at the outset that economic challenges were rising but not apparently that an economic meltdown was unfolding. Yet that was the state of affairs when the Martin-Florida report was released in January 2009.

Fundamentally the authors call for promotion of greater creativity throughout the workforce, mainly through improvements at all levels in the education and research systems. They see this as leading to more innovative and productive ways of doing things, giving Ontario a competitive advantage globally. The authors recognize that their recommendations are not a quick fix but nevertheless desirable in enhancing the way forward in the mid- to long-term.

The report is big-city-centred and does not adequately recognize the creativity already at work in various parts of Ontario, for example, among the computer-based high-tech farmers producing crops and livestock in the Carolinian or Huron Georegions. But then the report has little to say about agriculture, perhaps because the authors see the greatly increased efficiency of agriculture as a boon in allowing Ontario's farmers to meet "the basic consumption demands of Ontarians and untold numbers around the world with only 2 percent of our work force" (Martin and Florida 2009, 17). This drive to very high levels of productivity and efficiency through the increased use of powerful fertilizers and other technology has, however, left a sad legacy of water and air pollution, decimated woodlands and wetlands, and produced other environmental and social effects whose presence and solution are neglected in the Martin and Florida report. Furthermore, some of the challenges to Ontario are mainly political and would be difficult to erase based on creativity alone. An example is provided by the barriers erected by the United States to protect its wood producers.

But our main concern is the focused nature of the recommendations and their failure to reflect a sensitivity to Ontario's various georegions. Indeed, the report pushes Toronto-centred thinking by calling for growing linkages with the metropolis in an increasingly creative society. Other areas identified as possible growth nodes include Kitchener-Waterloo-Cambridge in the Carolinian Georegion and the City of Ottawa in the Ottawa Georegion. These cities are recognized because they already exhibit the high-tech innovative approach favoured by the authors. They pick winners based on their focused view of the best way forward, but they do not adequately recognize other places and regions and the challenges and opportunities facing them. One leading example would be Kingston and the Upper St Lawrence Georegion.

We need more precision in identifying what a creative urban-centred model would do for agricultural areas such as the Huron or the Carolinian Georegion. The culture here is often not big-city creative but rural and small-town inventive. What will the big-city creative model do for Sudbury, Thunder Bay, and Northeastern and Northwestern Ontario? Nickel and other mines and factories are closing in Sudbury as I write, leaving hundreds out of work, mainly because of a market decline and a glut of product. We need more specifics on what kind of creativity can reverse the situation. How specifically will creativity open up new lines of endeavour in these regions in the next one or two decades?

The report to the premier still views a growing Toronto metropolitan region as the economic engine and fundamental bellwether for Ontario. The Toronto Georegion is seen as central to a greatly expanded mega-region stretching from Montreal to Windsor. It relies here on distributive and trickle-down effects for the rest of the province, but, as noted earlier, they have not been highly effective and equitable in the past. Underlying a continued Toronto-centred approach is the image of growing traffic and travel to the big city by Ontario residents, increasing congestion, rising pollution, and escalating inequalities in and outside the metropolis.

In a recent analysis of economic and social changes in Toronto in the past forty years, University of Toronto professor David Hulchanski revealed dramatic changes in the distribution of income among people and districts in the city. The middle class is shrinking, and disparities are growing between upper- and lower-level groups, which are now distributed more widely throughout Toronto. Since 2001, fifteen middle-income neighbourhoods have vanished, the majority becoming low-income areas where individual earnings are 20 to 40 percent below the city average (*Toronto Sunday Star*, 8 February 2009, A3, A8, A9). Approaches to these challenges require basic changes in social and governance policies that seem to be held back more by ideology and politics than by creativity.

Land consumption is also a long-time problem following from continuing commitment to a Toronto-centred high-growth model. Even with efforts to control sprawl through intensification and high-volume transit in and out of the city, the consumption of surrounding lands will grow. In northern New Jersey, bordering the great metropolis of New York, more than 30 percent of the land is now dedicated to transport, resulting in great disruption to agricultural, small-town and big-city activities.

In reflecting on the Martin-Florida report, we continue to ponder what steps can be taken to lessen the deepening impacts of Ontario's economic decline. Many people seem to think a quick fix, or a set of quick fixes, will rejuvenate the development model that has been at work in the last four or more decades. These quick fixes involve changes in interest rates and monetary policy, as well as government-driven infrastructure and other stimulative fiscal measures. The quick fixes include efforts to rebuild a more efficient automobile industry and other industries, as well as forestry and comparable traditional large-scale enterprises. These quick fixes are generally being planned from the top down, with government support for banking and other financial institutions that are seen to be essential to development, growth, and job creation.

Some disparate voices have, however, suggested other approaches. One of these is the automobile company Toyota. With three major plants in Ontario, Toyota did not ask for bailout loans and other assistance but rather advocated that government support be directed to potential auto purchasers. They would then be in a position to choose which autos to buy, stimulating the economy and employment without unduly subsidizing the big car companies, which have not been successful competitors in the last decade and may very well not be in future.

Such bottom-up suggestions seem to be much more in line with the messages that the various chapters in *Beyond the Global City* send to government, business, and the citizenry. The chapters in this book reveal the complex history and varying environmental, economic, and social character of the different regions of Ontario, as well as the challenges they need to address in cooperation with the provincial and federal governments. Efforts need to be put into devising systems that will more effectively, efficiently, and equitably combine the needs of the regions with the power of senior governments.

We can deepen understanding of what we are advocating by referring to the ideas of Jeffrey D. Sachs, the well-known economist, author, and advisor to the United Nations, especially as set forth in his recent *Common Wealth: Economies for a Crowded Planet* (2009). Sachs is working at a much broader international scale than we are in this book. But the fundamental challenge is similar. He aims to identify, analyze, interpret, and respond to complex environmental, social, and economic variations among countries and regions around the world. Our task is to analyze, interpret, and at least suggest initial responses to the challenges facing the eleven different georegions we have identified in Ontario.

In calling for more effective and equitable approaches to challenges globally, Sachs sees the need for much greater cooperation among the diverse nations and regions of the world. He believes that "the problem can only be solved through an interactive approach that combines general principles with the details of a specific setting." For him, studies and programs "too often begin and end on the basis of general principles without due regard for ground level complexities" (2009, 13). He sees the challenge as akin to that faced by a physician who needs to understand the general theory and practice of medicine, as well as the detailed history and characteristics of the individual patient whose illness he or she wishes to deal with successfully.

These ideas of Sachs have been paralleled by the recent publication of a book by the well-known geographer, David Harvey. In *Cosmopolitanism and the Geographies of Freedom,* Harvey (2009) produces a tour de force on the importance of studying and understanding the particulars of places in some detail before applying sweeping general principles such as globalization or foreign aid. Among his salient examples are the very low level of understanding of the land, peoples, and cultures of the Middle East on the part of countries beginning wars in Iraq and Afghanistan.

Harvey is wary of applying broad principles and approaches to development and other issues given the diversity of histories, conditions, and cultures on the ground. Sachs calls for more cooperative institutions at the global level to facilitate the application of general principles. Our problem is comparable to that of Harvey and Sachs. We seek to modulate or modify the broad Toronto-centred approach in Ontario. Again, one promising way to begin would be through the establishment of an Ontario Georegional Consultative Committee or comparable cooperative mechanism. Such a body could commence building a shared approach to Ontario's complex conservation and development challenges, moving beyond the current great global-city model toward a more pluralistic future. Another useful way to move forward would be to encourage cross-sectoral georegional organizations such as the Carolinian Canada Coalition.

The final perspective we wish to explore very briefly is the possible effects of high-priced oil on the economic future of Toronto centred-planning in Ontario. In 2008 oil prices reached the high level of about $120.00 per barrel. Prices then plunged in the next few months to little more than $30.00 per barrel. This sharp drop occurred more or less in parallel with the emergence of problems with mortgages and related investment funds in US banks, as well as other financial institutions in North America, Europe, and Asia. These financial challenges were associated with a steep decline in the housing industry, major job losses, and economic meltdown globally.

In 2009, with massive infusions of government capital, some improvement in the financial and economic picture emerged. But the situation is far from secure, and further future bank losses anticipated on outstanding investment products, as well as continued economic insecurity and high unemployment. Furthermore by spring 2009 oil prices had climbed to about half their levels in spring 2008, and by fall 2010, to about two-thirds of those high levels. This can be seen as both a good sign and a threatening one, since economists like Jeff Rubin (2009) see oil prices as the major underlying cause of our current economic difficulties. In this scenario it is the high oil prices that undermine industrial and national capacities to ship goods long distances between continents, undercutting the world financial and trade system and the globalization model.

And in the early stages of the 2008 economic meltdown, some companies and countries did, indeed, begin to move away from increasingly costly international trade in steel, food, and other products and toward local sources of supply and markets. Rubin and others predict this will happen again as oil prices reach high levels in association with renewed economic growth, rising demand, and diminishing oil reserves. The point of all this is that the scenario of high oil and energy costs may be another very fundamental reason for moving from a Toronto-centred global planning model to one that takes greater cognizance of the needs and opportunities of Ontario's other georegions.

For further comments we return to the 2008–9 Annual Report of the Environmental Commissioner of Ontario, which calls for "building resilience" into Ontario planning and land use management. From an ecological perspective, resilience can be "described as the ability of an adaptive system to undergo change and reorganisation while maintaining its fundamental functions, processes and structures" (Environmental Commissioner of Ontario 2008–9, 17). The commissioner sees successful land use planning in a similar light. It should allow for change and reorganization while maintaining a region's fundamental functions, processes, and structures. This statement reflects the annual report's tendency to emphasize a regional approach to planning because of its greater sensitivity to local conditions and its capacity to provide more opportunity for local people to participate in determining their destiny.

A regional approach is also seen as involving what some observers call "redundancy" in provincial planning. The idea here is that a centralized, highly efficient system can be quite fragile. Unanticipated difficulties or failures can affect more people and a greater part of the well-being of Ontario than would a system based to a greater extent on decentralization and a regional or georegional approach, as advocated in *Beyond the Global City*. As an example of the risks in emphasizing efficiency and centralization, the commissioner notes the trend to build more and more dependence on water pipelines for a growing number of Ontario communities, thereby increasing the risks of breakdown or failure. In the end, then, the commissioner's *Annual Report 2008–2009* reflects much of the thinking in *Beyond the Global City*. The two documents call for a strengthened regional approach within generally applicable provincial directives and guidelines. A major way in which this could be accomplished is through more comprehensive applications of policies under the Provincial Policy Statement of Ontario.

Several relevant publications have come to our attention since the essential completion of *Beyond the Global City* in summer 2009. These publications are by leading authorities in fields such as economics, history, and the environment. They point in the same general directions as does *Beyond the Global City*. Among their major concerns are excessive reliance on a single approach to development, the environment, and ways of life. Another is the associated and increasingly salient concern about inequality or access to economic, social, cultural, and political opportunities.

Joseph Stiglitz (2007), the Nobel Prize winning economist, is concerned about the excessive focus on "free-market" economics that has dominated the globalization movement in the last several decades. This, in his view, has resulted in greater inequalities in income and other negative consequences for individuals and communities. He remains a strong advocate of capitalism but wants a larger and more balanced role for government in dealing not only with economic efficiency but with social justice, cultural diversity, the environment, access to health care, and consumer protection. Tony Judt (2010) expresses similar views based on his long historical studies of development and governance in Europe and North America.

In a recent *Wall Street Journal* report on the fortieth anniversary of Earth Day, 2010, William Ruckelshaus, administrator of the US Environmental Protection Agency in the early 1970s and again in the 1980s, sees a different approach to governance as necessary to meet climate change, non-point source pollution, and other environmental challenges of today as opposed to yesterday. In earlier years, according to Ruckelshaus, what was needed was a much stronger role for government, especially the central government, in developing and applying national standards to, for example, toxic emissions from industrial plants, lest they move operations to less well-regulated jurisdictions. This approach, to a considerable degree, is still needed. But so is a different one, a system that involves much wider understanding and commitment by decision makers and citizens to management of both development and environmental protection. Ruckelshaus sees this as necessary if we are to deal effectively and equitably with diffuse issues such as non-point pollution in the Great Lakes, for example.

Pollution from diverse and scattered sources such as excessive fertilizers from farms throughout different watersheds or domestic waste from numerous small towns is very difficult to control through centralized top-down governance systems, as is much of the carbon contributing to climate change from many small and dispersed sources. According to Ruckelshaus, such circumstances call for a much more wide-ranging understanding of, and involvement in, development and environmental stewardship. We need a broader democratic approach, better ways of getting more people involved in understanding and improving the process, and more accessible scientific, scholarly, and technical information for use by all those involved in the issues.

And behind all of this lies the image of a more locally responsible governance system that is close to communities and citizens, yet linked to higher levels of government and private-sector leaders with knowledge of and responsibilities for big geographic areas. Such a system would work collaboratively in dealing as effectively and equitably as possible with local and larger-scale needs. This is a very demanding task, one in which we are slowly getting involved. To move forward with vigor and confidence we require, in Ontario, a much greater commitment to understanding and involving the people of its many diverse places. This book is intended to make a strong contribution to that goal.

I would like to close with some personal reflections on the development history of Ontario's georegions, past, present, and future from a broad human ecological perspective. As indicated at various places in this book, the boundaries and outer parts of georegions have tended to change with technological, economic, social, and political changes. Cores of georegions have tended to persist even though their land use and landscape characteristics have changed, again in accordance with changes in technology, the economy, and the other forces mentioned above. Toronto, Peterborough, and Kingston and the Upper St Lawrence georegions are classic examples. For similar reasons new georegions have formed, for example, Northeastern and Northwestern Ontario have evolved from one North. The Niagara Escarpment Georegion has come into existence in the last two decades. And another georegion may be in the offing along the fringe of the Canadian Shield as The Land Between.

Social and political factors have been especially important in Niagara and The Land Between. The Niagara Escarpment evolved politically as a result of perceptions of the georegion by Toronto and the province, perceptions that often were not shared by local people. In contrast, The Land Between is the perception mainly of local people, who may force the province eventually to recognize and act on it. In this case, however, much of the local interest seems to derive from migrants into the area from the Toronto Georegion.

The Niagara Escarpment and The Land Between are the only new georegions to arise in Ontario in the last two or three decades. Both lie at the edges of the expanding Toronto metropolis, whose economic and social pressures have pushed them from essentially agricultural, forestry, and rural lands within other regions to become bands of discontinuously populated recreational lands primarily serving the expanding big city. Pressures from Toronto are also operating at an accelerating pace on the Carolinian Canada Georegion. Much of the Toronto Georegion was ecologically part of Carolinian Canada until its population, economic, social, technical, and other growth made it a separate georegion. The expansion of Toronto through the extension of road and rail systems and the creation of the green belt, smart growth, and other provincial policies is transforming or threatening to transform roughly the eastern third of Carolinian Canada.

The basic question is whether the citizens of the various georegions of Ontario want these Toronto-driven changes to continue or whether they wish to work on a different planning approach that challenges the big city's often unwanted land use and environmental and social effects and works to maintain a more diverse, equitable, sustainable, and resilient Ontario. Various possible ways of doing this have been suggested previously. What is ultimately required is a planning and decision-making system that is democratic in the sense that it offers a voice to the people living on the greater part of the land in the province, as well as to the millions of voters in the expanding Toronto Georegion.

To close the circle, we briefly return to the challenges facing the Toronto Georegion as an interactive part of Ontario and its other georegions. When examining those challenges we need to maintain a sense of the whole. As Lucy Sportza points out in chapter 4, Toronto is a lively city, rich in cultural diversity and economic opportunity, although not with a high degree of equity. It will likely continue to grow and sprawl, albeit in a way tempered by "smart city" and other measures to contain its advances into surrounding georegions. Its environmental, social, and political impacts will also continue to increase, in large part because of internal challenges relating to its density, layout, and environmental efficiencies, notably in terms of energy consumption and release, its carbon footprint and climate change.

In this respect it is instructive to think of David Owen's illuminating *Green Metropolis*, which among other things points out that the massive metropolis of New York is, to a considerable extent, a green city because its historically compact design and density make it a leader on a per capita basis in low levels of energy use, release of heat and carbon, and climate change. One of the keys to this situation is the relatively low use of cars in New York City, with the result that the main source of heat and carbon release is buildings. Owen does not, however, so carefully consider the loss of biodiversity, residents' access to nature, crowding, or other environmental and social effects of concern to the authors of this book. Nor does he so carefully consider the city in relation to its region and others beyond it. Nevertheless, the New York example is most useful when considering the future of the Toronto Georegion. Much more attention should be given to key factors such as transportation. Various options are before the city now, including the expansion of roads, greater use of streetcars, light rapid transit, subways, and rails. Decisions on these will have major effects on the environment, land use, and sustainability in Toronto and outlying lands. Much of the financing is expected to come from the province. The opportunity costs of its decisions will affect all georegions of Ontario.

In this context, it's possible to entertain two basic future images of the Toronto Georegion. One involves modified but continued emphasis on the great economic engine and global-city approach.

The other is a more dispersed set of cities, including greater growth in existing ones such as Kingston, Ottawa, Peterborough, and Waterloo-Kitchener. The realization of one or both of these images relies very much on the provincial government and its willingness to move away from its focus on the Toronto Georegion to planning for the georegions of the province as a whole. Inherent in all this are basic questions of population growth and a consumptive society. Much population growth results from immigration, whose costs and benefits are rising in the public mind and ultimately will have to be technically assessed and politically addressed by the province and the federal government.

Here it is useful to think of the report in the *Toronto Star*, 15 December 2010, by Professor David Hulchanski and colleagues at the University of Toronto's City Centre. In three vivid maps, Hulchanski shows dramatic changes from a 1970s Toronto composed mainly of middle-income owners, with small pockets of high and low incomes, to a 2005 Toronto in which middle-income earners have greatly receded, and high and low earners are segregated by neighbourhoods.

Projecting these trends to 2025 shows a Toronto in which high income earners are concentrated downtown and along subway lines, surrounded by predominantly low earners and some with middle incomes (Ann Mehler, *Globe and Mail*, 15 December 2010, 1). Little trickle-down effect is apparent here. This is definitely not the vision that planners and politicians saw arising from an emphasis on the great economic engine and the global-city approach. Beneath this planning rhetoric a number of factors such as immigration policy and social, economic, environmental, and land use policy and practice have been interacting to produce a set of challenges to the region, the province, and the federal government that are apparent inside and outside the great city. These need addressing directly by as yet unrecognized means. A key one, however, is a more regionally diverse understanding and planning of Ontario.

That this is an urgent and pressing need is vividly illustrated by the Ontario government decision to move ahead very quickly with the 2009 Green Energy Act. The intent was to unleash alternative energy sources such as wind and solar power, cut carbon emissions, and create jobs in a depressed economy. The act was passed with no white paper or comparable opportunity for careful preliminary comment by professionals and knowledgeable citizens who might identify issues that could arise and be addressed before passage. The act severely constrained the powers of regional or local governments to influence the location of wind towers, notably in relation to their current provincially approved official plans. The act overrode the existing provincial environmental-assessment procedures, putting the power to prepare project environmental assessments in the hands of the industry itself and placing judgments about their adequacy in a new provincial government office in Toronto. Citizen's comments on the assessments were only to be considered seriously if they could prove life- or health-threatening risks to humans or wildlife. The result has been strong widespread opposition to wind tower proposals, notably in southern Ontario.

With the Green Energy Act, the Ontario government overrode local knowledge, planning, and civic expectations in a top-down decision that likely would have been more effective and less disruptive if the development of wind or alternative sources of energy had been undertaken from a regional as well as a central perspective. The issue is not so much whether wind energy should be used as where, in the light of scientific and civic scrutiny of its environmental, social, and economic effects, it should be located. We can think back to the suggestion made earlier in this chapter for establishment of a regional body ultimately advising the premier and the government. Such a body would have been one useful source of guidance on the Green Energy Act, the pros and cons of wind power projects in various parts of Ontario, and the most appropriate choice of energy sources for these areas.

Concern appears to be growing about the Green Energy Act and its consequences, notably in Southwestern Ontario and Kingston and the Upper St Lawrence georegions. There may well have been better alternatives than the "one size fits all," or universal, approach of the province. It is impossible to say with certainty because of a lack of broad scientific and civic input and of monitoring and assessment of effects since the provincial decisions were made.

In the end, what is sorely needed in Ontario is more research and information on regional variations in environmental health, economic and social well-being, and governance. Largely forgotten, but deep regional differences in the province have been established in this book, as has a very strong case for more study, greater understanding, and improved recognition of regional needs and opportunities. These have been obscured by many decades of focus on

Toronto as a great global city in a globalizing world. Evidence is growing that this model has been overdone and needs to be balanced by more regional understanding, ideas, and policies.

REFERENCES

Adams, Peter, and Colin Taylor, eds. 1985. *Peterborough and the Kawarthas*. Peterborough, Ontario: Heritage Publications: A Division of Gould Graphic Services.

Attali, Jacques. 2006. *A Brief History of the Future*. New York: Arcade Publishing.

Courchene, Thomas J., and Colin R. Telmer. 1998. *From Heartland to North American Region State: The Social, Fiscal and Federal Evolution of Ontario*. Monograph Series on Public Policy, Centre for Public Management, Faculty of Management. Toronto: University of Toronto.

Dear, M.J., J.J. Drake, and L.G. Reeds, eds. 1987. *Steel City: Hamilton and Region*. Toronto, Buffalo, London: University of Toronto Press.

Environmental Commissioner of Ontario. 2009. *Building Resilience Annual Report 2008–2009*. Toronto, Ontario.

Hall, Roger, William Westfall, and Laurel Sifton MacDowell, eds. 1988. *Patterns of the Past: Interpreting Ontario's History*. Toronto and Oxford: Dundurn Press.

Hamil, Fred Coyne. 1951 (1973). *The Valley of the Lower Thames*. Toronto and Buffalo: University of Toronto Press.

Harvey, David. 2005. *A Brief History of Neoliberalism*. New York: Oxford University Press.

– 2009. *Cosmopolitanism and the Geographies of Freedom*. New York: Columbia University Press.

Johnson, Lorraine, ed. 2007. *The National Treasures of Carolinian Canada*. Toronto: Lorimer Press.

Judt, Tony. 2010. *Ill Fares the Land*. New York: Penguin Press.

Krugman, Paul. 2009. *The Conscience of a Liberal*. New York: W.H. Norton.

Martin, Roger, and Richard Florida. 2009. *Ontario in the Creative Age*. Toronto: Rotman School of Management, University of Toronto.

McIlwraith, Thomas F. 1999. *Looking for Old Ontario: Two Centuries of Landscape Change*. Toronto, Buffalo, London: University of Toronto Press.

Osborne, Brian, and Donald Swainson. 1988. *Kingston: Building on the Past*. Westport, ON: Butternut Press.

Owen, David. 2009. *Green Metropolis: Why Living Smaller, Living Closer, and Driving Less Are the Keys to Sustainability*. New York: Riverhead Books.

Paperny, Ann Mehler. 2010. *Globe and Mail*, 15 December, 1.

Roots, Betty L., Donald A. Chant, and Conrad E. Heidenreich, eds. 1999. *Special Places: the Changing Ecosystem of the Toronto Region*. Vancouver: UBC Press.

Rubin, Jeff. 1990 *Why Your World Is About to Get a Whole Lot Smaller*. Toronto: Random House Canada.

Ruckelshaus, William. 2010. "A New Shade of Green." *Wall Street Journal*, R1. Earth Day, Fortieth Anniversary Issue.

Sachs, Jeffry D. 2009. *Common Wealth: Economics for a Crowded Planet*. New York and Toronto: Penguin Books.

Sancton, Andrew. 2008. *The Limits of Boundaries: Why City Regions Cannot Be Self-governing*. Montreal and Kingston: McGill-Queen's University Press.

Saul, John Ralston. 2006. *The Collapse of Globalism*. Toronto: Penguin Canada.

Schumpeter, Joseph A. 2009 (1942). *Can Capitalism Survive? Creative Destruction and the Future of the Global Economy*. New York: Harperperennial, Harper Collins Publishers.

Sewell, John 1993. *The Shape of the City: Toronto Struggles with Modern Planning*, with a foreword by Jane Jacobs. Toronto: University of Toronto Press.

– 2009. *The Shape of the Suburbs: Understanding Toronto's Sprawl*. Toronto: University of Toronto Press.

Stiglitz, Joseph, E. 2007. *Making Globalization Work*. New York: Norton and Company.

Toronto Star. 2009. 8 February, A3, A8, A9.

– 2010. 10 December.

Trinklein, Michael J. 2010. Altered States. *Wall Street Journal*, 17–18 April, W1.

White, Randal. 1986. *Ontario 1610–1985: A Political and Economic History*. Toronto and London: Dundurn Press.

Wood, J. David. 2000. *Making Ontario: Agricultural Colonization and Landscape Re-Creation before the Railway*. Montreal and Kingston: McGill-Queen's University Press.

Wright, Ronald. 2009. *What Is America: A Short History of the New World Order*. Vintage Canada. Toronto: Random House of Canada.

Contributors

ALISON BAIN, associate professor of geography at York University in Toronto, studies contemporary Canadian urban and suburban culture. Her work examines the relationships between artists, cities, and suburbs, with attention to questions of identity formation and urban change. Her current research program focuses on cultural production and creative practice on the margins of Canada's largest metropolitan areas.

WAYNE CALDWELL, professor of rural planning and development, University of Guelph, has a long affiliation with the County of Huron. His focus continues to be on planning and change in rural and agricultural communities. He is a founding member and past chair of the Ontario Rural Council and a founding member of the Huron Stewardship Council and the Lake Huron Centre for Coastal Conservation. He was appointed by the Ontario government as chair of the provincial Nutrient Management Advisory Committee and in 2007–9 was president of the Ontario Professional Planners Institute. He is now president of the Association of Canadian University Planning Programs.

STEWART HILTS is professor emeritus, University of Guelph. His main research interests have been in rural planning and conservation. He was a founder of and active in land stewardship programs and land trusts in Ontario and has published numerous research and planning articles and monographs and works on conservation and planning in the Bruce Trail and Beaver Valley areas. Stewart Hilts recently was awarded the University of Guelph's Medal of Merit for his research and applied contributions to the university and to rural planning in Ontario.

MARGARET JOHNSTON works at Lakehead University in the School of Outdoor Recreation, Parks and Tourism. Her research examines the interactions of tourism change and climate change in the Arctic, the contributions of tourism to the social economy of the north, and risk management in recreation and tourism. She has also undertaken research on the outcomes associated with hosting special sporting events.

RHONDA KOSTER is director, the Instructional Development Centre, and associate professor with the School of Outdoor Recreation, Parks and Tourism at Lakehead University, Thunder Bay, Ontario. Her research focuses on the contribution of tourism to rural sustainability and includes building capacity for tourism development with First Nations communities, experiential tourism development, gateway communities and protected areas, rural tourism in the Canadian urban fringe, and frameworks for evaluating tourism as a community economic development endeavour.

PATRICK LAWRENCE is professor, the Department of Geography and Planning, University of Toledo, Ohio. His research includes environmental planning, water resources, coastal management, parks and protected areas, natural hazards, and the Great Lakes. He has published widely in geography, planning, GIS, remote sensing, and related topics and has worked with international, national, and regional organizations, including in leadership roles, with the Association of American Geographers, the International Association for Great Lakes Research, the Coastal Society, Maumee Great Lakes Area of Concern Public Advisory Committee, and Partners for Clean Streams Inc.

RAYNALD HARVEY LEMELIN is the LU-SSHRC research chair in Parks and Protected Areas at Lakehead University, Ontario. His research interests include northern parks and protected areas and their co-management with First Nations, aboriginal tourism, and wildlife tourism. Dr Lemelin is currently participating in a number of collaborative research projects with First Nations and Inuit communities located throughout Northern Canada.

NINA-MARIE LISTER, visiting associate professor of landscape architecture and affiliate faculty, master of design studies, Harvard

University, is associate professor of urban and regional planning at Ryerson University and a registered professional planner (MCIP, RPP). Her research, teaching, and practice focus on landscape infrastructure and ecological processes in metropolitan regions. She is co-editor of *The Ecosystem Approach: Complexity, Uncertainty, and Managing for Sustainability* (Columbia University Press 2008), with recent contributions in *Ecological Urbanism* (Harvard University with Lars Muller Publishers 2010), and *Large Parks* (Princeton Architectural Press 2008), winner of the J.B. Jackson Book Prize. Her work includes international exhibitions at the Canadian Centre for Architecture in Montreal and the Van Alen Institute in New York. Dr Lister grew up on the Trent-Severn River. Today her family farm is on the limestone plateau of Prince Edward County.

NIK LUKA is cross-appointed, School of Architecture and School of Urban Planning and is associate of the School of Environment, McGill University. His core interest is urban design as an interdisciplinary approach to better understanding the form, processes, uses, and meanings of space in everyday settings and how this understanding can enable us to develop sustainable yet strategic design and policy. An architect and planner, with a PHD in urban and cultural geography and long sojourns in Quebec City, Basel, and Sheffield, England, he serves on the board of directors and as scientific advisor for the Centre d'ecologie urbaine de Montreal. In his youth he spent family summers in Ontario cottage country: Manitoulin Island, Georgian Bay, and the Haliburton Highlands.

JOHN MARSH is emeritus professor of geography and director of the Trail Studies Unit at Trent University. He continues research and consulting on parks, trails, conservation, and heritage and tourism in Canada, Chile, and the United Kingdom. He has lived in the Peterborough area since 1971, has been president of the Trent Valley Archives and member of the Kawartha Highlands Park Advisory Board, and he was recently appointed to the new Smith-Ennismore-Lakefield Heritage Committee. He has been involved in numerous conservation, heritage, and planning issues in the Peterborough region.

STEPHEN D. MURPHY, professor and chair of the Department of Environment and Resource Studies, University of Waterloo, is chair, Centre for Ecosystem Resilience & Adaptation and Centre for Applied Science in Ontario Protected Areas. He is past chair, Society for Ecological Restoration Ontario; member of the Multi-Stakeholder Advisory Committee, Office of the Environment Commissioner of Ontario; and associate editor with *Restoration Ecology and Weed Science*. He has mentored and trained students in ecological integrity, ecological restoration, conservation ecology, ecological indicators, and management plans for private and public lands.

GORDON NELSON is a geographer and distinguished professor emeritus at the University of Waterloo. He continues his research in land use history, landscape and environmental change, conservation, and planning. He has published extensively and received awards from various scholarly and civic organizations, including the Massey Medal of the Royal Canadian Geographical Society and the Harkin Medal from the Canadian Parks and Wilderness Society. He remains active in a number of land use and heritage organizations such as the Carolinian Canada Coalition (chair), the Bruce National Park Advisory Committee and the Sources of Knowledge Subcommittee (Chair), the Ontario Parks Board of Directors, and local heritage and planning groups.

BRIAN OSBORNE is professor emeritus of geography at Queen's University, Kingston, and adjunct research professor at Carleton University, Ottawa. His past research areas include aboriginal history, settlement history, and the role of symbolic landscapes, monumentalism, and performed commemoration in building social cohesion and national identity. His research also relates to the commodification of heritage and culture in post-industrial societies and the impact of tourism as both an economic opportunity and a threat to sustainable communities. These themes are central to his chapter in this volume and to his most recent book, *Kingston: Building on the Past for the Future* (2011).

R.J. PAYNE is geographer and professor in the School of Outdoor Recreation, Parks and Tourism at Lakehead University, Thunder Bay, Ontario. He has expertise in parks and protected-areas management, nature-based recreation/tourism, and the human dimensions of natural resource management. He was a leader in the development of a parks degree program at Lakehead and served as a member of parks organizations such as the Parks Research Forum of Ontario. He has advised government organizations and published in relevant journals and proceedings.

SUSAN PRESTON's research and writing focus on culture-environment relationships. She is an adjunct scholar with McMaster University's Institute on Globalization and the Human Condition, where she completed a postdoctoral fellowship in conjunction with the Department of Anthropology. Her current work focuses on the development of ethically informed integrated approaches to understanding social, ecological, and economic values of nature and ecosystem services in support of environmental policy.

MARK SEASONS is associate professor with the School of Planning and has served as interim dean of the Faculty of Environment, University of Waterloo. His research interests include regional development/planning, program evaluation, and urban intensification. Seasons, a fellow with the Canadian Institute of Planners, is a native of Ottawa, Ontario, and knows the Ottawa Valley well.

ROBERT SHIPLEY, associate professor, School of Planning at the University of Waterloo, is chair of the University's Heritage Resources Centre and research fellow at Oxford Brookes University, Oxford, England. He is recognized as a leading international expert on the role of public participation in planning and is in the forefront of research on the economic aspects of heritage development. Studies of the financial values of heritage properties and the demolition of historic buildings have been key documents in improving heritage planning in Canada. He is a founding member of the Canadian Association of Heritage Professionals (CAHP) and has received the Sally Thorsen Award, the Margaret and Nicholas Hill Award, and the CAHP Award of Merit.

LUCY SPORTZA holds a PhD (Planning) from the University of Waterloo. Her research has involved conservation history, ecology, and urban planning. She has taught in the fields of environmental studies and urban geography at Universities of Guelph and Toronto, with a focus on the Toronto region, where she has lived for about a decade.

MICHAEL TROUGHTON was professor of geography, University of Western Ontario, and a leader in research on agricultural geography, rural planning, and conservation. He produced numerous research and policy papers and contributed substantially to Canadian and international geography and land use organizations. Michael was at the forefront in the establishment of the Thames as a Canadian Heritage River, serving as chair of the Management Committee. Before his death he was involved in the initial planning and first two chapters of this book and helped organize the first Canadian Association of Geographer's seminar from which it developed.

Index

ABC approach (abiotic, biotic, and culture method), 170, 171, 187
Aboriginal tourism, 269, 434
acidification, 347
acquifer, 403
adaptive reuse, 379
affordable housing, 213, 241
afforestation, 8, 63. *See also* conservation
aggregates (sands and gravels), 94, 322, 323
Agro-Climate Resource Index (ACRI), 109
Algoma Steel Corporation, 35
Algonquin Provincial Park, 30, 32, 42, 159, 173, 193, 338, 353, 394
Algonquin to Adirondack (A2A), 353
alienation, 271, 279, 284, 285, 287, 303, 400
alternative energy (wind or solar power), 138, 428
American Civil War, 39
American Fur Company, 12
American redstart (*Setophaga ruticilla*), 344
American Revolution, 14, 112, 207, 317

anglophone community, 235
Anishabe, 203, 255
Annual Report of the Environmental Commissioner of Ontario, 389, 423
anti-Catholic, 182
antiquarian movement, 368
anti-Semitism, 182
Arctic Edge Snowmobile Tour, 268
Areas of Concern (ACOs), 72
Areas of Natural and Scientific Interest (ANSIs), 92
Arena Societies, 50
artistic dividend, 214
Athens Charter, 369
Attawapiskat First Nation, 270
auto industry, 106, 417

Baby Boomers, 185
Beatty family, 40
Benefit Impact Agreement, 266
benefits, 165, 212, 214–15, 350, 356, 375, 398, 410, 414; economic, 136, 253, 285, 378, 379, 390, 397, 411, 413; environmental, 94, 348, 351; social, 147, 179, 214, 249, 266, 428
big-box retail, 238, 239

Big Picture, The, 393
biochemical oxygen demand (BOD), 64
biodiversity, 63, 104, 107, 109, 110, 191, 192, 392, 401, 403, 406, 427
biofuel, 355
biomass, 265
bioregions, 311
bioremediate, 350
biotechnology, 162, 238, 395
Blacks, 38, 39
blank slate, 103, 388, 390
boom and bust, 30, 35, 114, 177, 179, 182–4, 256, 261, 271, 278, 281, 284, 287, 399
Booth, J.R., 32, 33, 37, 38
borderlands, 18, 20, 42
bottom-up planning, 145, 147, 148, 408, 421
Boundary Waters Treaty, 71
break-in-bulk, 206
British North America Act, 72, 375
brownfields, 337, 347, 350, 351, 383, 407
Bruce National Park, 436
Bruce Power, 142
Bruce Trail Association (BTA), 327
Brûlé, Étienne, 10, 175
building code, 367
built heritage (cultural or human heritage), 50, 114, 361, 382, 385, 407, 408
Business Development Corporation, 143, 146
Buy Local, 137
by exception, 403
bylaw, definition, 348, 351, 371

Canada Central, 32

Canada Water Act, 72
Canadian Boreal Initiative (CBI), 289, 290
Canadian Environmental Protection Act (CEPA), 73
Canadian Heritage Rivers, 249
Canadian Inventory of Historic Buildings, 376
Canadian Pacific Railway, 33, 49, 114, 256, 281
canals, 21, 22, 23, 28, 30, 31, 43, 65, 67, 156, 179, 206, 318, 396, 397
capitalism, 188, 196, 415, 424
carbon and global warming, 265, 342, 425, 427, 428
Carolinian Canada Coalition (CCC), 102, 392, 393, 409, 422, 436
central-planning (corporate, or top-down planning), 76
Centre of Excellence in Mining Innovation, 265
Champlain Canoe Circuit, 249
Champlain, Samuel de, 8, 9, 10, 38, 156, 175, 269
Chapleau Game Preserve, 4, 268, 270
Chatham-Kent, 410
Citizens Opposed to Paving the Escarpment (COPE), 326
city beautiful, 85
city-region, 3
Civic Guild of Art, 88
civic input, 405
civil society, 322
Clay Belt, 3
clear-cutting, 59
climate change, 55, 72, 74–5, 92, 98, 137, 253, 265, 268, 270, 272, 280, 352–7, 391–2, 406–7, 415, 424–5, 427, 434

Coalition on the Niagara Escarpment (CONE), 322
coastal planning, 391
Cobourg, 20, 163, 396
co-housing, 352
collaborative community planning, 408, 409
Collingwood, 27, 28, 37, 38, 178, 320, 395
colonization, 4, 20, 23, 27, 33, 39, 170, 176, 187, 188, 189, 193, 209, 317, 395; history, 317; railways, 209
Committee on the Status of Endangered Wildlife in Canada (COSEWIC), 343
community-based land use planning, 293
community development, 145, 147, 148, 408, 421
Community Futures Development Corporation, 145, 146, 148, 150
community improvement plans, 373, 382, 408
Community Matters, 146
commutershed, 320
comprehensive-planning, 89. *See also* ABC approach
connectivity, 92, 164. *See also* biodiversity, corridors
conservation, 102, 105, 118, 119, 145, 327, 392, 433. *See also* husbandry; stewardship
conservation authorities (CAS), 117, 143, 144, 145, 158, 322, 323, 339, 340, 356, 406
conservation easements, 118, 119
contaminants, 59, 60, 73, 343
conurbation (sprawl), 4, 115, 323. *See also* metropolis

cooperative planning, 392
core-periphery, 303
corridors, 13, 36, 94, 166, 176, 240, 346, 349, 411. *See also* connectivity
cosmopolitanizing, 219
cost-price squeeze, 136
cottage country, 169, 175, 193
Crawford Lake Conservation Area, 317
creative cities, 214, 217, 220
creative destruction (Schumpeter), 413, 414
Cree, 253, 254, 255, 258, 272
Crown reserves, 47
culinary tourism, 219
cultural crossroad, 242
cultural diversity, 83, 424, 427
cultural-mapping, 163
cumulative impacts, 149, 336, 338, 342, 343, 348, 353
cycle of dependency, 271
cycles, 256, 261, 281. *See also* boom and bust

De Beers, 266, 270
dechannelized, 337
deforestation, 3, 339
demolition by neglect, 373
Denhez, Marc, 366, 378
Department of Fisheries and Oceans, 73, 75
Department of Government Services, 73
depopulation, 140, 277
Design for Development: The Toronto Centred Region, 323
design guidelines, 371
designation, definition, 371
Diamond Royalty Tax, 266

disparity, 187
diversification, 125, 256, 261, 262, 265, 267, 271, 286
downloading, 416
dust bowls, 339

ecological services, 393, 406
economic engine, 98, 201, 390, 417, 428
economic multipliers, 212
economic shadow, 239
ecoregions, 52, 229, 230, 240, 311, 409
ecosystem integrity, 339
ecotone, 171, 343
ecotourism, 291, 320
ecumene (settlement), 20, 21, 22
edge effects, 344
elitist, 215, 377
embodied energy, 380
employment, 131, 134, 136, 149, 211, 213, 410; First Nations, 258, 270; loss, 283, 414, 417; older workers, 411; regional, 98, 133, 140, 165, 233, 236, 238, 239, 256, 261, 286, 396; urban and rural, 126; youth, 411. *See also* jobs; mining; sustainability
empowerment, 271, 272, 286, 287
endangered species, 102, 103, 108, 115, 316, 343. *See also* biodiversity; species at risk
Endangered Species Act, 339
energy efficiency, 348, 349, 362, 379, 408
Environment Canada, 2, 72, 75, 144, 229, 230, 235
environment, definition of, 359, 360
environmental assessment, 241, 288, 326, 429

Environmental Commissioner of Ontario, 389, 392, 403, 423
Environmental Farm Planning program, 118
Environmentally Sensitive Landscapes (ESLS), 92, 340, 354
Environmentally Sensitive Policy Areas (ESPAS), 340
Environmentally Significant Areas (ESAS), 91
equalization, 417
equity, 50, 99, 149, 420, 421, 426
ethnic heritage, 3, 48, 52, 86, 97, 256, 363, 399, 400
ethnoburbs, 97
eutrophic, 50, 64. *See also* mesotrophic; oligotrophic
eutrophication, 64, 116
exclusivity, 182, 332
exotics, 66, 73, 76, 343, 345, 346, 348, 391. *See also* biodiversity; invasive species

facilitation, 393
Far North, 4, 270, 301, 302, 400
Far Northern Georegion, 400
Federation of Ontario Naturalists, 327, 339
Finns, 37, 256
fire, 7, 116, 253, 270, 364
First Nations Land Policy, 266
fisheries, 42, 73, 75, 176, 182, 395
flat-earth, 103, 388, 390. *See also* blank slate
flooding, 55, 68, 70, 75
Florida, Richard, 2, 4, 215, 414, 418, 419, 420
fly-in tourism, 282
Fort Frontenac, 10, 12, 13, 14, 15, 206

Fort William, 13, 24, 35, 37
forwarding, 45, 46. *See also* transshipment
fragmented landscape, 115, 117, 344, 348, 411. *See also* biodiversity; species at risk
Fram, Mark, 371
free market, 416. *See also* globalization, neoliberalism
Friends of the Escarpment (FOE), 327
Fugitive Slaves Act, 39
Fuller, Tony, 23, 113
functional region, 161

Galt, John, 18
Gananoque, 14, 15, 23, 27, 397
garden city, 85
Gatineau, 30, 33, 230, 236, 237, 240, 398
Georegional Consultative Committee, 412
Georgian Bay, 3, 9, 13, 32, 42, 169, 170, 182, 346, 398, 414
German Company Block, 47
Gertler Report, 322, 323, 330
getaway country, 159
glaciers (glacial drift, glacial lake), 40, 60, 61, 111, 173, 280. *See also* moraine
global cities, 390, 413. *See also* globalization; world city
global city model, 390
globalization, 52, 81, 97, 98, 121, 203, 217, 220, 265, 424; costs of, 97, 357; regional, 217; theory of, 388, 390, 394, 413, 422, 423; Toronto, 95. *See also* free market; global cities; neoliberalism; world city

glocalization, 2, 201, 211, 217, 220
GO Transit, 95
gobi (also goby), 397
Golden Horseshoe, 80, 81, 82, 83, 95, 98, 99, 104, 188, 201, 322, 394
Grand River Canal, 23
Grand River, 20, 47, 48
Grand Trunk Railway, 33
grassroots, 218. *See also* bottom-up planning; community development
Great Arc, 313, 314
Great Depression, 181
Great Lakes Binational Toxics Strategy, 71
Great Lakes Charter Annex, 72
Great Lakes Commission, 66, 76
Great Lakes Regional Collaboration, 72
Great Lakes–St Lawrence Seaway Study (GLSLS), 74
Great Lakes–St Lawrence Water Resources Compact, 74
Great Lakes United, 73
Great Lakes Water Quality Agreement, 71, 72, 73, 74, 116
Great Western Railway, 45, 125
Greater Golden Horseshoe, 60, 81, 82, 88, 98, 99, 322
Greater Peterborough Area Economic Development Corporation, 155, 160
Greater Toronto Area (GTA), 80, 81, 82, 83, 93, 94, 95, 96, 104, 105, 106, 110, 115, 120, 121, 123, 124, 131, 169, 231, 234, 299, 301, 321, 326, 327, 332
Greater Toronto Bioregion, 93
Gréber Plan, 236

green city, 238, 427
Green Energy Act, 429
green infrastructure, 401, 411
Greenbelt, 95, 120, 123, 124, 236, 310, 326, 353, 354, 356
Greenbelt Plan, 322, 326
green-field, 212
greenways, 352
greyfields, 350
groundwater, 66, 106, 143, 342, 343, 353, 355, 403
growth poles, 242, 293, 294, 295
Guelph, 18, 19, 48, 95, 99, 144, 162, 312, 353, 368, 433, 437

Haliburton Highlands, 158, 159, 163, 169, 170, 171, 173, 182, 184, 435
Harris, Mike, 297, 325, 415, 416
Harvey, David, 160, 202, 413, 422
have-not cities, 356
health costs, 415
heartland era, 415
heritage conservation district, 384, 385
Heritage Impact Assessment (HIA), 372, 408
heritage landscapes, 142, 235, 237, 376, 382
Heritage Resources Centre, 437
heritage sites, 372, 375, 376
high-tech, 355, 419
high-tech farmers, 419
Highway 401, 89, 104, 111, 113, 115, 121, 124, 160, 164, 322
Historical Atlas for Wentworth County, 330
home place, 52
homelessness, 213
Homestead Act, 208

homogenization, 336
Hudson Bay, 37, 251, 252, 253, 255, 266, 279
Hudson Bay Lowlands, 248, 249, 250, 270, 280
Hudson River, 10, 22
Hudson's Bay Company, 12, 247
Hulchanski, David, 420, 428
Huron Business Development Corporation, 143, 146, 150
Huron Clean Water Project, 144
Huron Community Development Corporation, 131
Huron County Department of Planning and Development, 134, 144
Huron Road, 18
Huron Tract, 18, 113, 176, 208
Huron Water Protection Steering Committee, 143
Huronia, 8, 11, 175
husbandry, 337. *See also* conservation stewardship

immigrants, 3, 18, 35, 59, 83, 85, 86, 137, 233, 234, 240, 241, 282
Indian Reserve, 35, 317
indigenous people, 7, 11, 38, 51, 111, 227, 316
industrial era, 4, 320
Industrial Revolution, 15, 366, 368, 369
inequalities, 420, 424
infill development, 212, 351
innovation(s), 111, 123, 161, 215, 301, 349, 350, 351, 392, 415
integrated pest management, 348
integrity (cultural, ecological), 5, 94, 189, 190, 339, 364, 436

intensification, 138, 185, 241, 341, 351, 365, 376, 407, 420, 437. *See also* urban planning
International Joint Commission (IJC), 392
International Union for Conservation of Nature (IUCN), 339
invasive (loss of biodiversity), 65, 72, 116, 356, 391
Iroquois Confederacy, 13

Jacobs, Jane, 90, 214
Jewish people, 36, 182
jobs, 284, 378; agricultural, 122; First Nations, 288; gain, 294, 406, 428; industrial, 122, 282; loss, 137, 261, 282, 284, 389, 405, 410; middle-income, 213; regional, 133, 137, 262, 267, 271, 298, 301; service, 134, 284, 410; technology, 162. *See also* employment
Judt, Tony, 389, 424

Kawartha Lakes, 31, 99, 155, 156, 157, 158, 159, 160, 171, 173, 178, 179, 181, 395
Kenora district, 282, 283, 284
King's Highway, 20, 21, 396
Kitchener, 1, 47, 48, 130, 403, 410

La Rivière aux Dindes, 16
La Salle, 10
Lake Huron Centre for Coastal Conservation, 144, 433
lake levels, 68, 70, 72, 391
Lake Michigan, 12, 55, 71
Lake Nipissing, 9, 32, 229, 245, 269

Lake of the Woods, 12, 35
Lake Simcoe, 8, 10, 38, 39, 176, 178, 182
Lake Simcoe Protection Act, 403
Lake St Clair, 16, 112
Lake Superior National Marine Conservation Area, 291
Lake Temagami, 11
Lake Temiskaming, 34, 31
Lakehead (Fort William, Port Arthur), 24
Lakewide Management Plans, 72
Land Between, The, 154, 401, 425
land clearing, 3, 339
land trusts, 118, 119, 433
land use complex, 82, 83, 86, 88, 95
landowner contact, 117
Lands for Life, 12, 287, 296, 297, 298, 299, 302, 304
Laurentian University, 265
Leadership in Energy and Environmental Design (LEED), 380
Leading Edge, 312
League of Slow Cities, 217
leapfrog development (sprawl), 96
level playing field, 416
living countryside, 219
living machines, 350
Local Architectural Conservancy Advisory Committees (LACAC), 375
local dependency, 294
local knowledge, 429
locally responsible, 425
logging, 19, 20, 24, 30, 31, 33, 36, 38, 41, 59, 63, 177, 180, 182, 280, 291. *See also* lumbering
London, 18, 20, 27, 29, 39, 353, 391, 392, 394, 407

Loyalists, 14, 16, 25, 29, 112, 207, 316, 317
lumbering, 4, 20, 24, 29, 31, 35, 36, 37, 40, 41, 42, 51, 182, 209, 399, 401. *See also* logging

Macdonald, Sir John A., 216
Manitoulin Island, 9, 35, 37, 38, 313, 317, 331, 435
Manitowaning, 35, 38
Martin-Florida report, 418, 420
Marxian, 190
mass transit, 94, 99, 354
McGuinty Liberal government, 97, 325
McMaster University, 312, 410
mechanization, 265, 336, 337
megalopolis, 52, 99, 261. *See also* conurbation, metropolis
mega-projects, 147
mega-region, 420
Mennonites (Plain Folk), 16, 47
mercantile capitalism, 188
mesotrophic, 64. *See also* eutrophic; oligotrophic
Métis, 254
Metrolink, 402, 405
metropolis, 4, 89, 115, 169, 171, 183, 186, 323, 390, 395, 404, 405, 419, 420, 426, 427. *See also* global city; megalopolis
Metropolitan Toronto, 80, 81, 86, 413, 417
Metropolitan Transportation and Region Transportation Study, 323
middle-income, 213, 420, 428
Mid-Peninsula Transportation Corridor, 325
Miller, David, 98

Mine Development and Closure Act, 265
mining, 4, 28, 33, 34, 46, 51, 256, 269, 337; First Nations, 34, 247, 255, 258, 266, 286, 289, 303, 399; legislation, 266, 270, 286; regional, 4, 35–7, 42, 209, 245, 253–4, 256, 258, 262, 266–7, 271–2, 278, 282–3, 304, 355, 399; technology, 261, 265
Ministry of Agriculture, Food, and Rural Affairs, 142
Mississauga, 3, 157, 208
modernism, 363, 366
modernity, 202, 217
Mohawks, 15
monitoring (research, planning), 75, 76, 249, 429
moraine (ice shields), 154, 314, 403, 405. *See also* glaciers
municipal cultural planning, 163
Municipal Heritage Committee, 376
Mushkegowuk Council of Chiefs, 255

National Capital Commission (NCC), 237
National Farmers Union, 219
National Historic Site, 39, 156, 216
national parks, 69, 103, 325
National Research Council (NRC), 238, 398
National Wildlife Refuges, 70
natural areas, 4, 68, 110, 116, 117, 121, 141, 339, 344, 346, 349, 392, 393
natural decay, 363, 364
Natural Heritage Stewardship Program, 117

Nature Conservancy, 63, 339
nature conservation, 42, 315, 393
neo-liberalism, 388, 390, 391
Neutral people (indigenous people), 8, 317
new economy hubs, 272
New Ontario, 245, 247, 272
new urbanism, 212, 213, 215, 216, 352
New York City, 22, 43
Niagara Escarpment Commission (NEC), 312, 316, 324, 325
Niagara Escarpment Parks and Open Space System, 327
Niagara Escarpment Plan, 310, 312, 322, 326, 327, 329, 331, 332
Niagara Fruit Belt, 110, 115
Niagara-on-the-Lake, 15
Niagara to GTA Corridor, 327
nickel belt (Sudbury), 37
NIMBYism, 191
non-government organizations (NGO), 102, 116, 125, 290, 409
non-point source pollution, 116, 424
non-profit governance, 161
non-regulatory approaches, 145, 325
non-timber forest products, 265
North American Free Trade Agreement (NAFTA), 416
Northern Ontario Sustainable Communities Partnership, 265
Northern Railway, 28, 41
Northwestern Ontario Municipal Association (NOMA), 285
Nunavut, 284, 400
nutrient management, 138

Oak Ridges Moraine, 93, 94, 123, 169, 310, 322, 335, 336, 344, 356, 403, 406
oak savannah, 108
Ohio Lake Erie Commission, 76
oil prices, 422, 423
Ojibwa (indigenous people), 38, 175, 176
old-growth forests, 177
oligotrophic, 64. *See also* eutrophic; mesotrophic
one-industry towns, 394, 399
one-season places, 394
Ontario biotechnology corridor, 163
Ontario Farmland Trust, 124
Ontario Federation of Anglers and Hunters, 246
Ontario Federation of Snowmobile Clubs (OFSC), 268
Ontario Forest Industries Association, 296
Ontario Forestry Coalition, 271
Ontario Georegional Commissioner, 412, 413
Ontario Georegional Forum, 412
Ontario Healthy Communities Coalition, 360
Ontario Heritage Act (OHA), 370, 373, 375, 378, 407
Ontario Heritage Trust (OHT), 327
Ontario Hydro, 255, 270
Ontario Ministry of Environment and Energy (MOEE), 72, 73
Ontario Ministry of Natural Resources (OMNR), 73, 158, 161, 294, 295, 303, 312, 393, 400
Ontario Ministry of the Environment, 75, 360
Ontario Municipal Board, 353

Open Society, 49, 50
open space, 85, 212, 216, 327, 408. *See also* natural areas
opportunity costs, 145, 427
oral tradition, 175
Otonabee Region Conservation Authority, 158
Ottawa, Arnprior, and Parry Sound Railway, 181
ozone (o$_3$), 342

Paleo-Indians, 5, 280
Parkway Belt, 324
Parry Sound, 32, 33, 37, 40, 171, 173, 181
particulates, 341
Partnership for Public Lands, 304
Payment for Ecological Goods and Services (PEGS), 121, 145
PCBs (polychlorinated biphenyls), 59, 64
peak oil, 138
Penetanguishene, 9, 20, 38, 394
penitentiaries, 210, 216
perceptions and land use, 166, 309, 311, 324, 329, 408, 426
peregrine falcon, 347
permafrost, 251, 270, 279
pesticides, 338, 342
Peterborough County, 154, 155, 156, 157, 158, 159, 160
Peterborough Liftlock, 156
phytoremediation, 351
Pikangikum First Nation, 283, 287, 289, 290
place-entrepreneurs, 3
Places to Grow Act, 299, 301, 322
Plain Folk (Mennonites and Quakers), 16
planned obsolescence, 378

pluralistic society, 422
point sources, 116. *See also* pollution
Polar Bear Conservation and Education Habitat, 268
Polar Bear Provincial Park, 248, 249, 253, 270
pollution, 47, 49, 55, 66, 72, 341, 343, 354, 360, 419, 420, 425; air, 217, 349; biological, 42; water, 64, 66, 68, 116, 144, 192, 355, 394. *See also* eutrophication; point sources; non-point source pollution
Port Arthur (Lakehead), 24, 32, 35, 37
postmodern, 214
powerlessness, 294
prairies, 4, 32, 63
Precambrian Shield, 209
primary-resource extraction, 247
private-public partnerships, 162, 212
property rights, 362, 373, 375
protected areas, 69, 80, 103, 249, 253, 269, 272, 298, 315, 326–7, 336–8, 354, 356, 406; First Nations, 249, 255, 434; place or regional, 247–9, 270, 302, 327, 331, 356; public input, 297; urban, 91, 340. *See also* national parks; nature conservation; provincial parks
provincial parks, 182, 249, 338–9
Provincial Policy Statement (PPS), 123, 298, 299, 382, 403, 408, 424
Provincially Significant Wetlands, 92
public hearings, 325

Public Lands Act, 176
Public Lands and Colonization Act, 208
public participation, 352, 437
public transit, 96, 212, 213, 238, 241, 244, 365

Quakers (Plain Folk), 16
quality of life, 139, 162, 201, 210, 211, 214, 218, 238, 241, 293
quality of place, 211, 214, 217
Queen Elizabeth Way (QEW), 104, 114, 394, 402, 405
Queen's Bush, 38
Queen's University, 397, 436
quick fix, 418, 420

recession, 417
recharge areas, 342
reciprocity treaty, 177
reconciliation, 337, 347, 349
Red Hill Creek Valley, 110
reforestation, 63. See also conservation
Regional Municipality of Metropolitan Toronto, 81
Regional Municipality of Ottawa-Carleton, 241
Regional Municipality of Waterloo, 340
regional planning imperative, 188
regional state, 415, 416
regional transit authority, 96. See also Metrolink
regulatory, 145, 266, 401
Remedial Action Plans (RAPs), 72
remediation, 337, 347, 351, 406
renewable energy, 272, 300, 350
renovation, 367, 378, 379, 408

reserves (First Nations), 16, 38, 176, 216, 289, 338, 353. See also conservation
resilience, 290, 406, 423
resort industry, 178, 181, 184
resource depletion, 261
resource hinterland, 188, 258, 399
resource use, 55, 75, 324, 338, 352
resource-dependent communities, 281, 183
retirees, 137, 232, 233, 395, 401, 411
re-urbanization, 365
Rideau Canal, 19, 22, 24, 29, 45, 216, 228, 237, 395
Robinson-Superior Treaty, 255
Room to Grow, 298
Rouge Valley, 336, 406
Royal Commission on the Future of the Toronto Waterfront, 93
Royal Military College, 397
Rubin, Jeff, 423
Ruckelshaus, William, 389, 424, 425
rural and small-town inventive culture, 419
Rural Revolution, 142
Ruskin, John, 368

Sachs, Jeffrey D., 421, 422
salt, 342. See also pollution; resource use
Sancton, Andrew, 404
Sault Ste Marie Canal, 28
savannahs, 3, 8, 108, 337
Schumpeter, Joseph, 413, 414, 415
Science North, 268
Second World War, 88, 181, 378
seigneurial, 15
self-employed persons, 137, 262

self-help, 147, 290
self-organizing system, 347, 348
self-sustaining system, 347
senior citizens, 233
sense d'appartenance, 272
sense of entitlement, 271
sense of place, 201, 202, 361, 379
Septic System Re-inspection Program, 144
Sewell, John, 90, 404
Shoreline Management Program, 72, 73
Short Distance Society, 24, 27, 113
shovel-ready, 410
single-family, 88
single-sector towns, 277, 283
Slow City Manifesto, 217
Slow Food, 217, 218. *See also* quality of life
small-farm survival, 219
smart growth, 212, 298, 300, 301
smog, 174, 342
snowmobiles, 268, 269
social construct, 165
social justice, 213, 218, 424
social-learning, 348
social-support programs, 411
Society for the Protection of Ancient Buildings (SPAB), 368
softwood lumber, 262, 285
Sources of Knowledge Subcommittee, 436
source-water protection, 158
Spadina Expressway, 89, 90
species at risk, 83, 108, 343, 392
St Lawrence Islands National Park, 397
stagflation, 185
State of the Lakes Ecosystem Conferences (SOLEC), 71

stewardship, 102, 105, 116, 117, 118, 119, 145, 148, 327, 381, 392, 393, 425, 433. *See also* conservation; husbandry
stewardship councils, 392–3, 433
Stewardship Manual for Cottagers and Non-farm Rural Residents, 144
Stiglitz, Joseph, 389, 424
stormwater, 342, 346, 347, 350
strategic planning, 145, 147
subsidies, 121, 123, 125
suicide rate, 289
sunk energy, 408
surveillance, 392
sustainability, 92, 93, 98, 190, 191, 205, 213, 217, 218, 219, 242, 287, 290, 291, 359, 360, 380, 397, 399, 427

tax breaks, 372
tax policy, 367
Temagami, 11, 254, 269, 336
Thames (River), 8, 10, 16, 17, 18, 19, 20, 24, 29, 38, 48, 49, 111, 437
Thomson, Tom, 181, 189
Thousand Islands, 205, 208, 209, 216
tile drainage, 208
top-down planning, 295, 297, 301, 425, 429. *See also* bottom-up planning
Toronto Carrying Place, 175, 176
Toronto-centred concepts, 91, 120, 124, 169, 170, 183, 188, 189, 388, 397, 404, 419, 420, 422, 423
Toronto Islands, 43

Index 451

tourism, 52, 65, 143, 159, 183, 193, 249, 267, 320, 398, 434, 436; agro-tourism, 218; benefits, 97, 137, 161, 180, 182, 184, 267–8; eco-tourism (nature-based tourism), 291, 293, 295, 320, 434, 436; environmental effects, 220, 269, 291–2, 362, 434; First Nations, 205, 249, 272, 434; heritage or cultural or rural, 163, 210, 214–16, 219, 239, 435; mass, 170, 209; place or regional, 31, 35, 36, 41–2, 44, 143, 155–7, 174, 186–7, 238–9, 242, 245, 249, 262, 269, 272, 282, 290, 297, 395
toxins, 350, 351
Toyota, 110, 120, 421
trans-boundary activities, 46, 75, 325
transfer payments, 417
transshipment, 27, 35, 45, 46, 51, 206
Treaty of Paris, 14, 206
Trent Conservation Coalition, 158
Trent River, 31, 152, 155, 156
Trent-Severn, 31, 155, 156, 169, 435
trickle-down effects, 294, 390, 413, 417, 420, 428

Ukrainians, 37
unemployment, 258, 260, 262, 264, 411, 414, 415, 423
Union of Ontario Indians, 255
United Nations Education, Social and Cultural Organization (UNESCO), 369
University of Guelph, 144, 433, 437

University of Waterloo, 19, 410, 436, 437
Upper Canada, 16, 17, 19, 22, 28, 112, 158, 176, 177, 247, 318
uranium, 267
urban ecology, 349, 352, 361, 407
urban ecosystems, 341, 342, 344
urban heat-island, 349, 350
urban planning, 84, 85, 211, 341, 437
urban policy, 407
urban renewal, 214, 365, 366, 367
urban sprawl, 69, 80, 91, 120, 124, 212, 341, 349, 353, 365, 392
urbanizing region, 128
US National Trust for Historic Preservation, 380
utility corridors, 346

value-added, 262, 265
values, 76, 372, 403; disagreements, 105, 115, 120, 292, 311, 362, 381; economic, 109, 120–1, 185–6, 332, 437; environmental, 401; First Nations, 205; philosophical, 110, 437; regional, 102–3, 123, 271, 321, 333; social, 94, 215, 220, 374
Venice Charter, 364, 369, 377
Victorian, 33, 114, 346, 368, 369
vision, 45, 128, 139, 149, 162, 236, 265, 291, 357, 390, 404, 406, 413, 415, 417, 418, 428

War of 1812–14, 16, 17, 111, 112, 227
water control, 62
water pipelines, 391, 424
Waterloo, 19, 47, 48, 82, 128, 130, 340, 354, 363, 391, 403

Waterloo Moraine, 340
watersheds, 17, 32, 46, 48, 55, 56, 143, 152, 158, 255, 311, 344, 351, 391, 399, 406, 425
welfare, 136, 293
Welland Canal, 22, 23, 43, 58, 66, 113, 207, 318, 396
West Nile virus, 344
Whitefeather Forest Proposal, 289, 290, 303, 304
wilderness, 8, 21, 67, 192, 193, 207, 361
wildlands, 21, 23, 24, 51, 247, 399
Wildlands League, 296
wildlife, 51, 60, 73, 270, 277, 281, 295, 429, 434
wildlife habitat, 55, 399
wind towers, 141, 406, 429
wind turbines, 138
Windsor–Quebec City corridor, 160

wise use, 337
withdrawals, 55, 66, 392
Wood, David, 20
World Biosphere Reserve, 70, 309, 312, 322, 326, 327, 328, 329, 331
world city, 81, 97, 99n5. *See* global cities
World Heritage Convention (WHC), 369
World Heritage Site, 216

Yi Fu Tuan, 202
York, 42, 80, 82, 83, 84, 99, 113, 354
Youth Matters, 146

zebra mussels, 65, 392
zero-till, 116